Structural engineering design in practice

PROJECT DESIGN EXAMPLES

General Editor: Colin Bassett, Bsc, FCIOB, FFB

Related titles

Basic Soil Mechanics 3rd Edition, *Roy Whitlow*

Structural engineering design in practice

PROJECT DESIGN EXAMPLES

Roger Westbrook B.Sc., C.Eng., MICE, MIStructE
Lecturer in Structural Engineering, University of Bradford

and

Derek Walker M.Sc., DipTP, C.Eng., FIStructE, MICE, MIQA
Senior Structural Engineer, TBV Consult

Third edition

LONGMAN

Addison Wesley Longman Limited
Edinburgh Gate
Harlow, Essex CM20 2JE
England
and Associated Companies throughout the world

First published 1984
Second edition 1988
Third edition 1996
Reprinted 1997 (twice)

British Library Cataloguing-in-Publication Data
A catalogue entry for this title is available from the British Library.

ISBN 0 582 23630 4

Set by 7 in Times 10/12

Produced by Longman Singapore Publishers (Pte) Ltd.
Printed in Singapore

Contents

Foreword

I was encouraged by the response to 'Structural Design in Practice – 1st & 2nd Editions' to publish a 3rd Edition, but with a difference.

Although still based on calculations and drawings, this time, the emphasis is on different types of structures, rather than different Structural Codes of Practice.

The structures are fictitious buildings and bridges, and have been chosen to provide a wide range of design techniques and solutions similar to those encountered in a modern design office.

As this task is more ambitious than before, I have a co-author in Derek Walker, who has contributed to the design philosophy and concepts, checked the calculations and prepared the drawings and details.

Derek is a part-time lecturer in Structural Design at the University of Bradford and has been a tutor on the Institution of Structural Engineers Part 3 Preparation Course, run by the University, for some years. Formerly with PSA Projects, he is now a senior professional civil and structural engineer with TBV Consult, which is a multi-disciplinary design consultancy within the Tarmac Professional Services Group.

The calculations and drawings presented in this edition are by no means complete, but should provide Civil and Structural Engineering students, Postgraduate MSc students, HND and Engineering Council Part 2 students with a good guide as to how to approach real structural design projects in practice.

For students preparing for the Part 3 professional examination of The Institution of Structural Engineers, the design examples should give useful assistance on how to present calculations and drawings for a variety of building types and structures. However, it must be emphasised that the examples given are not presented as definitive solutions or model answers.

The book should also provide a comprehensive reference for practising engineers who do not have the opportunity to design the full range of buildings and structures covered in the text.

Finally, my thanks go to those Engineers who bought the 1st and 2nd Editions (or both), and who took the trouble to write and suggest this further edition.

July 1995

Roger Westbrook,
University of Bradford

Suggested Procedure for preparing structural calculations

1.0 The purpose of the calculations are to:-

i) Show that the design is in accordance with good structural practice, and, where appropriate, comply with the current and relevant National Standards and Building Regulations.
ii) Demonstrate that the design is adequate in relation to stability, strength and serviceability requirements.
iii) To aid, instruct and assist the draughtsman preparing the general arrangement and detailed drawings.
iv) Provide a permanent record for future reference.

There is a requirement for the calculations to be comprehensive, self-explanatory and well presented, and in order to achieve these qualities, they must conform to:-

v) a standard method of documentation
vi) a logical sequence of presentation
vii) a consistent set of units and notation

2.0 Structural Form

Structural stability should be achieved by making an appropriate choice of the structural form and arrangement. The use of additional and expensive features to achieve structural stability should be avoided. Benefits are gained if the structural engineer is involved in the design from the earliest possible moment, and the method of preventing disproportionate collapse established.

3.0 Health and Safety at Work Act and C.D.M. Regulations

These regulations render every designer legally responsible for the safety of structures under his/her control, and require a risk assessment to be carried out on every project.

4.0 Designing for Stability

The following is a good guide in the process of designing for stability:-

i) The overall stability of the structure in the transverse and longitudinal directions should be checked using wind loads or notional forces.
ii) Check that the most adverse combination of unfactored loads do not exceed the safe bearing capacity of the soil.
iii) Any floor or roof acting as a horizontal diaphragm should not have any discontinuities, such as expansion joints, which will nullify the design assumptions.
iv) Shear walls, bracing, etc., should be continuous throughout the height of the building, and any points of high stress concentrations (i.e. openings) should be checked.
v) Check for adequate resistance to torsional moments on asymmetrical buildings by calculating the shear centre of application of wind or other horizontal loading.
vi) Position ties in accordance with the relevant National Standards for reinforced concrete, structural steelwork, masonry and timber.
vii) At bearings, or junctions of structural members, and in particular, pre-cast concrete, check that:-
   (a) There is adequate overlap of steel reinforcement
   (b) Positive lateral and torsional restraints
   (c) Adequate restraint on the longitudinal axis
viii) Check the site history for any special features such as made-up ground, tunnels, gas mains, mining, contamination, etc.
ix) Ensure that the stability of the structure during construction is adequate.
x) Consider the effect of expansion and construction joints on the stability of the structure, and ensure that they are provided where necessary.
xi) When designing stability walls, rigid frames etc., for horizontal forces, ensure that the following criteria are met:-
   (a) Check stability against local overturning to ensure that there is an adequate factor of safety
   (b) Any horizontal displacements at the top of the building, relative to the base, should be restricted to 1/500 of the overall height. Inter storey distortion should be limited to 1/250 of the storey height under consideration
   (c) In certain cases, particularly tall steel framed buildings and steel chimneys, it may be necessary to check for wind-induced oscillations and provide damping systems
xii) Consider the following factors in the design stage:-
   (a) Fire resistance required, and the effects of damage by fire
   (b) Compatibility of structural materials used with the required design life of the building
   (c) For cladding and facing materials ensure that the fixings can accommodate movement

due to thermal effects and moisture, etc. The fixings should be of a mechanical 'fail-safe' type, and be resistant to corrosion

(d) For large structures, where different sections of the work are designed and detailed by different designers, ensure that there are no deficiencies or discontinuities at the interface

(e) Carry out a risk assessment and ensure that the proposed solution complies with the C.D.M. Regulations.

Chapter 1 Road bridge over railway

A Local Authority propose to develop a site at the far side of an existing overpass and existing railway from the nearest trunk road. In order to assess the feasibility of developing the site (for Industrial purposes), you have been approached, as Consultant, to provide a preliminary design for the main span beams and slab. The proposed scheme is to be composite steel plate girders at 3 m. centres and a reinforced concrete deck, as shown in Fig. 1/001. The bridge is **not** to be continuous because the piles necessary for the abutments and piers are in weak/stiff clays and the predicted settlement of the supports is significant (say, exceeding span/1000) (Ref 1/1). To avoid overstress, the structure should be made statically determinate by use of simply-supported spans. The bridge is to be 3 lanes, with a 2 m. pathway each side, as shown in Fig. 1/001, and the proposed scheme as shown has a minimum clearance of 5.7 m. under the existing viaduct, and has a minimum clearance under of 5.0 m. to the railway tracks.

Loading – live

Traffic loading to be taken as 10 kN/m^2 for preliminary design. An alternative load is to be considered, for element design, of 100 kN wheel load on a 0.3m. x 0.3m. square contact area. Footway loading is to be 5.0 kN/m^2

Loading – wind

The Local Authority do not require a check for wind loading at this time.

Loading – impact

It can be assumed that the leaf piers and abutments shown on Fig. 1/001 are sufficiently far back (4.5 m.), and protected from the railway tracks for accidental impact not to be a consideration.

Loading – superdead

Take average characteristic loading of surfacing as 2.4 kN/m^2.

Loading – dead

Take unit weight of concrete as 24 kN/m^3

Site conditions

Fill, alluvium and weak clay down to 12.0 m., where stiff clay (undrained Cu = 75 kN/m^2) was encountered to a large depth.

Design Code of Practice

BS 5400 Parts 3 & 5 – Composite Steel Bridges

Materials

Steel plates – EN 10025 Grade Fe 510B – fy = 355 N/mm^2
Concrete deck – fcu = 40 N/mm^2 and for reinforcement – fry = 460 N/mm^2

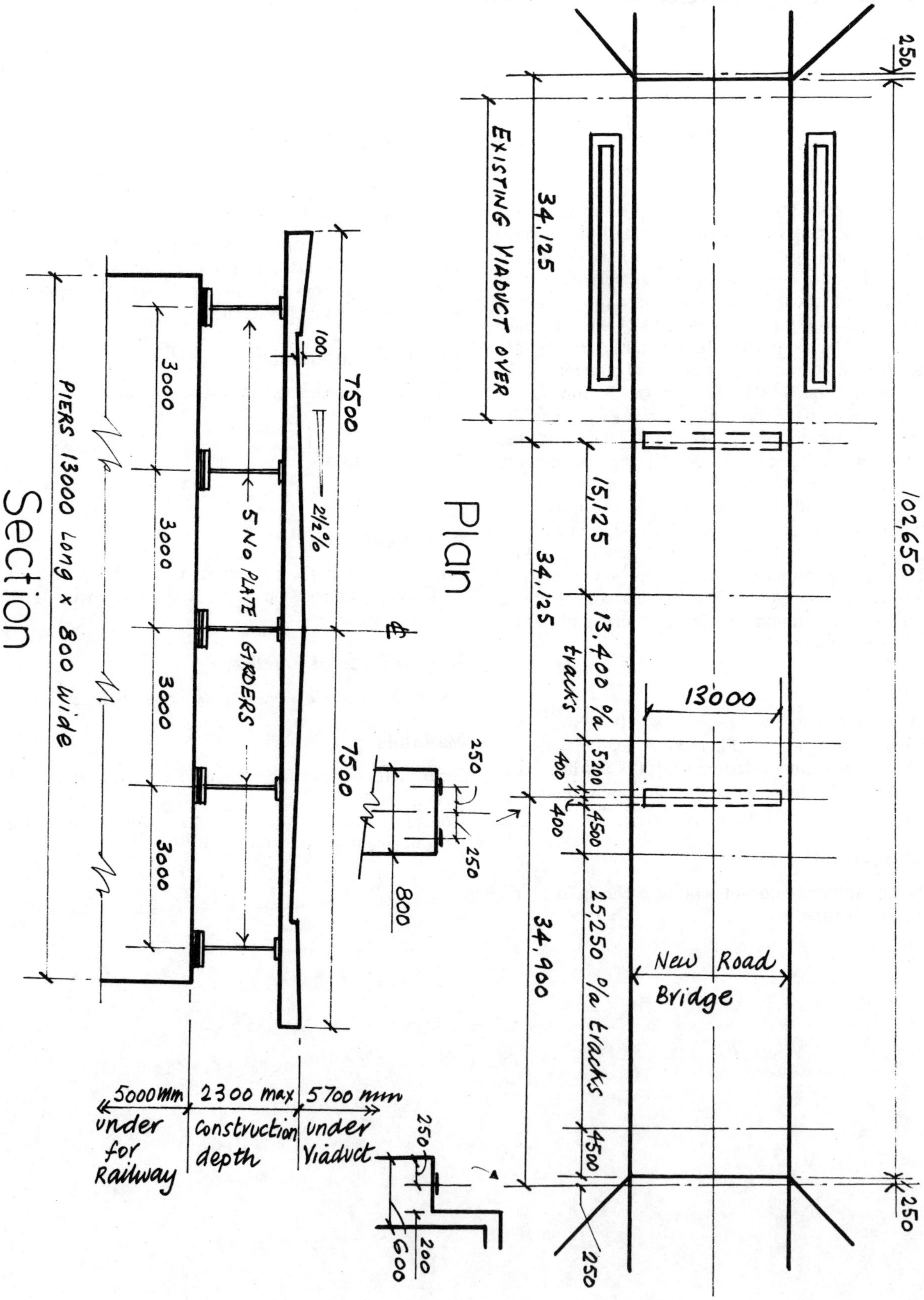
Plan
Section
EXISTING VIADUCT OVER
34,125
34,125
34,900
15,125
13,400 c/a tracks
5200
4500
25,250 c/a tracks
4500
102,650
250
250
250
13000
New Road Bridge
PIERS 13000 LONG x 800 WIDE
3000
3000
3000
3000
5 NO PLATE GIRDERS
7500
7500
100
2½%
250
250
800
5000mm under for Railway
2300 max construction depth
5700 mm under Viaduct
600
200

Fig.1/001.

Ref. (BS5400)	Initial girder size.
	Span over rail tracks, c/c of bearings = 34900 − 250 = 34650 mm centres of 5 girders = 3000 mm
Ref 1/2	(Practical parameter: centres usually in range 2500 mm → 3800 mm.)
	Simply-supported spans, no haunch, spliced in middle.
Ref 1/2	(Practical parameter: plates not fabricated greater than 23 m. long).
	Deck slab 230 mm. average thickness
Ref 1/2	(Practical parameter: min. thickness = 220 mm. on basis of cover & crack width criteria) From fig 1/001, "construction depth" − beam + slab + surfacing ≯ 2300 mm
Ref 1/1	Using the above criteria & consulting a chart of flange & web sizes in Ref 1/1 for simply supported bridges, assuming:− 1) Deck slab 230 mm av. thickness (5.52 kN/m²) 2) Super dead loads (surfacing) equivalent to 100 mm. of finishes (2.40 kN/m²) 3) Steel: EN 10013 grade Fe 50B: py = 385 N/mm²
Ref 1/2	4) span-to-girder depth ratios 20 → 30 (1732.5 → 1155 mm.). Note! for continuous, span-to-girder depth usually 20 → 25 5) Webs have stiffeners at approx. 2.0 m. crs. where such stiffening is required. 6) Elastic stress analysis. 7) Steel beams "unpropped". 8) Lateral torsional buckling not a problem due to sufficient cross-bracing. Initial overall depth of beam governed by construction depth. A 1600 mm beam will leave 2300 − 1600 = 700 mm for slab + finishes ✓ o.k. Also within condition (4)

Ref. (BS 5400)

From chart (Ref. 1/1),
for L = 34.65m ; $L/D = \frac{34650}{1600} = \underline{21.65}$,
and assuming HA loading as being closest to the preliminary design loading:

$\underline{A_{ft} = 0.018\ m^2}$; $\underline{A_{fb} = 0.025\ m^2}$

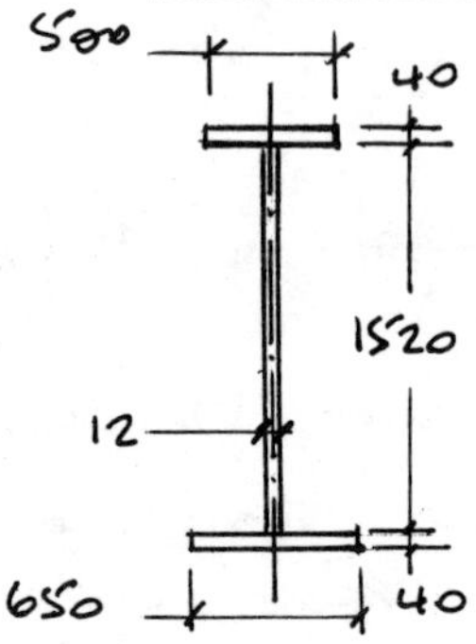

Fig. 1/002
Proposed section

Try a top flange 500 x 40 mm
$\therefore A_{ft} = \frac{500 \times 40}{10^6} = 0.02 m^2$ ($> 0.018 m^2$)

Try a bottom flange 650 x 40 mm
$\therefore A_{fb} = \frac{650 \times 40}{10^6} = 0.026 m^2$ ($> 0.025 m^2$)

Also $\underline{t_w \simeq 12 mm.}$

$d/t_w = 1520/12 = 127$ mm
& $A_w = \frac{1520 \times 12}{10^6} = \underline{0.018\ m^2}$

Total area = 0.020 + 0.026 + 0.018
= $\underline{0.064\ m^2}$

$\therefore$ Self-weight = 77 x 0.064 = $\underline{5.0\ kN/m.}$

Design parameters:

Steel: $E_s = 205 \times 10^3\ N/mm^2$: $\sigma_y = 355\ N/mm^2$

Concrete: $f_{cu} = 40\ N/mm^2$: $E_{cs} = 31 \times 10^3\ N/mm^2$
$E_{cL} = 15.5 \times 10^3\ N/mm^2$

Reinforcement : $f_{ry} = 460\ N/mm^2$
min. cover = 35 mm (top)
& 40 mm (bottom)

Pt.3 Table 2 — Material factor, steel, $\gamma_m = 1.05$

Section properties for inner beam

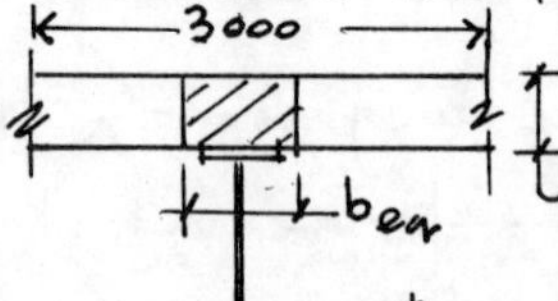

Fig. 1/003
Equivalent section

Modular ratios :

Short term, $E = \frac{205}{31} = \underline{6.6}$

Long term, $E = \frac{205}{15.5} = \underline{13.2}$

Equivalent slab width, bea
Short = 3000 / 6.6 = 454 mm.
Long = 3000 / 13.2 = 227 mm.

Ref.
(BS5400)

Steel only		A (mm^2)	y (mm)	Ay ($\times 10^3 mm^3$)	Ay^2 ($10^6 mm^4$)	I ($10^6 mm^4$)
T_f	500 x 40	20000	1580	31600	49928	3
W	12 x 1520	18240	800	14592	11674	3512
Bf	650 x 40	26000	20	520	10	4

Table 1/1 $\Sigma = 64240$ $\Sigma = 46712$ $\Sigma = 61612$ $\Sigma = 3519$

$$I = 61612 + 3519 - \left(\frac{46712^2}{64240}\right) = 31164 \times 10^6 mm^4$$

$\therefore I = 31.2 \times 10^9 mm^4$

$$y_{na} \text{ (see Fig. 1/003)} = \frac{46712 \times 10^3}{64240} = 727 mm.$$

Section classification:

Web: compact if depth of neutral axis from compression end of web $\ngtr 28 t_w \sqrt{\frac{355}{\sigma_w}}$

Depth = 1560 − 727 = 833 mm.
Limit = $28 \times 12 \times \sqrt{\frac{355}{355}}$ = 336 mm.
} Non-compact.

Top flange: (compression)

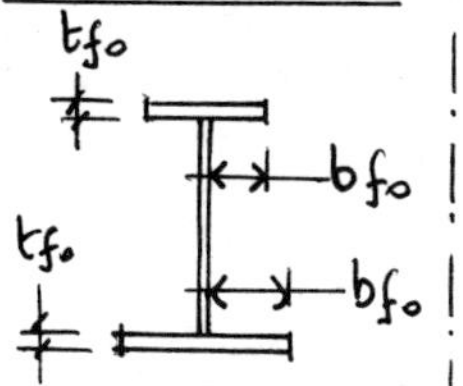

Fig. 1/004.

$\frac{b_{fo}}{t_{fo}} \ngtr 12 \sqrt{\frac{355}{\sigma_f}}$ for compact.

$b_{fo}/t_{fo} = 244/40 = 6.1 < 12$ - 'compact'

Bottom flange (tension)

$\frac{b_{fo}}{t_{fo}} \ngtr 16$ for compact.

$b_{fo}/t_{fo} = 319/40 = 8 < 16$ - 'compact'

$\therefore$ Section 'non-compact'

Composite (* Values from Table 1/1)

short term	A (mm^2)	y (mm)	Ay ($10^3 mm^3$)	Ay^2 ($10^6 mm^4$)	I ($10^6 mm^4$)
slab 454 x 230	104420	1715	179080	307123	460
beam *	64240	727	46712	61612	3519

Table 1/2 $\Sigma = 168660$ $\Sigma = 225792$ $\Sigma = 368735$ $\Sigma = 3979$

$$I = 368735 + 3979 - \left(\frac{225792^2}{168660}\right) = 70437 \times 10^6$$

$\therefore I = 70.44 \times 10^9 mm^4$

$$y_{na} = \frac{225792 \times 10^3}{168660} = 1339 mm.$$

Ref.	Long Term	A (mm²)	y (mm)	Ay (10^3 mm³)	Ay² (10^6 mm⁴)	I (10^6 mm⁴)
(BS5400)	Slab 227×230	52210	1715	89540	153561	230
	beam	64240	727	46712	61612	3519
	Table 1/3	Σ = 116450		Σ = 136252	Σ = 215173	Σ = 3749

$$I = 215173 + 3749 - \left(\frac{136252^2}{116450}\right) = 59501 \times 10^6 \text{ mm}^4$$

$$\therefore \; I = 59.5 \times 10^9 \text{ mm}^4$$

$$y_{na} = \frac{136252 \times 10^3}{116450} = 1170 \text{ mm.}$$

Section modulii :–

Steel only :

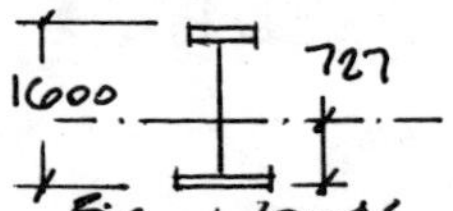

Fig. 1/005

$$Z_{top} = \frac{31.2 \times 10^9}{(1600 - 727)} = 35.7 \times 10^6 \text{ mm}^3$$

$$Z_{bot} = \frac{31.2 \times 10^9}{727} = 42.9 \times 10^6 \text{ mm}^3$$

Composite (short) :

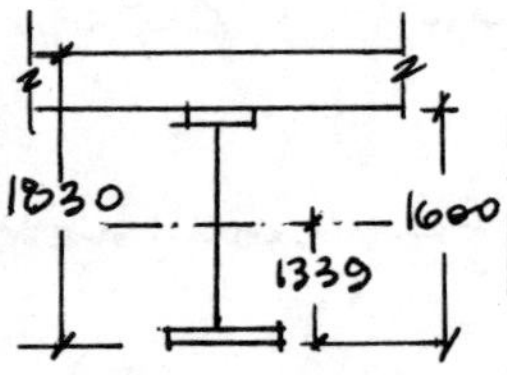

Fig. 1/006

$$Z \text{ (top of slab)} = \frac{70.44 \times 10^9}{(1830 - 1339)} = 143.5 \times 10^6 \text{ mm}^3$$

$$Z \text{ (u/s of slab)} = \frac{70.44 \times 10^9}{(1600 - 1339)} = 270 \times 10^6 \text{ mm}^3$$

$$Z \text{ (beam bot.)} = \frac{70.44 \times 10^9}{1339} = 52.6 \times 10^6 \text{ mm}^3$$

Composite (long) :

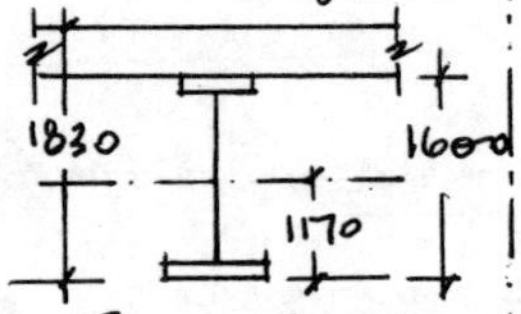

Fig. 1/007

$$Z \text{ (top of slab)} = \frac{59.5 \times 10^9}{(1830 - 1170)} = 90.1 \times 10^6 \text{ mm}^3$$

$$Z \text{ (u/s of slab)} = \frac{59.5 \times 10^9}{(1600 - 1170)} = 138.4 \times 10^6 \text{ mm}^3$$

$$Z \text{ (beam bot.)} = \frac{59.5 \times 10^9}{1170} = 50.8 \times 10^6 \text{ mm}^3$$

Limiting stresses.

Plate girder compression (top) flange is continuously restrained by the concrete deck slab :–

Allowable compression & tension in steel beam $= \dfrac{\sigma_y}{\gamma_m \cdot \gamma_{f3}} = \dfrac{355}{(1.05 \times 1.1)} = 307 \text{ N/mm}^2$

Allowable concrete stress

$$= \frac{0.5\, f_{cu}}{\gamma_{f3}} = \frac{0.5 \times 40}{1.1} = 18.2 \text{ N/mm}^2$$

Ref.	
(BS 5400)	**Global analysis of bending moments.**
	In normal circumstances, a plate analysis would be undertaken on the deck slab + beams. For this preliminary design, the conservative approach of taking one beam + supported slab will be used. (3m. width)
	S.S. Span = 34.65 m.
	Steel beam: self-weight = 5.0 kN/m (calculated previously)
	∴ U.L.S. moment = $\frac{1.05 \times 5.0 \times 34.65^2}{8}$ = 788 kN.m. (γ_{fL} = 1.05)
	Wet concrete: s.wt. = 0.23 × 3 × 24 = 16.6 kN/m.
	∴ U.L.S. moment = $\frac{1.15 \times 16.6 \times 34.65^2}{8}$ = 2858 kN.m. (γ_{fL} = 1.15)
	Super dead load: surfacing = 2.40 kN/m^2 – see previous calcs.
	∴ U.L.S. moment = $\frac{1.75 \times (2.4 \times 3) \times 34.65^2}{8}$ = 1891 kN.m (γ_{fL} = 1.75)
	Live load: 10 kN/m^2 (see brief)
	∴ U.L.S. moment = $\frac{1.5 \times (10 \times 3) \times 34.65^2}{8}$ = 6753 kN.m. (γ_{fL} = 1.50)
	Total combination 1:
	Steel beam: total steel + concrete = 788 + 2858 = 3646 kN.m.
	Top flange stress = $\frac{3646 \times 10^6}{35.7 \times 10^6}$ = 102.1 N/mm^2
	Bottom flange stress = $\frac{3646 \times 10^6}{42.9 \times 10^6}$ = 85.0 N/mm^2
	Super dead load (long term)
	Stress in top of slab (concrete units) = $\frac{1891 \times 10^6}{90.1 \times 10^6 \times 13.2}$ = 1.6 N/mm^2
	Stress in top of beam (steel units) = $\frac{1891 \times 10^6}{138.4 \times 10^6}$ = 13.7 N/mm^2
	Stress in bottom of slab (concrete units) = 13.7 / 13.2 = 1.0 N/mm^2

Ref. (BS5400)	
	Stress in bottom of beam (steel units) $= \frac{1891 \times 10^6}{50.8 \times 10^6} = \underline{37.3\ N/mm^2}$
	<u>Live load</u> (short term)
	Stress in top of slab (concrete units) $= \frac{6753 \times 10^6}{152.8 \times 10^6 \times 6.6} = \underline{6.7\ N/mm^2}$
	Stress in top of beam (steel units) $= \frac{6753 \times 10^6}{270 \times 10^6} = \underline{25.0\ N/mm^2}$
	Stress in bottom of slab (concrete units) $= 25/6.6 = \underline{3.8\ N/mm^2}$
	Stress in bottom of beam (steel units) $= \frac{6753 \times 10^6}{52.6 \times 10^6} = \underline{128.4\ N/mm^2}$

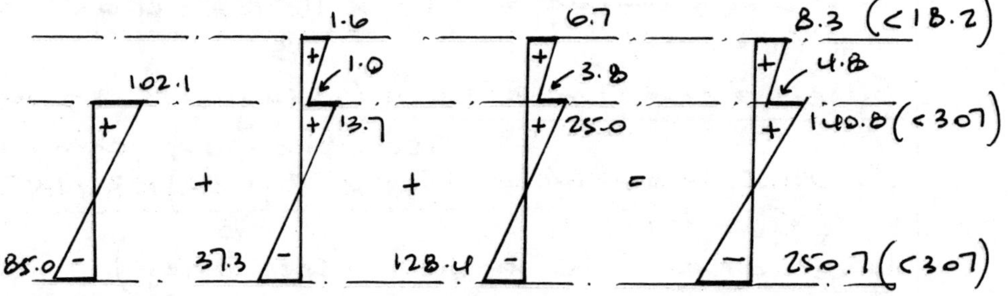

<u>Beam + Concrete</u> + <u>Long Term</u> + <u>Short Term</u> = <u>Final stresses</u>

<u>Fig. 1/008</u>

∴ <u>Plate girder o.k. in longitudinal bending</u>

	<u>Stiffened web shear resistance at U.L.S.</u>
Cl. 3/9.9.2.1	Allowing 50 mm. clearance at each end of the girder, from Fig 1/001, O/all beam length = 34900 − 50 + 200 − 50 = 35000 mm = <u>35.0 m.</u>
	<u>Steel beam</u>: $V_{max} = \frac{1.05 \times 5.0 \times 35}{2} = \underline{92\ kN}$. ($\gamma_{fL} = 1.05$)
	<u>Concrete</u>: $V_{max} = \frac{1.15 \times 16.6 \times 35}{2} = \underline{334\ kN}$. ($\gamma_{fL} = 1.15$)
	<u>Superdeadload</u>: $V_{max} = 1.75 \times \left(\frac{2.4 \times 3}{2}\right) \times 35 = \underline{221\ kN}$. ($\gamma_{fL} = 1.75$)
	<u>Live load</u>: $V_{max} = 1.5 \times \left(\frac{10 \times 3}{2}\right) \times 35 = 788\ kN$ ($\gamma_{fL} = 1.50$)
	$\Sigma = \underline{1435\ kN}$.

Ref.	
(BS5400) Cl. 3/9.9.2.2	$\phi = \dfrac{a}{d_{we}} = \dfrac{1500}{1520} = \underline{1.0}$; $\lambda = \dfrac{1520}{12}\sqrt{1.0} = \underline{127}$
	$\tau_y = \dfrac{355}{\sqrt{3}} = \underline{205\ N/mm^2}$
Pt.3 Fig 11	For $m_{fw} = 0$, $\dfrac{\tau_e}{\tau_y} = 0.66$ $\{\lambda = 127, \phi = 1.0$
	$\therefore \tau_e = 0.66 \times 205 = \underline{135.3\ N/mm^2}$
	$\therefore V_D = \dfrac{12 \times 1520}{1.05 \times 1.10} \times \dfrac{135.3}{10^3} = \underline{2137\ kN.}$
	$\therefore \underline{V_D > V_{max.}}$ ✓ o.k.
	(2137) (1435)

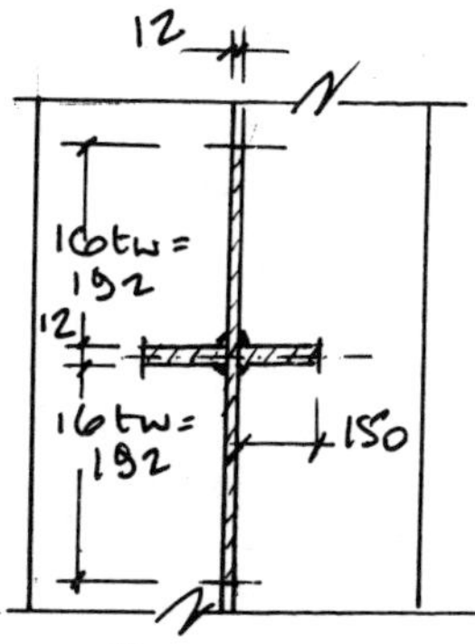

Plan on stiffener.

Fig. 1/009

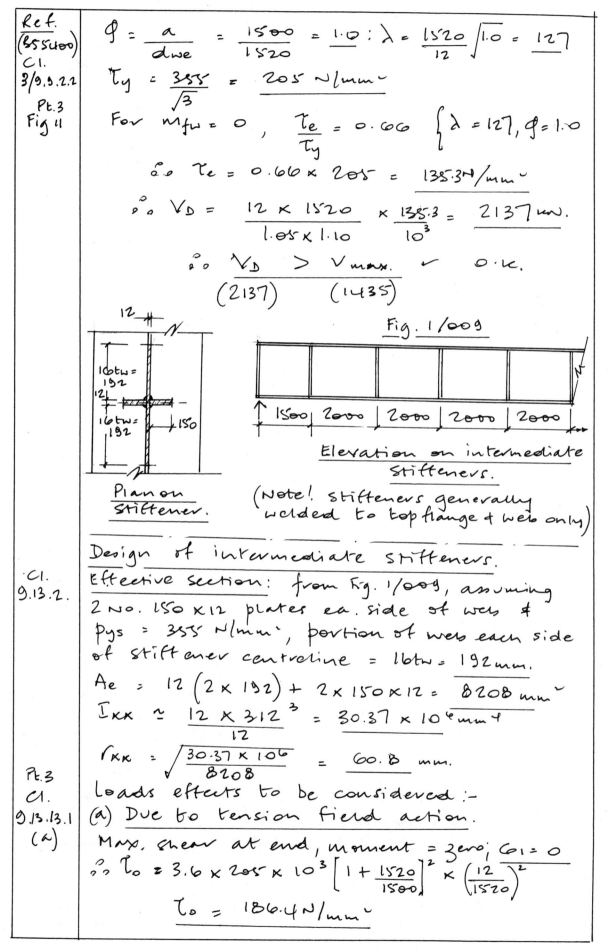

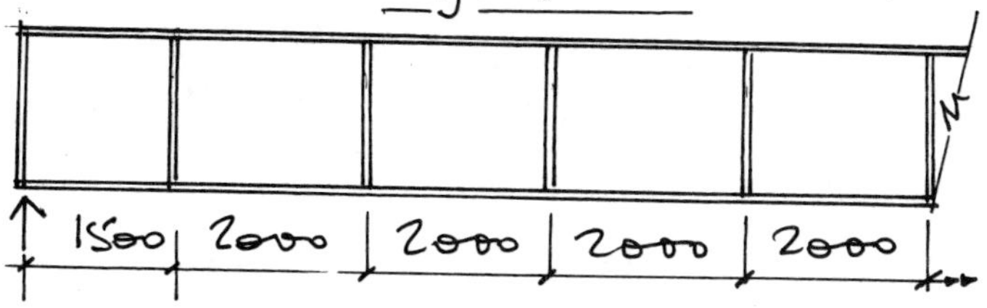

Elevation on intermediate stiffeners.

(Note! stiffeners generally welded to top flange + web only)

	Design of intermediate stiffeners.
Cl. 9.13.2.	Effective section: from Fig. 1/009, assuming 2 No. 150 × 12 plates ea. side of web + p_{ys} = 355 N/mm², portion of web each side of stiffener centreline = $16t_w$ = 192 mm.
	$A_e = 12(2 \times 192) + 2 \times 150 \times 12 = \underline{8208\ mm^2}$
	$I_{xx} \simeq \dfrac{12 \times 312^3}{12} = \underline{30.37 \times 10^6\ mm^4}$
	$r_{xx} = \sqrt{\dfrac{30.37 \times 10^6}{8208}} = \underline{60.8}\ mm.$
Pt.3 Cl. 9.13.13.1 (a)	Loads effects to be considered :-
	(a) Due to tension field action.
	Max. shear at end, moment = zero; $G_1 = 0$
	$\therefore \tau_o = 3.6 \times 205 \times 10^3 \left[1 + \dfrac{1520}{1500}\right]^2 \times \left(\dfrac{12}{1520}\right)^2$
	$\underline{\tau_o = 186.4\ N/mm^2}$

Ref. (BS5400) (Pt. 3)

Average shear stress in web, $\tau = \dfrac{143 \times 10^3}{1520 \times 12} = 78.6\ N/mm^2$

$\therefore$ as $\tau \not> \tau_o$, no force from tension field action in stiffener.

(b) axial force from destabilizing action –

Cl. 9.13.13.3 & Cl. 9.13.6.

as only the force in (b) in action, design to 9.13.6.

Stiffeners O.K. if $\sigma_R \leq \dfrac{A_{se} . a_{max} . \sigma_{ls}}{K_s . t_w . l_s^2 . \gamma_m . \gamma_{f3}}$

in which, $\sigma_R = \tau_R = 78.6\ N/mm^2$ (τ)

$A_{se} = 8208\ mm^2$ (see previous calcs)

$a_{max} = 1500\ mm.$

Cl. 9.13.5.3

σ_{ls} is buckling stress of stiffener:

$\lambda = \dfrac{l_s}{r_{se}} = \dfrac{1520}{60.8} = 25$

Pt. 3 Fig. 23

$\therefore$ From Fig. 23, $\sigma_{ls} = 323.05\ N/mm^2$

K_s, from Fig. 23, $= 0.075$

$R.H.S = \dfrac{8208 \times 1500 \times 323.05}{0.075 \times 12 \times 1520^2 \times 1.05 \times 1.10} = 1656\ N/mm^2$

$\therefore$ L.H.S < R.H.S.

(78.6) (1656)

Stiffeners 150 x 12 x 2 No. OK.

Design of shear connectors.

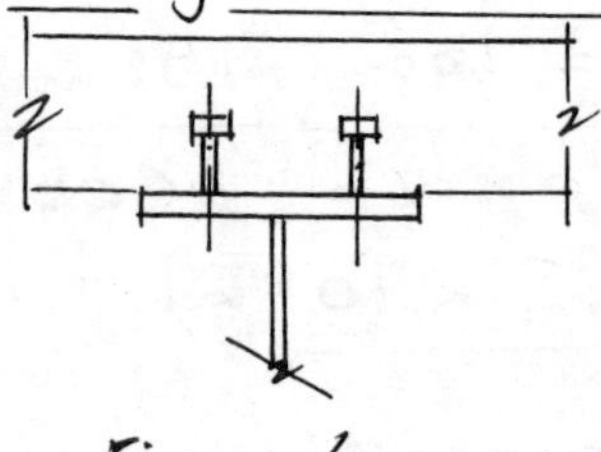

Fig. 1/010 (shear studs)

Try 2 rows of 19 mm ϕ headed studs x 100 mm. lg.

Nominal strength / stud

Pt. 5 Table 7

$P_u = 109\ kN$

Design strength / p_r

$= 2 \times 0.55 \times 109 = 120\ kN$.

Cl. 5/6.3.4

Shear connectors need not be considered at U.L.S, since the section is not compact, nor is there uplift.

Ref.	
(BS 5400 Pt. 3)	**S.L.S. shear forces (end of beam)**

Steel beam – long term: $V_{max} = \frac{1.0 \times 5.0 \times 35}{2} = 87.5\,kN$

$(\gamma_{fL} = 1.0)$

Concrete – long term: $V_{max} = \frac{1.0 \times 16.6 \times 35}{2} = 290.5\,kN$

$(\gamma_{fL} = 1.0)$

Super dead – long term: $V_{max} = 1.2\left(\frac{2.4 \times 3}{2}\right) \times 35 = 151.2\,kN$

$(\gamma_{fL} = 1.20)$

Total V_{max} (long term) $= 529.2\,kN$.

Live load – short term: $V_{max} = 1.2 \times \left(\frac{10 \times 3}{2}\right) \times 35 = 630\,kN$

$(\gamma_{fL} = 1.20)$

Total V_{max} (short term) $= 630\,kN$.

Considering uncracked sections, and conservatively ignoring shear lag :–

$\frac{A\bar{y}}{I}$ (long term) $= \frac{52210}{59.5 \times 10^9}(1715 - 1170) = 0.48 \times 10^{-3}$ (mm^{-1})

$\frac{A\bar{y}}{I}$ (short term) $= \frac{104420}{70.44 \times 10^9}(1715 - 1339) = 0.56 \times 10^{-3}$ (mm^{-1})

$\therefore$ Shear flow $= \left(\frac{529.2 \times 0.48}{10^3} + \frac{630 \times 0.56}{10^3}\right) \times 10^3$

$= 607\,N/mm$.

Try pairs of studs at 150 mm. crs:

Shear per group $= \frac{607 \times 150}{10^3} = 91\,kN$.

From previous page, design strength $= 120\,kN$ ✓ o.k.

$\therefore$ use 19 mm studs, 100 lg. min @ 150 crs.

Flange to web welds:

These depend on the horizontal shear forces in the joints between flanges and web. For the top flange one has to consider the shear due to dead load on the girder alone and the shear due to composite action:

$A\bar{y}/I$ coefficients:–

a) Girder alone.

A_s = area of top flange $= 500 \times 40 = 20000\,mm^2$

Ref. (BS5400 Pt. 3)

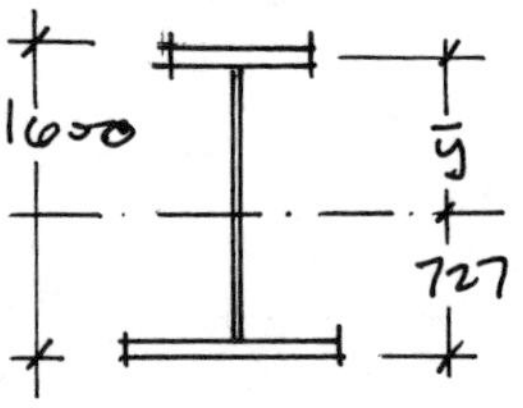

Fig. 1/011

From Fig. 1/011,

$\bar{y} = 1600 - 727 - 20 = \underline{853\,mm}$.

$\therefore \frac{A\bar{y}}{I} = \frac{20000 \times 853}{31.2 \times 10^9} = \frac{0.55 \times 10^{-3}}{(mm^{-1})}$

(b) composite beam

From Fig. 1/012, $S_t = A_e . d_c$

For slab, coefft = S_t / I_t

For the steel flange, coefft $= \frac{A_s}{I_t}\left(d_c - \frac{230}{2} - \frac{40}{2}\right)$

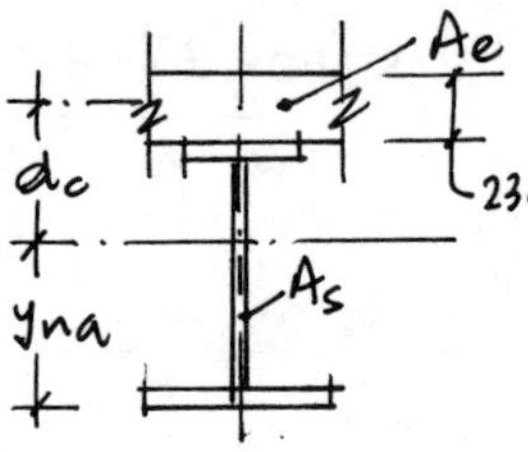

Fig. 1/012

	Short (m = 6.6)	Long (m = 13.2)
$d_c = 1600 - \frac{230}{2} - y_{na}$	1446mm	315mm
$I_t =$	70.44×10^9	59.5×10^9
$S_t = A_e \times d_c =$	15.24×10^6	16.45×10^6
$d_c - \frac{230}{2} - \frac{40}{2} =$	11 mm	180 mm.
$S_t / I_t =$	0.220×10^{-3}	0.280×10^{-3}
$\frac{A_s}{I_t}\left(d_c - \frac{230}{2} - \frac{40}{2}\right) =$	0.003×10^{-3}	0.006×10^{-3}
Summing =	0.223×10^{-3}	0.286×10^{-3}

Shear flows, using U.L.S. shears calculated previously :–

Steel beam $= 92 \times 10^3 \times 0.55 \times 10^{-3} = 50.6\,N/mm$

Wet concrete $= 334 \times 10^3 \times 0.55 \times 10^{-3} = 183.7\,N/mm$.

Super dead (long) $= 220 \times 10^3 \times 0.286 \times 10^{-3} = 62.9\,N/mm$.

Live load (short) $= 787 \times 10^3 \times 0.223 \times 10^{-3} = 175.5\,N/mm$

$\Sigma = 472.7\,N/mm$

Cl. 14.6.3.4 Using simple method of assessing fillet weld capacity: $\tau_D = \frac{k\,(\sigma_y + 455)}{\gamma_m . \gamma_{f3} . 2\sqrt{3}}$

For side welds, $k = 0.9$: $\tau_D = \frac{0.9\,(355 + 455)}{1.1 \times 1.1 \times 2\sqrt{3}} = 174\,N/mm^2$

min. throat thickness (2 welds) $= \frac{472.7 \times 0.5}{174} = 1.35\,mm$

min. leg length $= 1.35 / 0.707 = 1.91\,mm$.

$\therefore$ 6mm fillet welds O.K.

Ref.	
BS5400 (Pt 3) Cl. 5/6.2.3 (DTp version)	**Deck slab (U.L.S)**

Strength of reinforcement $= \frac{0.87 \times 460}{1.1\ (\gamma_{f3})} = 364\ \text{N/mm}^2$

Try slab reinforcement as in Fig. 1/013.

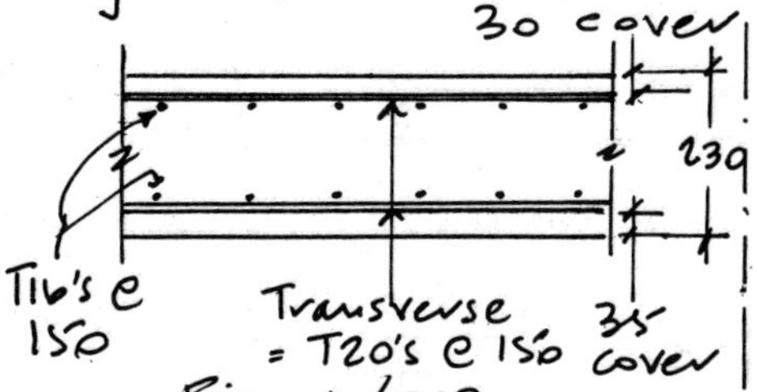

Fig. 1/013

($f_{cu} = 40\ \text{N/mm}^2$).

Using bending moment coefficients from Ref 1/5, using the Load Case loadings as in Fig. 1/014 & Fig. 1/015

L.C. I

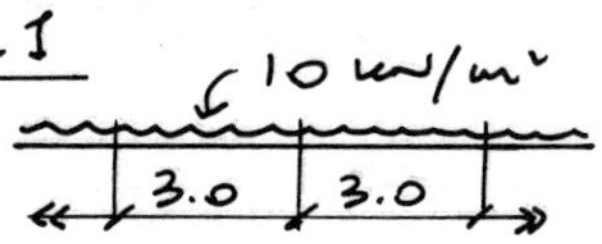

Fig. 1/014

L.C. II (see brief)

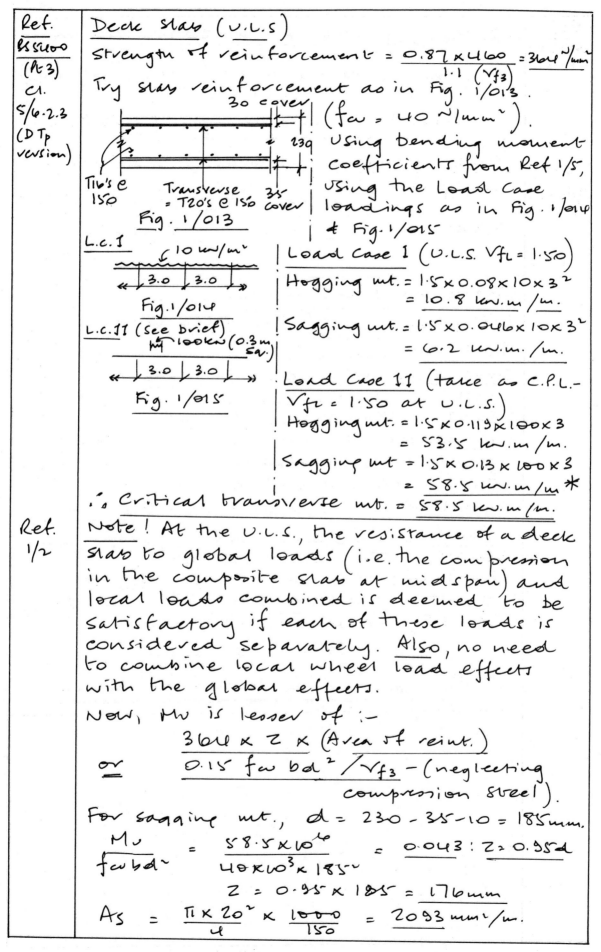

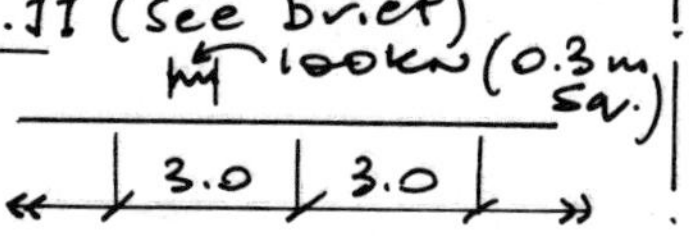

Fig. 1/015

Load Case I (U.L.S. $\gamma_{fL} = 1.50$)

Hogging mt. $= 1.5 \times 0.08 \times 10 \times 3^2$
$= 10.8$ kN.m/m.

Sagging mt. $= 1.5 \times 0.046 \times 10 \times 3^2$
$= 6.2$ kN.m./m.

Load Case II (take as C.P.L. – $\gamma_{fL} = 1.50$ at U.L.S.)

Hogging mt. $= 1.5 \times 0.119 \times 100 \times 3$
$= 53.5$ kN.m/m.

Sagging mt $= 1.5 \times 0.13 \times 100 \times 3$
$= 58.5$ kN.m/m *

∴ Critical transverse mt. $= 58.5$ kN.m/m.

Ref. 1/2	Note! At the U.L.S., the resistance of a deck slab to global loads (i.e. the compression in the composite slab at midspan) and local loads combined is deemed to be satisfactory if each of these loads is considered separately. Also, no need to combine local wheel load effects with the global effects.

Now, M_u is lesser of :–

$364 \times z \times$ (Area of reint.)

or

$0.15\ f_{cu}\ b d^2 / \gamma_{f3}$ – (neglecting compression steel).

For sagging mt., $d = 230 - 35 - 10 = 185$ mm.

$$\frac{M_u}{f_{cu} b d^2} = \frac{58.5 \times 10^6}{40 \times 10^3 \times 185^2} = 0.043 : z = 0.95d$$

$$z = 0.95 \times 185 = 176\ \text{mm}$$

$$A_s = \frac{\pi \times 20^2}{4} \times \frac{1000}{150} = 2093\ \text{mm}^2/\text{m}.$$

Ref. (BS 5400 Pt. 3)

$\therefore M_u\ (\text{steel}) = \dfrac{364 \times 176 \times 2093}{10^6} = 134$ kN.m/m

$M_u\ (\text{conc}) = \dfrac{0.15 \times 40 \times 10^3 \times 185^2}{1.1 \times 10^6} = 187$ kN.m/m

$\therefore M_u = 134$ kN.m/m (> 58.5 ✓ o.k.)

Cl 5/6.3.3.4 — Check minimum transverse reinforcement:

Area $\not< 0.8\, s h_c / f_{ry} = \dfrac{0.8 \times 230}{460} = 0.4$ mm²/m

Actual area $= \dfrac{2 \times 2093}{10^3} = 4.2$ mm²/m ✓ o.k.

Check on beam camber.

Deflections $\left(\dfrac{5wL^4}{384EI}\right)$ are calculated using S.L.S. γ_{fL} values :→

Steel beam; $w = 1.0 \times 5.0 = 5$ kN/m.

$\dfrac{5wL^4}{384EI} = \dfrac{5 \times 5 \times (34650)^4}{284 \times 205 \times 10^3 \times 31.2 \times 10^9} = 14.7$ mm.

Wet concrete; $w = 1.0 \times 16.6 = 16.6$ kN/m.

$\dfrac{5wL^4}{384EI} = \dfrac{5 \times 16.6 \times (34650)^4}{384 \times 205 \times 10^3 \times 31.2 \times 10^9} = 48.8$ mm.

Superdead; $w = 1.2 \times 2.4 \times 3 = 8.64$ kN/m.

$\dfrac{5wL^4}{384EI} = \dfrac{5 \times 8.64 \times (34650)^4}{384 \times 205 \times 10^3 \times 59.5 \times 10^9} = 13.3$ mm.

Live load; $w = 1.2 \times 10 \times 3 = 36$ kN/m.

$\dfrac{5wL^4}{384EI} = \dfrac{5 \times 36 \times (34650)^4}{384 \times 205 \times 10^3 \times 70.44 \times 10^9} = 46.8$ mm.

Total = 123.6 mm.

$\therefore$ Put camber in beams of 150mm.

Chapter 2 Highway roundabout overpass beams

A Highway Authority requires an elevated 4-lane dual carriageway road access, to and from an Industrial Estate, to be built on a level site. The scheme is shown in Figure 2/001, and consists of two ramps (maximum gradient 1:20), together with a flat overpass. The height to the carriageway of the overpass, shown on Figure 2/001, is to give a clearance of 5.1 m. to the underside of the structural members. The structure itself comprises of 16 No. Class 1 M9 beams, with a composite deck slab 250 mm. thick, together with an average of 165 mm. of surfacing. The beams-on-piers scheme has been chosen to minimise disruption to the busy carriageway and roundabout underneath. Pier centres of 21m. are ideal for ramp slopes, and for clearance to the sides of, and in the middle of, the roundabout. The client requires a preliminary design of the bridge beams, and has allowed the following assumptions to be used: (i) Simply supported spans, with continuity for shrinkage and temperature stresses in a final design; (ii) An isolated internal beam to be designed before final analysis is carried out using a finite element plate program; (iii) checks are satisfactory for composite beam flexure in SLS, shear connector design, shear checks in ULS, and beam flexure check in ULS using the initial SLS design results.

Loading

The client requires either (but not both) of:-

(a) Type HA loading.

(b) The load from a line of 32 tonne vehicles, shown on Figure 2/001, which may form queues either entering, or leaving, the Industrial Estate.

Site Conditions

Site investigations revealed that there is 0.5 m. of made ground lying over 2.5 m. of loose sand (N=8) over a substantial depth of dense sand (N=40). It is proposed to use displacement piles, founded in the dense sand, supporting pile caps under the piers for the final design.

Design Code of Practice

The Bridge is to be designed to BS 5400, Part 2 (Loading), Part 4 (Concrete Bridges) and Part 5 (Composite Bridges).

Materials

Concrete in pre-cast beams (Class 1) – fci = 50 N/mm^2, fcu = 60 N/mm^2.

In-situ concrete – fcu = 50 N/mm^2.

Prestressing wires: area = 181mm^2; fpu = 1657 N/mm^2; initial prestress per wire = 210 kN.

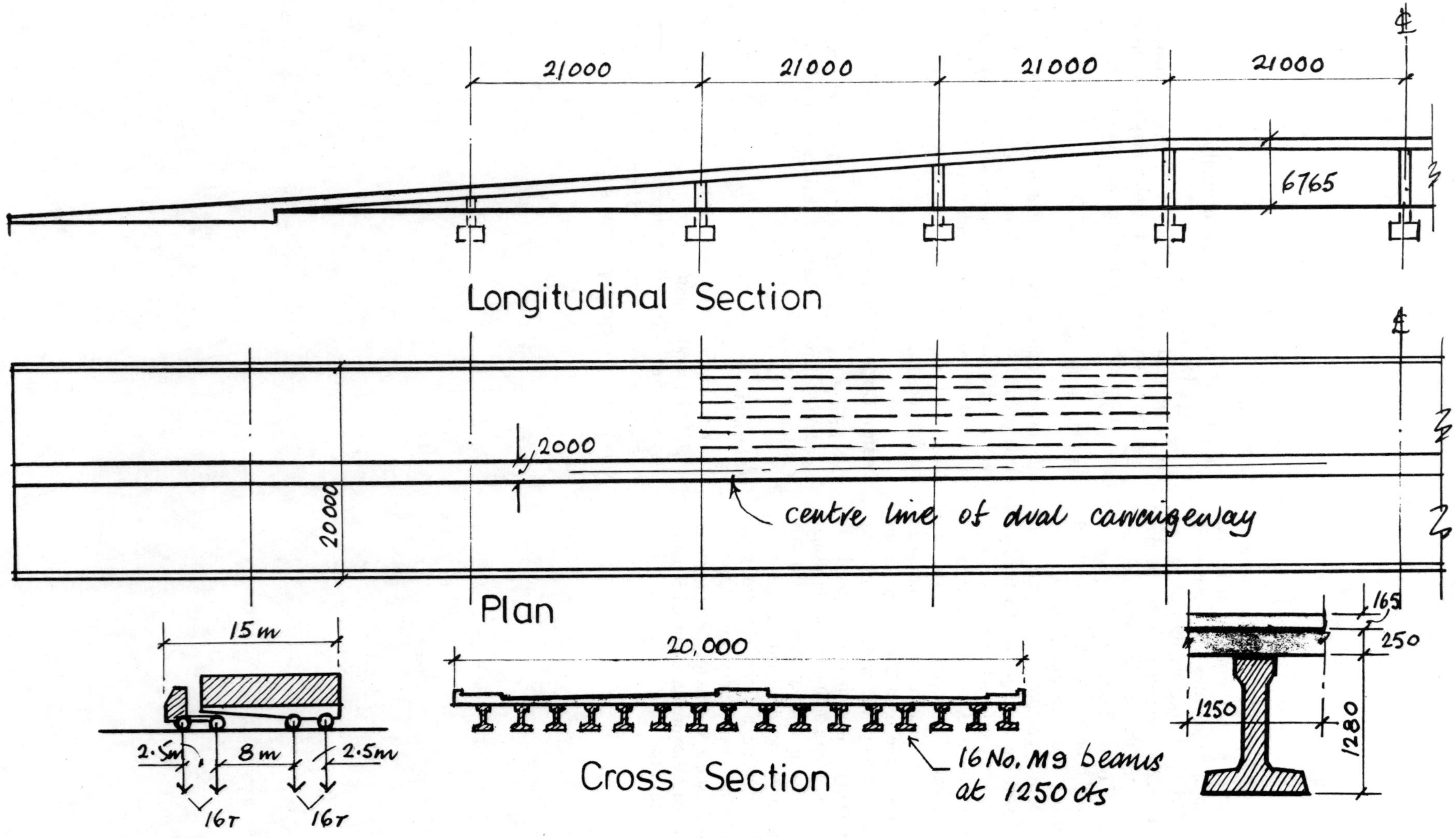
21000
21000
21000
21000
6765
Longitudinal Section
2000
20000
centre line of dual carriageway
Plan
15m
2·5m
8m
2·5m
16T
16T
20,000
Cross Section
16 No. M9 beams at 1250 cts
165
250
1250
1280

Fig. 2/001

Ref.	
(BS 5400 Pt. 4)	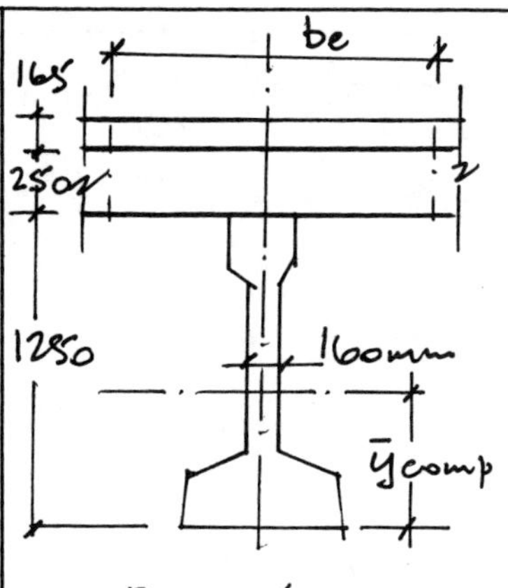 Fig. 2/002 From manufacturer's data: M9 Beam properties are:- Area = 425450 mm^2 self-wt. = 10.04 kN/m. y_b (neutral axis depth from u/s of beam) = 512 mm. $I_{beam} = 8.29 \times 10^{10}$ mm^4 $Z_{bot} = 161.96 \times 10^6$ mm^3 $Z_{top} = 108.09 \times 10^6$ mm^3
	Slab properties: for preliminary design take modular ratio between pre-cast & in-situ concrete = 1.0.
Cl. 5.3.1.2	Effective breadth: for U.L.S = 1250 mm (beam centres). For S.L.S., $be = \frac{span}{10} + web = \frac{21000}{10} + 160$ $be = 2260$ mm ∴ Take be = beam centres = 1250 mm. ∴ From Fig 2/002, Slab area = 250 × 1250 = 312500 mm^2 $I = \frac{1250 \times 250^3}{12} = 1.63 \times 10^9$ mm^4 Composite section (see Fig. 2/002) $A_{comp} = 425450 + 312500 = 737950$ mm^2 $\bar{y}_{comp}$ (see Fig 2/002): taking moments of area about beam soffit:- $\bar{y}_{comp} = \frac{(425450 \times 512) + (312500 \times 1375)}{737950}$ $\bar{y}_{comp} = 877.4$ mm. $I_{comp} = 8.29 \times 10^{10} + 425450(877.4 - 512)^2 + 1.63 \times 10^9 + 312500(1375 - 877.4)^2$ ∴ $I_{comp} = 2.19 \times 10^{11}$ mm^4 ∴ $Z_{b\,comp} = 2.19 \times 10^{11} / 877.4 = 249.6 \times 10^6$ mm^3 $Z_{t\,beam} = 2.19 \times 10^{11} / (1250 - 877.4) = 587.8 \times 10^6$ mm^3 $Z_{t\,slab} = 2.19 \times 10^{11} / (1500 - 877.4) = 351.7 \times 10^6$ mm^3

Ref.	
(BS5400 Pt.4)	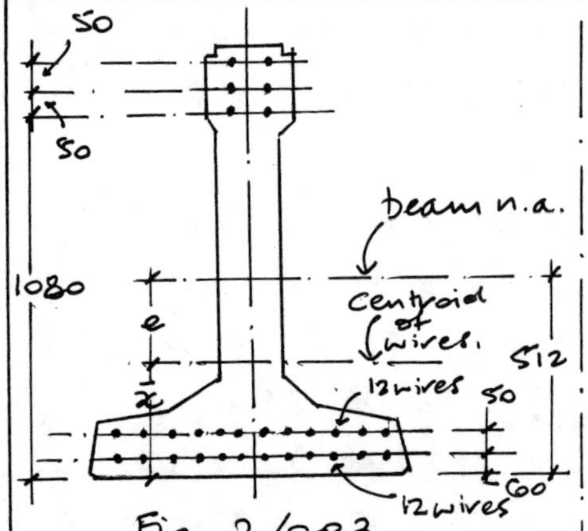 Fig. 2/003 (Tendon layout) **Calculation of initial tendon force, 'P_i' & eccentricity of tendons, 'e' :-** Try 30 No. wires at P_i = 210 kN / wire (see brief) $\therefore P_i = 30 \times 210$ $P_i = 6300$ kN. Taking moments of area about soffit of beam $30\bar{x} = 12 \times 60 + 12 \times 110 + 2(1080 + 1130 + 1180)$ $\therefore \bar{x} = 294$ mm. $\therefore e = y_b - \bar{x} = 512 - 294 = 218$ mm
Cl. 5.3.1.1	**S.L.S. Bending Moment & Shear Force Values.** 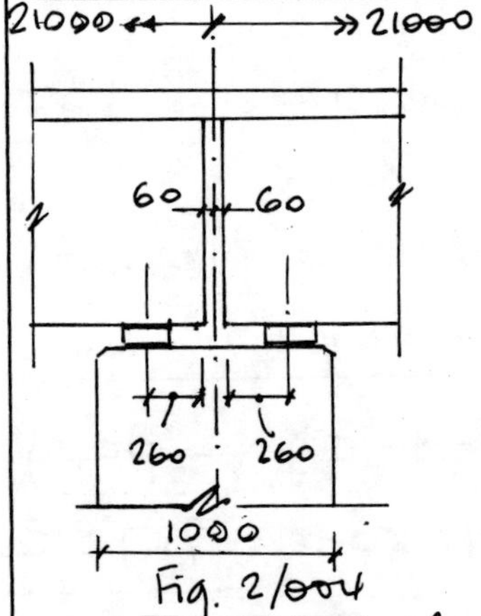 Fig. 2/004 (sketch at top of pier) 'Effective span' is distance between centres of bearings. From Fig. 2/004 :- $L_e = 21000 - 2(260+60)$ $= 20360$ mm or 20.36 m. Length (overall) of beam $= 21000 - 2 \times 60 = 20880$ mm or 20.88 m. Preliminary design - take 1.25 m of slab / beam. (crs.) Precast beam (swt = 10.04 kN/m) :- S.F. (max) $= \frac{10.04 \times 20.88}{2} = 104.8$ kN. ($\gamma_{fL} = 1.0$) B.M. (max) $= \frac{10.04 \times (20.36)^2}{8} = 520.2$ kN.m. Deck slab: U.D.L $= 0.25 \times 24 \times 1.0 \times 1.25 = 7.5$ kN/m S.F. (max) $= \frac{7.5 \times 20.88}{2} = 78.3$ kN. ($\gamma_{fL} = 1.0$) B.M. (max) $= \frac{7.5 \times (20.36)^2}{8} = 388.6$ kN.m. ($\gamma_{fL} = 1.0$)

Ref.	
(BS 5400 Pt. 4)	**Super dead load (surfacing)** ($\gamma = 23$ kN/m³)

($\gamma_{fL} = 1.20$) U.D.L $= 1.25 \times 1.2 \times 0.165 \times 23 = 5.7$ kN/m

S.F. (max) $= \dfrac{5.7 \times 20.88}{2} = 59.4$ kN.

B.M. (max) $= \dfrac{5.7}{8} \times (20.36)^2 = 295.3$ kN.m.

Live load

a) HA acting alone. $\gamma_f = 1.20$

U.D.L = 30 kN/m. + K.E.L. of 120 kN per lane, anywhere on span.

Lane width, 3.65 m: K.E.L/m. width

$= 120/3.65 = 32.9$ kN/m.

S.F. max. (K.E.L. at support)

$= 1.25 \times 1.2 \times 32.9 + 1.25 \times 1.2 \times 30 \times \dfrac{20.88}{2}$

$= 519.15$ kN.

B.M. max. (K.E.L. at midspan)

$= \dfrac{1.25 \times 1.2 \times 32.9 \times 20.36}{4} + \dfrac{1.25 \times 1.2 \times 30 \times (20.36)^2}{8}$

$= 2427.5$ kN.

b) HB loading acting alone. $\gamma_f = 1.10$.

32 tonne vehicle, load/wheel on beam

$= 32/8 = 4$ tonne $= 40$ kN. ($\times \gamma_f$)

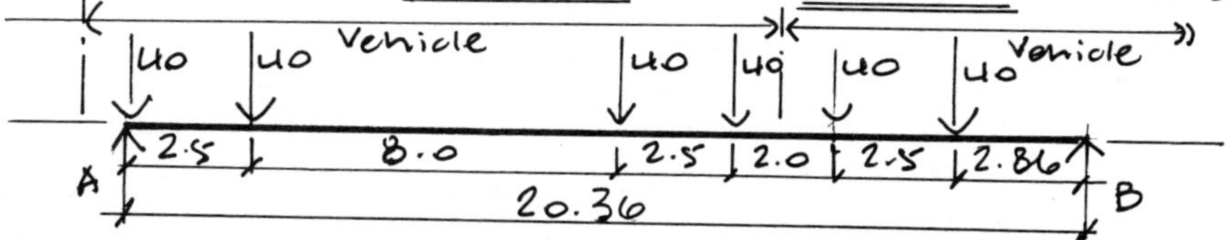

Fig. 2/005 Positions of vehicles for max. shear (loads in kN's).

$R_A \times 20.36 = 40(20.36 + 17.86 + 9.86 + 7.36 + 5.36 + 2.86) \times 1.10$

$R_A = 137.6$ kN (< HA + K.E.L.)

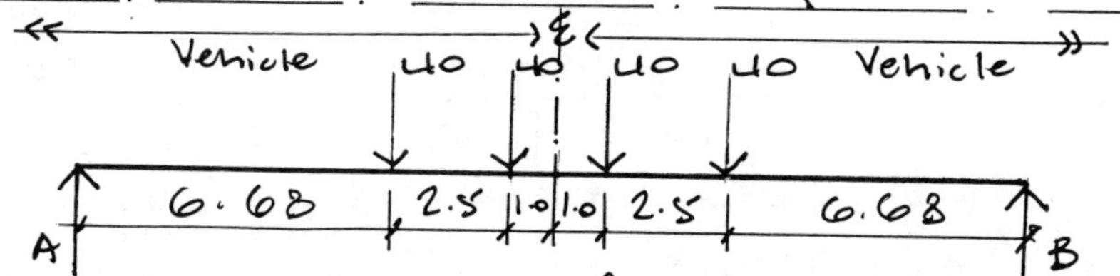

Fig 2/006 Positions for max. B.M.

Ref. (BS 5400 Pt 4)

$R_A = R_B = (4 \times 40)/2 = \underline{80\,\text{kN}}$

$M_{MAX} = (80 \times 10.18 - 40 \times 3.5 - 40 \times 1.0) \times 1.10$

$M_{MAX} = \underline{697.8}\ \text{kN.m.}\ (< \text{H.A.} + \text{K.E.L.})$

∴ HA + K.E.L. Shears & moments critical.

S.L.S. B.M.'s :-

At transfer: M_1 = beam only = $\underline{520.2\ \text{kN.m.}}$

Working I : M_2 = beam + wet slab

$= 520.2 + 388.6 = \underline{908.8}\ \text{kN.m.}$

Working II : M_3 = S.D.L. + Live load

$= 295.3 + 2427.5 = \underline{2723}\ \text{kN.m}$

Check transfer stresses: Class 1 beam −

Cl. 6.3.2.4 (b) & Cl. 6.3.2.2 (b) & Table 23

$f'_{min} = -1.0$ & $f'_{max} = 0.5 f_{ci} = 25\ \text{N/mm}^2$ or $0.4 f_{cu} = \underline{24\ \text{N/mm}^2}$, whichever is least.

Assuming 5% losses, $\alpha = \underline{0.95}$

Top fibre stress :-

$$+\alpha P_i/A - \alpha P_i.e/Z_t + M_1/Z_t > -1.0$$

$$\frac{0.95 \times 6300 \times 10^3}{425450} - \frac{0.95 \times 6300 \times 10^3 \times 218}{108.09 \times 10^6} + \frac{520.2 \times 10^6}{108.09 \times 10^6}$$

$= +14.07 - 12.07 + 4.81 = \underline{6.81} \not< -1.0$ ✓ o.k.

Bottom fibre stress :-

$$+\alpha P_i/A + \alpha P_i.e/Z_b - M_1/Z_b < 24.0$$

$$\frac{0.95 \times 6300 \times 10^3}{425450} + \frac{0.95 \times 6300 \times 10^3 \times 218}{161.96 \times 10^6} - \frac{520.2 \times 10^6}{161.96 \times 10^6}$$

$= +14.07 + 8.06 - 3.21 = \underline{18.92} < 24$ ✓ o.k.

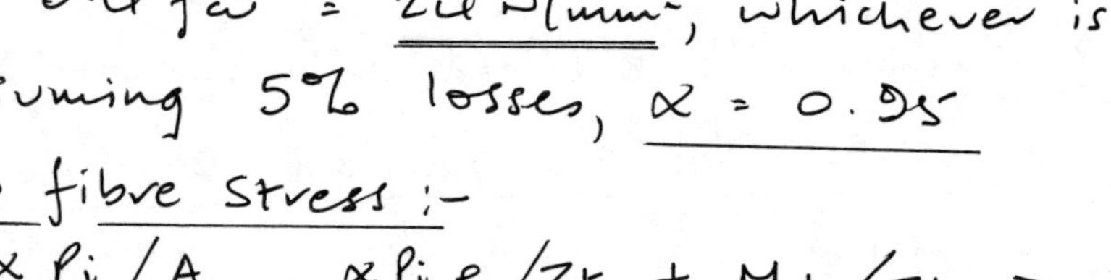

Fig. 2/007 Transfer Stresses.

Check working I stresses at service. Class 1 beam − f_{min} = zero & $f_{max} = 0.4 f_{cu} = 24\ \text{N/mm}^2$

Cl. 6.3.2.4 (a)(1) & Table 22

Assuming 20% losses, $\beta = \underline{0.80}$

Top fibre stress :-

$$+\beta P_i/A - \beta P_i.e/Z_t + M_2/Z_t > 0.$$

Ref. (BS 5400 Pt. 4)

$$+\frac{0.80\times 6300\times 10^3}{425450} - \frac{0.80\times 6300\times 10^3\times 218}{108.09\times 10^6} + \frac{908.8\times 10^6}{108.09\times 10^6}$$

$$+11.85 \quad - 10.16 \quad + 8.41 = \underline{\underline{10.1}} \not< 0.0 \quad \checkmark \text{ o.k.}$$

Bottom fibre stress:

$$+\beta P_i / A + \beta P_i e / Z_b - M_2 / Z_b < 24$$

$$+\frac{0.80\times 6300\times 10^3}{425450} + \frac{0.80\times 6300\times 10^3\times 218}{161.96\times 10^6} - \frac{908.8\times 10^6}{161.96\times 10^6}$$

$$+11.85 \quad + 6.79 \quad - 5.61 = \underline{\underline{13.03}} < 24 \quad \checkmark \text{ o.k.}$$

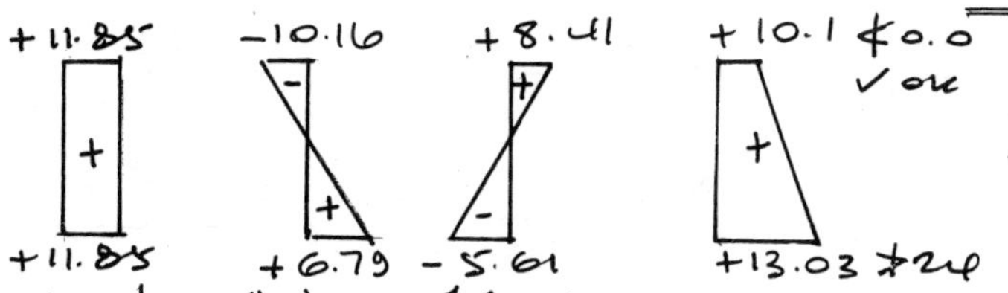

Fig. 2/008 Working 1 Stresses.

Check Working II stresses at service.

Cl. 7.4.1 — As f_{cu} for precast beam ≯ f_{cu} for in-situ slab by more than 10 N/mm², then limits as for working 1. (f_{cu} precast = 60 N/mm², f_{cu} in-situ = 50 N/mm²). Add to working 1

Composite section stresses:–

stresses in slab, top fibre $= \dfrac{2723\times 10^6}{351.7\times 10^6} = \underline{7.74}$ N/mm²

stress in beam, top fibre $= \dfrac{2723\times 10^6}{587.8\times 10^6} = \underline{4.63}$ N/mm²

stress in beam, bottom f. $= \dfrac{-2723\times 10^6}{249.6\times 10^6} = \underline{-10.91}$ N/mm²

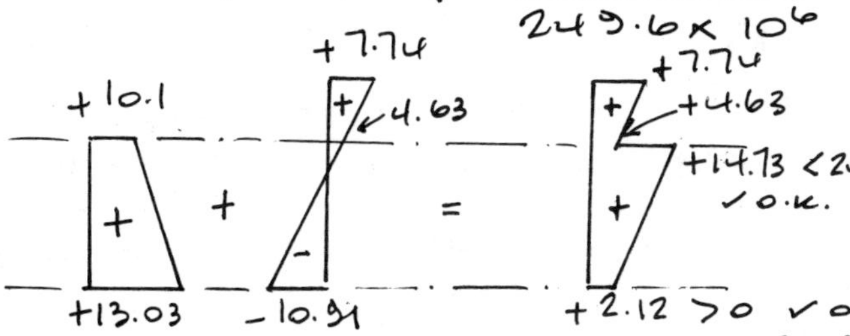

Fig. 2/009 Final Stresses.

∴ Beam & slab o.k. in S.L.S. of flexure.

Check that beam bearings of 1000 kN (100 tonne) S.L.S. capacity are o.k.

S.F. values (previous calcs.)

Beam only	=	104.8 kN.
Deck slab	=	78.3 kN.
Superdead	=	59.4 kN.
Live load	=	519.15 kN.
Total	=	761.65 < 1000 kN ✓ o.k.

Bearing satisfactory

Ref.	
(BS5400 Pt.4) Cl.7.4.2.2 (a) (1)	Check shear resistance of beam.
	Precast beam takes shear (Cl. 7.4.2.2 (a)(1))
	Reactions are at U.L.S.: factors are $\gamma_{fL} \times \gamma_{f3}$ (see Cl. 4.2.3), where $\gamma_{f3} = 1.1$
	Shear forces at U.L.S.:–
	Beam only $(\gamma_{fL} = 1.15)$ $= 104.8 \times \frac{1.15 \times 1.1}{1.0} = 132.6$ kN.
	Deck slabs $(\gamma_{fL} = 1.15)$ $= 78.3 \times \frac{1.15 \times 1.1}{1.0} = 99.0$ kN.
	Superdead $(\gamma_{fL} = 1.75)$ $= 59.4 \times \frac{1.75 \times 1.1}{1.20} = 95.3$ kN.
	Live load $(\gamma_{fL} = 1.50)$ $= 519.15 \times \frac{1.50 \times 1.1}{1.20} = 713.8$ kN.
	Total $= 1041$ kN.
Cl. 6.3.4.1	Design check – V_c to be considered for both uncracked & cracked in flexure sections.
Cl 6.3.4.2	Uncracked: $V_{co} = 0.67bh\sqrt{(f_t^2 + f_{cp} \cdot f_t)}$ – Eq. 28
	where $f_t = 0.24\sqrt{f_{cu}} = 0.24 \times \sqrt{60} = 1.86\ \text{N/mm}^2$
Cl. 4.2.3	$f_{cp} = \frac{0.8 \times 6300 \times 10^3 \times 0.87}{425450} = 10.3\ \text{N/mm}^2$ $(\gamma_{fL} = 0.87)$
	$b = b_w = 160$ mm (see Fig. 2/002)
	& h = overall depth = 1500 mm (see Fig. 2/002)
	$\therefore V_{co} = \frac{0.67 \times 160 \times 1500}{10^3}\sqrt{1.86^2 + 10.3 \times 1.86}$
	$V_{co} = 764.7$ kN.
Cl. 6.3.4.3 Eq. 29	Cracked: for class 1 beams,
	$V_{cr} = 0.037bd\sqrt{f_{cu}} + \frac{M_{cr}}{M} \cdot V \nless 0.1bd\sqrt{f_{cu}}$
	where $b = b_w = 160$ mm (see Fig. 2/002)
	$d = 1250 - \bar{x} = 1250 - 220.3 = 1030$ mm (see Fig. 2/003)
	$M_{cr} = (0.37\sqrt{f_{cu}} + f_{pt})\, I/y$
	in which, $f_{pt} = \beta P_i / A + (\beta P_i . e . y / I)$
	$f_{pt} = \left[\frac{0.8 \times 6300 \times 10^3}{425450} + \frac{0.8 \times 6300 \times 10^3 \times 291.7 \times 512}{8.29 \times 10^{10}}\right] 0.87$
	$f_{pt} = 18.2\ \text{N/mm}^2$: $M_{cr} = \frac{(0.37\sqrt{60} + 18.2)\, 8.29 \times 10^{10}}{512 \times 10^6} = 3411$ kN.m

Try cracked check at beam quarter point – calculation of M & V at 1/4 pt.

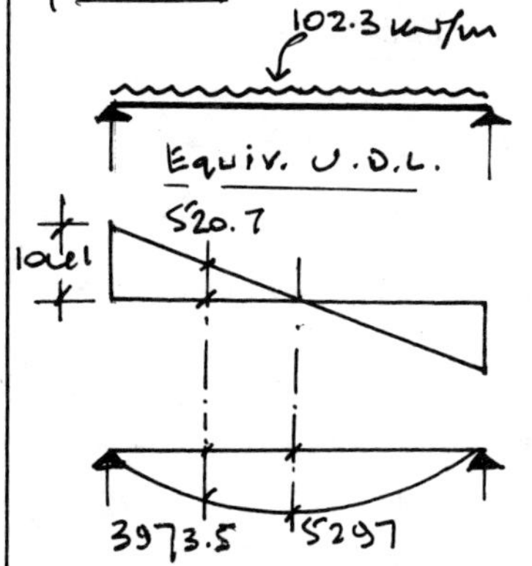

Fig. 2/010
Moments & Shears at U.L.S.

At U.L.S., end reaction = 1041 kN (see previous page)

∴ Total beam load $= 1041 \times 2 = 2082$ kN.

∴ U.D.L $= \frac{2082}{20.36} = \underline{102.3}$ /m.

∴ U.L.S. moment at midspan

$= 1041 \times 10.18 - 102.3 \times \frac{10.18^2}{2} = \underline{5297}$ kNm

U.L.S. moment at 1/4 pt

$= 1041 \times 5.09 - 102.3 \times \frac{5.09^2}{2} = 3973.5$ kN.m.

U.L.S. Shear at 1/4 pt. = 520.7 kN.

∴ $M = 3973.5$ kN.m : $V = 520.7$ kN.

∴ $V_{cr} = 0.037 \left(\frac{\times 160 \times 1030\sqrt{60}}{10^3} + \frac{3411 \times 520.7}{3973.5}\right)$

$\underline{V_{cr} = 488.7 \text{ kN.}}$

But $V_{cr} \not< \frac{0.1 \times 160 \times 1030\sqrt{60}}{10^3} = 127.6$ kN. ✓ ok

∴ $V_c = V_{cr} = 488.7$ kN. at 1/4 pt

& $V_c = V_{co} = 764.7$ kN. at end.

In both cases, $V > V_c$ (reinf. reqd.)

Cl. 6.3.4.4. For 'M' beams – use T10 links : $A_{sv} = 157$ mm² for both legs: at supports :–

∴ $S_v = \frac{A_{sv} \times f_{yv} \times d_t}{V + 0.4\, b\, d_t - V_c}$

∴ $S_v = \frac{157 \times 460 \times (1500 - 291.7)}{1041 \times 10^3 + 0.4 \times 160 \times (1500 - 291.7) - 764.7 \times 10^3}$

$\underline{S_v = 247 \text{ mm.}}$

∴ Use T10 links @ 200 mm crs. throughout.

Cl. 7.4.2.3

Fig 2/011

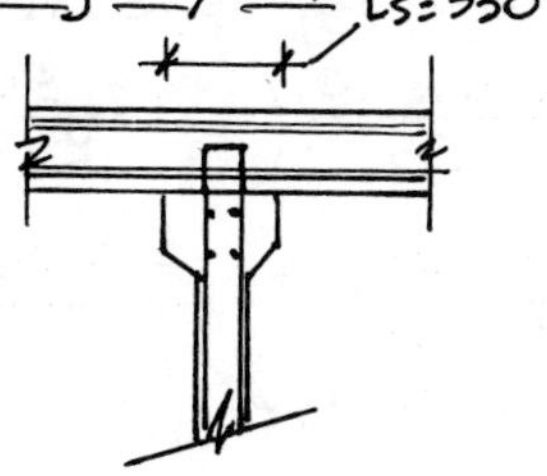

Check horizontal shear at precast / in-situ interface.

At midspan, total force in slab $= 0.45 \times f_{cu} \times b_e \times h_{slab}$

$= 0.45 \times 50 \times 1250 \times 250 / 10^3$

$= \underline{7031}$ kN.

Ref.	Calculations
(BS5400 Pt4)	V_1 = Horizontal shear / (span/2) $= \dfrac{7031 \times 10^3}{(20.62/2 \times 10^3)} = \underline{682}$ N/mm.
	V_1 should not exceed lesser of :-
Table 31	(a) $k_1 . f_{cu} . L_s$: where $k_1 = 0.15$ for Type 1 surface, L_s = 330 mm, as shown in Fig 2/011
	$\therefore k_1 . f_{cu} . L_s = 0.15 \times 50 \times 330 = \underline{2475}$ N/mm
	or (b) $v_1 \times L_s + 0.7 \times A_e \times f_y$; where A_e = area of links /mm $= 157 \times \dfrac{10^3}{200} \times \dfrac{1}{10^3} = \underline{0.785}$ mm²/m.
Table 31	For Type 1 surface, $v_1 = 0.8$ (Table 31);
	$\therefore v_1 L_s + 0.7 A_e f_y = 0.8 \times 330 + 0.7 \times 0.785 \times 460 = \underline{517}$ N/mm.
	$\therefore$ (b) < V_1 <u>NOT</u> <u>OK</u>.
	Decrease spacing of links towards support to 100 c/s. $A_e = 157 \times \dfrac{10^3}{100} \times \dfrac{1}{10^3} = \underline{1.57}$ mm²/m.
	$\therefore$ (b) $= 0.8 \times 330 + 0.7 \times 1.57 \times 460 = \underline{769}$ N/mm ✓ <u>O.K.</u>

<u>Check bending in U.L.S. using strain compatibility method</u>

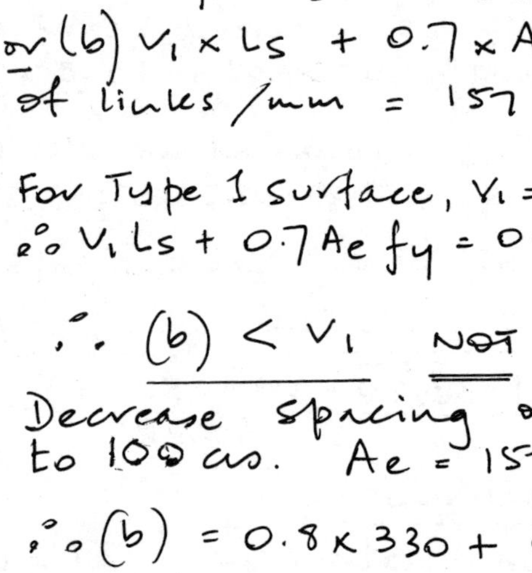

Fig. 2/012

Assume depth of N.A. at U.L.S. = 170 mm. from top of composite section.

Cl. 9.3.3.1 — Design to cl. 9.3.3.1.

<u>Step 1</u> Calculate the strain, ε_u, at tendon centroid from Fig. 2/012.

Depth of tendon centroid = 1500 − 294 = $\underline{1206}$ mm

By similar triangles: $\dfrac{\varepsilon_u}{1036} = \dfrac{0.0035}{170}$

$\underline{\varepsilon_u = 0.0213}$

<u>Step 2</u> Calculate the strain in the concrete at tendon level, ε_e, caused by prestress

$$f = \frac{\beta P_i}{A}\left(1 + \frac{Ae^2}{I}\right) = \frac{0.8 \times 6300 \times 10^3}{425450}\left(1 + \frac{425450 \times 218^2}{8.29 \times 10^{10}}\right)$$

$f = \underline{14.73 \text{ N/mm}^2}$: $\varepsilon_e = \dfrac{14.73}{15.5 \times 10^3} = \underline{0.00095}$

<u>Step 3</u> Total strain, $\varepsilon_a = \varepsilon_u + \varepsilon_e$

$\varepsilon_a = 0.0213 + 0.00095 = \underline{0.0223}$

Ref.	
(BS5400 Pt. 4)	**Step 4.** Calculate the strain in the tendons, ε_{pe}, after losses.

$$f_{pe} = \frac{\beta . P_i \times 10^3}{(\gamma_m \times A_{ps} \times \text{No. of wires})} = \frac{0.8 \times 6300 \times 10^3}{1.15 \times 181 \times 30}$$

(A_{ps} = area of wire – see brief)

$\therefore f_{pe} = 807.1$ N/mm² : $\varepsilon_{pe} = \dfrac{807.1}{205 \times 10^3} = 0.0039$

Step 5. Calculate total strain in the tendons,

$\varepsilon_{pb} = \varepsilon_a + \varepsilon_{pe} = 0.0223 + 0.0039 = 0.0262$

Step 6 From BS 5400, Pt 4, Fig 1, calculate stress in tendons, f_{pb}, at Ultimate. (Ref.: Fig. 1)

From brief, $f_{pu} = 1657$ N/mm² : $\gamma_m = 1.15$ (Cl. 9.3.3.1)

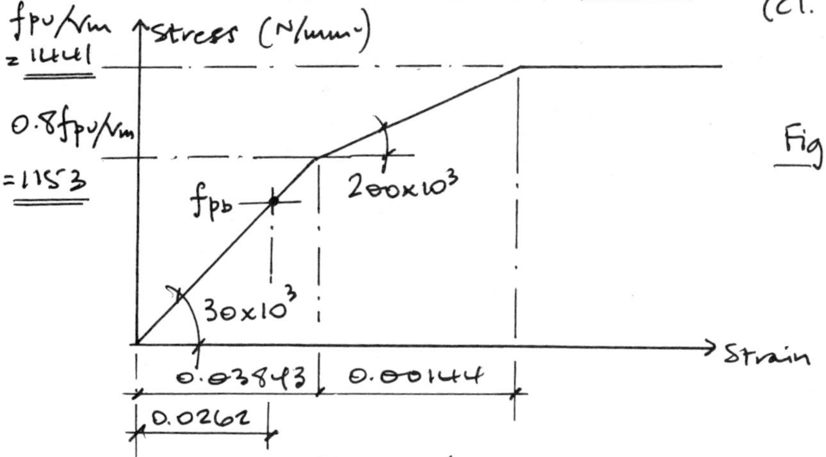

Fig. 2/013

Stress/strain diagram.

From calculation of the remaining strain,
$\varepsilon = 0.0262$, $f_{pb} = 0.0262 \times 30 \times 10^3 = 786$ N/mm².

Step 7 $T = f_{pb} \times A_{ps} \times$ (No. of wires)

$T = 786 \times 181 \times 30 / 10^3 = 4268$ kN.

Step 8 C = compression block

$= \dfrac{0.45 \times 50}{10^3}(170 \times 1250)$

$= 4781$ kN.

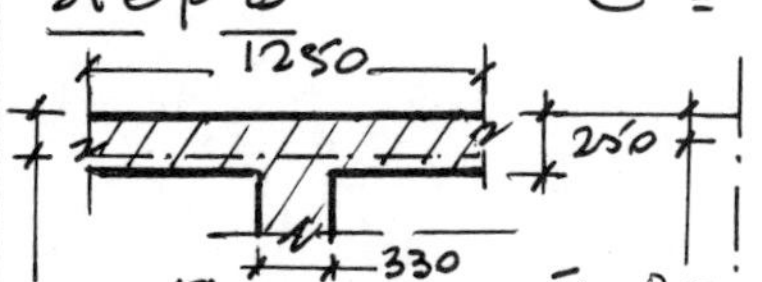

170 (x) Fig. 2/014 $\bar{y} = 85$

Difference between 'T' & 'C' = 10 % ✓ o.k.

Step 9. Lever arm = 1500 – 294 – 85
= 1121 mm (see Fig. 2/014)

so $M_u = 4268 \times 1121 / 10^3 = 4784$ kN.m

Ref.	Step 10 – Calculate Bending Moment at U.L.S.
	Taking B.M. values at S.L.S. & applying partial load factors, we have :–
	Beam : $M_{ULT} = 520.2 \times \frac{1.15}{1.0} = 598.23$
	Deck : $M_{ULT} = 388.6 \times \frac{1.15}{1.0} = 446.89$
	Superdead : $M_{ULT} = 295.3 \times \frac{1.75}{1.20} = 430.64$
	Live : $M_{ULT} = 2427.5 \times \frac{1.50}{1.20} = 3034.37$
	Total = 4510.13
	∴ $M_U > M_{ULT}$ ✓ O.K. (4784) (4510.13)
	Composite beam O.K.

Chapter 3 Factory for the production and storage of polythene

A multi-national polythene manufacturer requires a large, 60 m. square, single-storey factory, on a 'green-field' site, as shown in Fig. 3/001. The client's Architect has stipulated that one central column only is to be allowed in the factory, which is clad with profiled steel sheet sandwich panels (50% roof lighting, no windows in the sides of the factory). Apart from small personnel doors, Fig. 3/001 also shows the requirement for one large sliding door opening, 5 m. high, and a minimum clearance inside the building of 6m. Only the major structural members are offered as calculations in this example, and only selected important connections are designed. Also, several checks normally included in a full design, such as secondary bending and local buckling of the truss member walls, have been omitted.

Loading – live qk

Roof loading (offices and factory)	– 0.75 kN/m^2
Factory roof services	– 0.65 kN/m^2

Loading – wind wk

Basic wind speed 46m/sec (outskirts of a town – level site)
Note! In this example not all load cases are considered, and wind uplift on the roof trusses has been ignored.

Soil safe bearing capacity

Stiff clay, safe bearing 250 kN/m^2

Design Code of Practice

The building is to be designed in accordance with BS 5950, Structural Steelwork

Other Design Aids

The roof purlins have been taken from the manufacturer's design tables without publication in this book.
Roof purlins – from the 'Zeta' overlap tables, 20016 'Zeta' purlins at 3.0 m. centres will span 6 m. between roof trusses, and will carry an unfactored total load of 1.05 kN/m^2 (cladding + snow).

Materials

All structural steelwork to be EN10025 – S275.

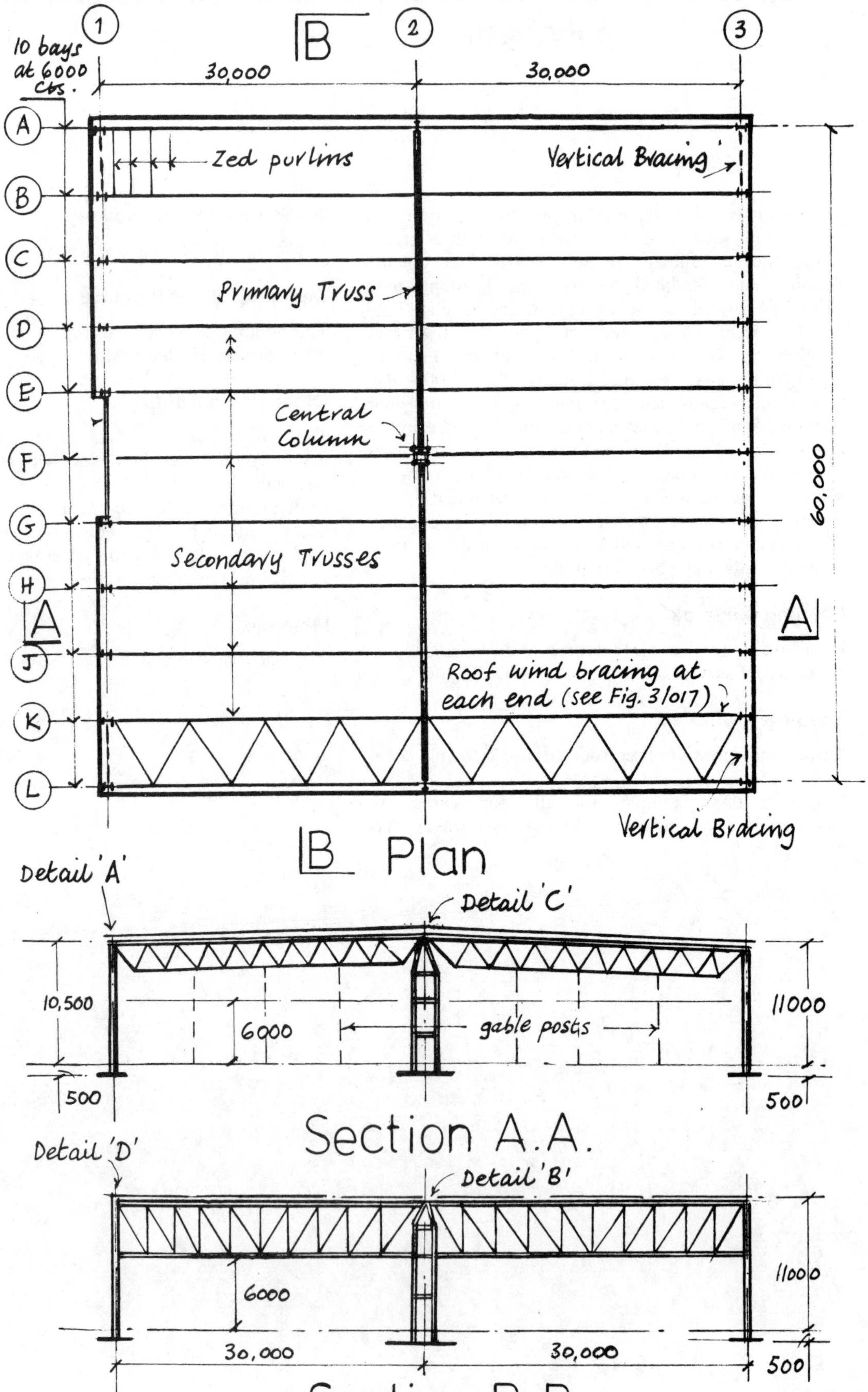
1
2
3
B
10 bays at 6000 Ctrs.
30,000
30,000
A
B
C
D
E
F
G
H
J
K
L
Zed purlins
Vertical Bracing
Primary Truss
Central Column
Secondary Trusses
60,000
A
A
Roof wind bracing at each end (see Fig. 3/017)
Vertical Bracing
B Plan
Detail 'A'
Detail 'C'
10,500
6000
gable posts
11000
500
500
Section A.A.
Detail 'D'
Detail 'B'
6000
11000
30,000
30,000
500
Section B.B.

Fig. 3/001

Ref.	
All BS5950 U.N.O.	**Step 1 Design of secondary roof trusses:** Span 30 m. : trusses at 6 m. centres. One load case only :- 1.6 qk + 1.4 gk. (wind uplift ignored)
	Design loading :
	q_k = 0.75 kN/m² (snow) + 0.65 kN/m² (services) = 1.4 kN/m²
	g_k = 0.30 kN/m² (cladding) + 0.20 kN/m² (steel) = 0.5 kN/m²
Table 2	∴ Design loading = 1.6 × 1.4 + 1.4 × 0.5 = 2.94 kN/m² (say) 3.0 kN/m²
	Proposed layout (all S.H.S.) Fig. 3/002

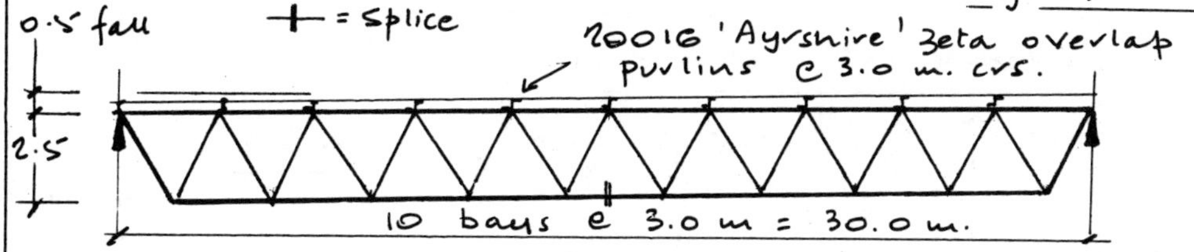

U.D.L. on truss = 3.0 × 6 = 18.0 kN/m.

Purlin point load = 18.0 × 3 = 54 kN.

Reaction, R = 18 × 30 / 2 = 270 kN.

Fast analysis :-

a) at end support :-

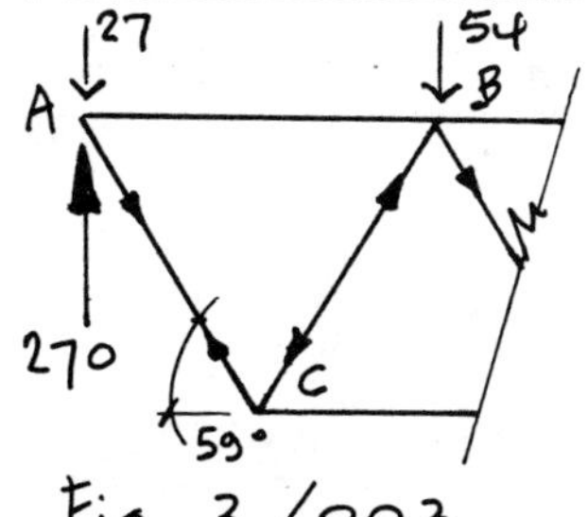

Fig. 3/003

At A: F_{AC} = max. internal tensile force.

$F_{AC} \sin 59° + 27 = 270$

F_{AC} = 282.5 kN.

At C: F_{BC} = max. internal compressive force.

$F_{AC} \sin 59° = F_{CB} \sin 59°$

∴ F_{BC} = 282.5 kN.

b) at centre-line :-

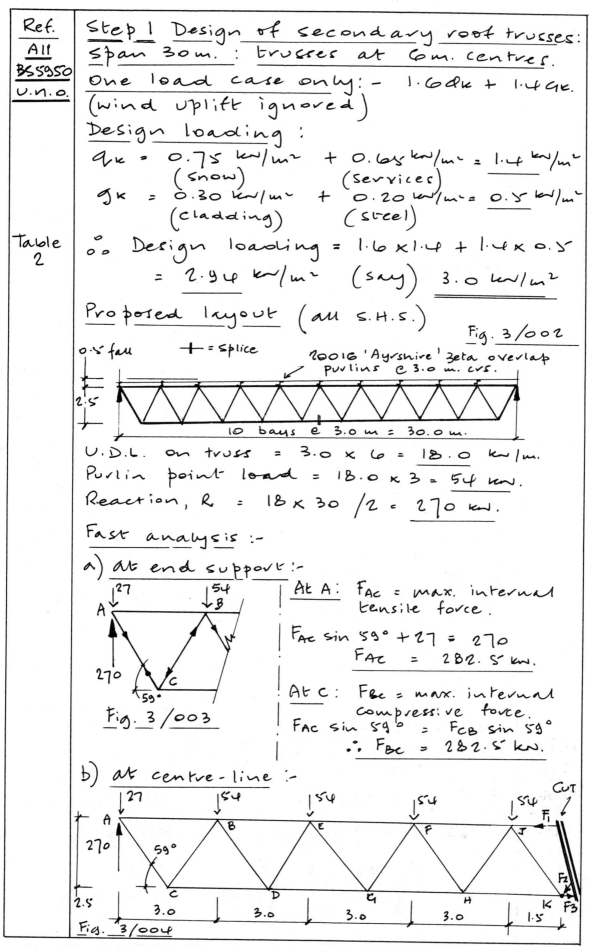

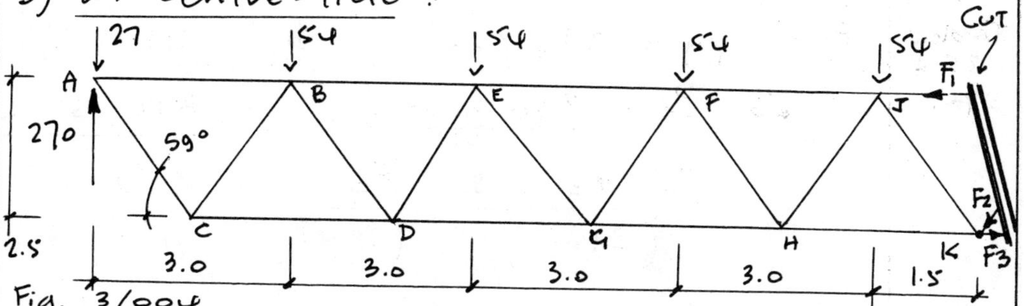

Fig. 3/004

Ref.	
	From Fig. 3/004 :-
	Taking moments about K.
	$F_1 \times 2.5 + 27 \times 13.5 + 54[10.5 + 7.5 + 4.5 + 1.5]$
	$= 270 \times 13.5$
	$\therefore F_1 = 794$ kN. (comp.)
	Taking moments about C.
	$794 \times 2.5 + 27 \times 1.5 = 270 \times 1.5$
	$+ 54[1.5 + 4.5 + 7.5 + 10.5] + F_2 \times 12 \sin 59°$
	$\therefore F_2 = 31.5$ kN (comp.)
	Taking moments about A.
	$F_3 \times 2.5 = 54(3 + 6 + 9 + 12) + 31.5 \times 13.5 \sin 59°$
	$\therefore F_3 = 794$ kN. (tens.)
	Check at centreline :-
	↑ $2 \times 31.5 \sin 59° = 54$ kN. } o.k.
	↓ 54 kN.
	Member Design:
	Top boom : Max. force = 794 kN. (comp.)
	Try a 120 x 120 x 8.0 S.H.S ($p_y = 275$ N/mm²)
	Check classification :
Table 7	$\frac{b}{T} = \frac{d}{t} = \frac{B - 3T}{T} = \frac{d - 3t}{t} = \frac{120 - 3 \times 8.0}{8.0}$
	$= 12$
	$\varepsilon = \sqrt{\frac{275}{275}} = 1.0 \;;\; 39\varepsilon = 39$
	12 < 39 : section "not slender"
	$L_E = 1.0L = 3000$ mm.
4.7.3. 2(a)	$\lambda = \frac{3000}{45.6} = 66$ (< 180 ✓ o.k.)
Tables 25 & 27(a)	For $p_y = 275$ N/mm², $p_c = 230$ N/mm²
	$\therefore P_c = \frac{230 \times 3550}{10^3} = 816.5$ kN.
	$\therefore P_c > F_c$ ✓ o.k.
	(816.5) (794)

Ref.	
4.6.1	**Bottom boom : Max. force = 794 kN (tens.)**
	Area required $= \frac{794 \times 10^3}{275} = 2887\text{mm}^2$
	Try a 120 × 120 × 8.0 SHS ($A = 3550\text{mm}^2$)
	$P_t = \frac{275 \times 3550}{10^3} = 976 \text{ kN}$.
	$\therefore P_t > F_t$ ✓ o.k.
	(976) (794)
	Compression diagonal BC (see Fig. 3/004)
	Max. force = 282.5 kN.
	Try an 80 × 80 × 5 S.H.S. ($p_y = 275 \text{ N/mm}^2$)
	Check classification :-
Table 7	$\frac{b}{T} = \frac{d}{t} = \frac{80 - 3 \times 5}{5} = 13$
	13 < 39 : Section 'not slender'
	Length = 2913 mm : $L_E = 0.85L$ (both ends welded) $= 0.85 \times 2913 = 2476\text{mm}$.
Table 27(a)	$\lambda = \frac{2476}{30.5} = 81$: $p_c = 201 \text{ N/mm}^2$
	$\therefore P_c = \frac{201 \times 1490}{10^3} = 299.5 \text{ kN}$.
	$\therefore P_c > F_c$ ✓ o.k.
	(299.5) (282.5)
	Tension diagonal AC (see Fig. 3/004)
	Max force = 282.5 kN.
	Try an 80 × 80 × 5 S.H.S ($p_y = 275 \text{ N/mm}^2$)
Cl. 4.6.1	$P_t = \frac{275 \times 1490}{10^3} = 410 \text{ kN}$.
	$\therefore P_t > F_t$ ✓ o.k.
	(410) (282.5)
	Member sizes :-
	Booms - 120 × 120 × 8 S.H.S.
	Diagonals - 80 × 80 × 5 S.H.S.

Ref.

Check central deflection (approx. method)

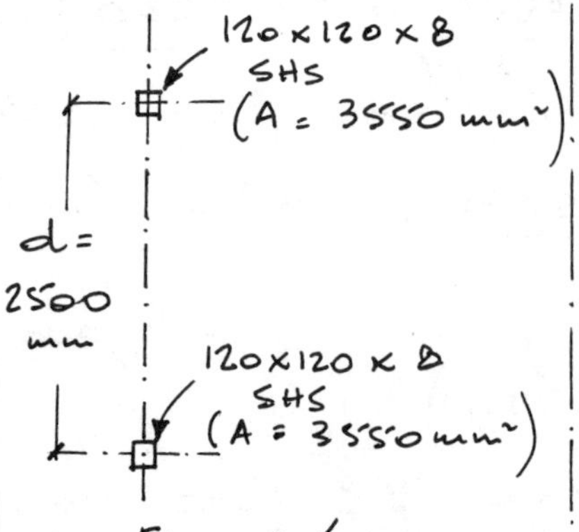

Fig. 3/005

$$I_{approx} = \frac{A_f \times d^2}{2}$$

(A_f = area of boom, d = lever arm)

$$\therefore I = \frac{3550 \times 2500^2}{2}$$

$$I = 11.1 \times 10^9 \text{ mm}^4$$

Service U.D.L. on truss = 1.4 x 6 = 8.4 kN/m.

$$\therefore \Delta = \frac{5}{384} \times \frac{8.4 \times (30000)^4}{205 \times 10^3 \times 11.1 \times 10^9} = 39 \text{ mm}$$

Table 5 — Allowable = 30000 / 360 = 83.3 mm

Allowable (83.3) > Actual (39) ✓ o.k.

Secondary Truss - Connections.

Detail 'A' - see Figs. 3/001 & 3/018

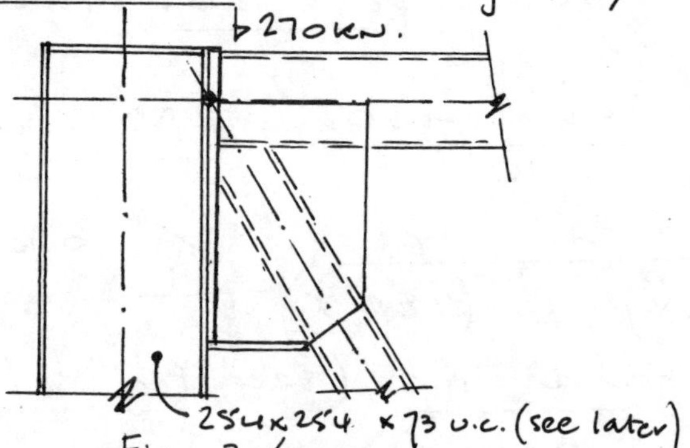

Fig. 3/006

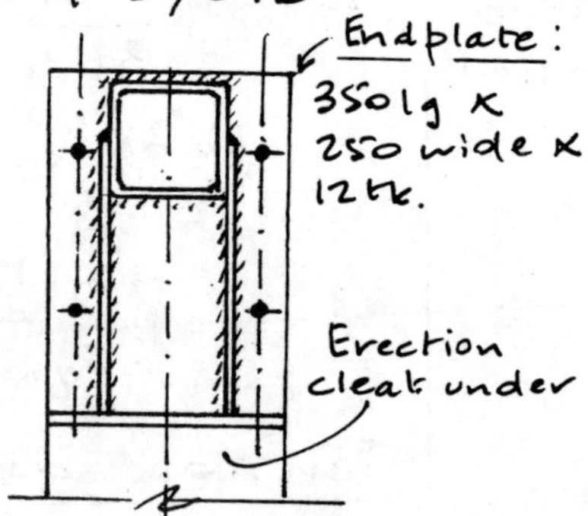

Fig. 3/007 (A-A)

Ref 3/1 — No. of bolts required:

Shear = 270 kN: try M20 8.8 bolt, single shear = 91.9 kN.

4 No. M20 8.8 bolts = 367.6 kN ✓ o.k.

Fillet welds:

Length of weld (see Fig. 3/007) ≏ 800 mm.

Ref 3/1 — For a 6 mm ◺, shear resistance = 0.9 kN/mm

∴ Shear capacity = 0.9 x 800 = 720 kN.

∴ Use 6 mm ◺ welds: 4 No. M20 8.8 bolts.

Ref.	

Site splice – see Fig. 3/002

Splice at centre-line, tension boom, Force = 794 kN.

Ref. 3/1

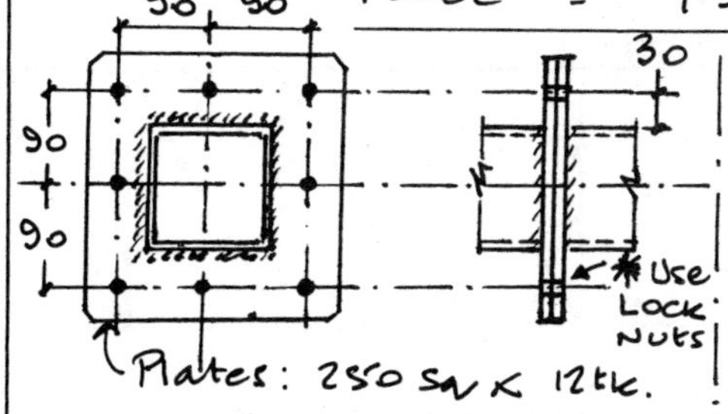

Fig. 3/008.

Bolts: For one M20 8.8 bolt, $P_t = 110$ kN.

8 No. M20 8.8 bolts $= 8 \times 110 = 880$ kN.

$\therefore$ $P_t > F_t$ (880) (794) ✓ o.k.

Check plate prying action.

Moment $= \dfrac{P_t \times L'}{2}$, where L' is the distance between the bolt head & the section.

For M20 bolt, head width = 30 mm.

$\therefore L' = 30 - 30/2 = 15$ mm. (see 3/008)

$\therefore M = \dfrac{110 \times 15}{2 \times 10^3} = 0.82$ kN.m.

Thickness of plate, $t = \sqrt{\dfrac{4M}{B \times p_y}}$, where B is spacing of bolts.

$\therefore t = \sqrt{\dfrac{4 \times 0.82 \times 10^6}{90 \times 275}} = 11.5$ mm.

$\therefore$ Use 12 mm tk. plates.

Typical Internal 'K' joint (at C)

Ref 3/2

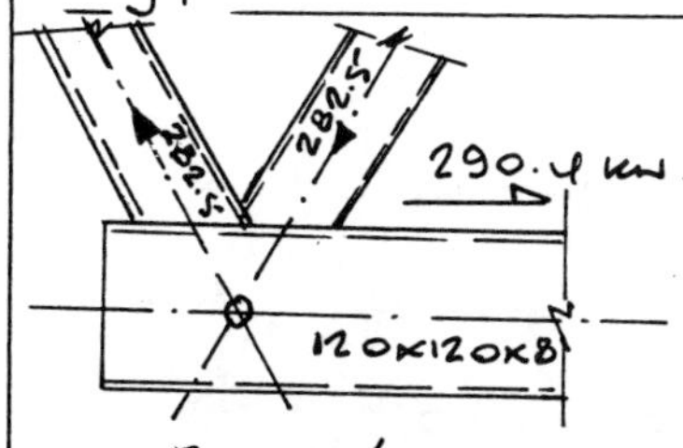

Fig. 3/009

For full design of a 'K' overlap joint, see Ref 3/2 or E.C.3 (European Code for Steel), inc. web buckling

From Fig 3/010, length of weld $= 2 \times 170 + 2 \times 80 = 500$ mm.

Ref. 3/1

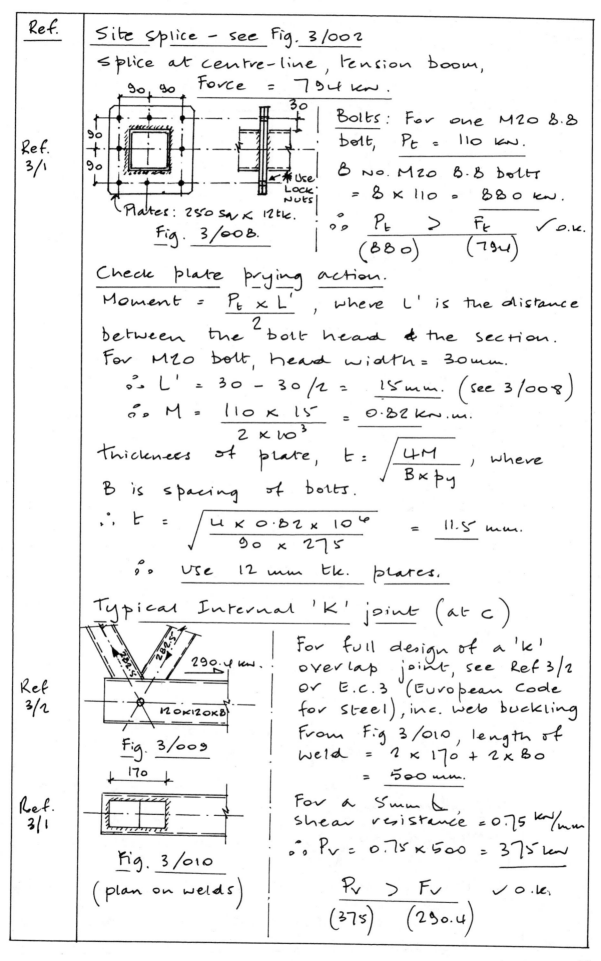

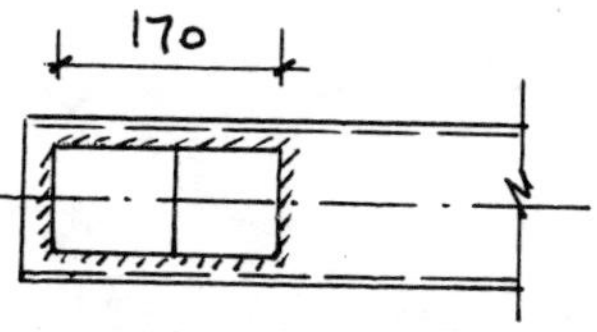

Fig. 3/010 (plan on welds)

For a 5 mm ⊾, shear resistance = 0.75 kN/mm

$\therefore P_v = 0.75 \times 500 = 375$ kN

$P_v > F_v$ (375) (290.4) ✓ o.k.

Ref.

Step 2 Design of primary trusses - span 30m.

One load case only :- $1.6Q_k + 1.4G_k$.

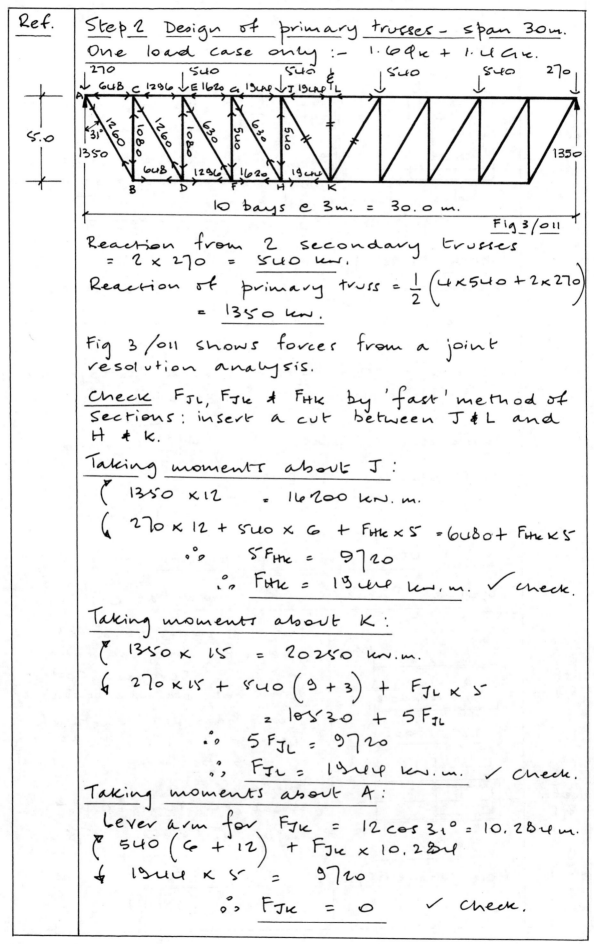

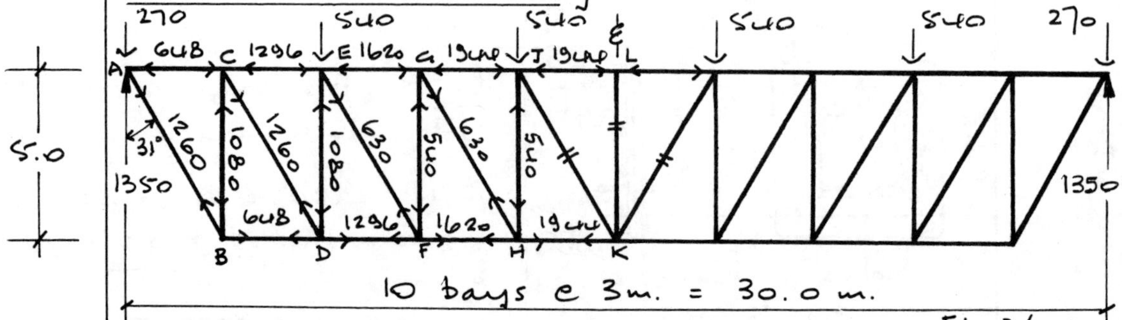

Fig 3/011

Reaction from 2 secondary trusses
= 2 × 270 = 540 kN.

Reaction of primary truss = $\frac{1}{2}(4 \times 540 + 2 \times 270)$
= 1350 kN.

Fig 3/011 shows forces from a joint resolution analysis.

Check F_{JL}, F_{JK} & F_{HK} by 'fast' method of sections: insert a cut between J & L and H & K.

Taking moments about J:

↻ $1350 \times 12 = 16200$ kN.m.

↺ $270 \times 12 + 540 \times 6 + F_{HK} \times 5 = 6480 + F_{HK} \times 5$

∴ $5F_{HK} = 9720$

∴ $F_{HK} = 1944$ kN.m. ✓ check.

Taking moments about K:

↻ $1350 \times 15 = 20250$ kN.m.

↺ $270 \times 15 + 540(9 + 3) + F_{JL} \times 5$

$= 10530 + 5F_{JL}$

∴ $5F_{JL} = 9720$

∴ $F_{JL} = 1944$ kN.m. ✓ check.

Taking moments about A:

Lever arm for $F_{JK} = 12 \cos 31° = 10.284$ m.

↻ $540(6 + 12) + F_{JK} \times 10.284$

↺ $1944 \times 5 = 9720$

∴ $F_{JK} = 0$ ✓ check.

Ref.	Member design (using max. forces only).
	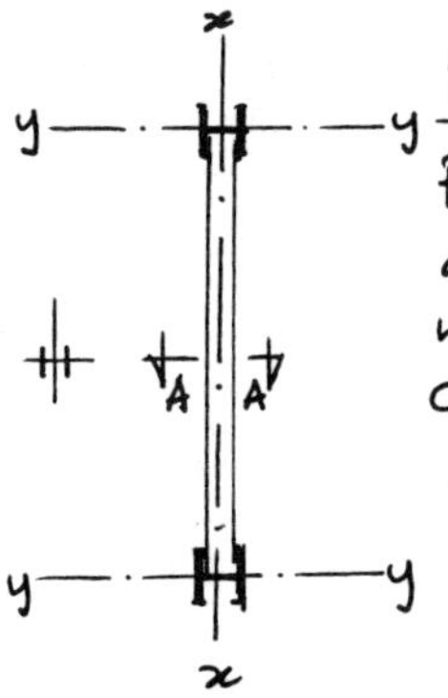 Fig. 3/012 Proposed truss - all U.C.'s - welded gusset connections
	Top boom: max. comp. force = 1944 kN.
	$L_{ex} = 0.85 \times 6 = 5.10$ m.
	$L_{ey} = 1.0 \times 3 = 3.0$ m.
	Try a 203 × 203 × 86 U.C. ($p_y = 265$ N/mm²)
	Check classification:–
Table 7	Flange: $b/T = 5.09 < 15\varepsilon$ } 'Not slender'
	Web: $d/t = 12.4 < 39\varepsilon$ } 'Not slender'
Table 27(b)	$\left(\frac{L_e}{r}\right)_x = \frac{5100}{92.7} = 55$: $p_{cx} = 221.5$ N/mm²
Table 27(c)	$\left(\frac{L_e}{r}\right)_y = \frac{3000}{53.2} = 56$: $p_{cy} = 202^*$ N/mm²
	* critical: $p_c = 202$ N/mm²
	$\therefore P_c = \frac{202 \times 11000}{10^3} = 2222$ kN $> F_c$ ✓ o.k.
	Bottom boom: max. tensile force = 1944 kN
4.6.1	Area req^d = $\frac{1944 \times 10^3}{275} = 7069$ mm²
	Try a 203 × 203 × 60 U.C ($p_y = 275$ N/mm²)
	$P_t = \frac{275 \times 7580}{10^3} = 2084$ kN.
	$\therefore P_t > F_t$ ✓ o.k. (2084) (1944)
	Internal compression members (vertical)
	Welded gussetted connections: $L_e = 0.85L$
	$\therefore L_e = 0.85 \times 5 = 4.25$ m.
	Max force = 1080 kN (comp.) (B-C)
	Try a 203 × 203 × 60 U.C. ($p_y = 275$ N/mm²)
	Check classification:–
Table 7	Flange: $b/T = 7.23 < 15\varepsilon$ } 'Not slender'
	Web: $d/t = 17.3 < 39\varepsilon$ } 'Not slender'

Ref.

Table 27(c) $\left(\frac{l_e}{r}\right)_y = \frac{4250}{51.9} = \underline{82}$; $p_{cy} = \underline{157\ N/mm^2}$

$\therefore P_c = \frac{157 \times 7580}{10^3} = \underline{1190\ kN}$.

$\therefore \underline{P_{cy}}\ \underline{>\ F_c}$ ✓ o.k.

(1190) (1080)

Internal Tension Members (diagonal)

Max force = 1260 kN (A – B)

Area required = $\frac{1260 \times 10^3}{275} = \underline{4582\ mm^2}$

Try a 203 × 203 × 46 U.C. ($p_y = 275\ N/mm^2$)

$P_t = \frac{275 \times 5880}{10^3} = \underline{1617\ kN}$.

$\therefore \underline{P_t\ >\ F_t}$ ✓ o.k.

(1617) (1260)

Primary truss members :–

Top boom : 203 × 203 × 86 U.C.
Bottom boom : 203 × 203 × 60 U.C.
Verticals : 203 × 203 × 60 U.C.
Diagonals : 203 × 203 × 46 U.C.

Typical gusset connection (B)

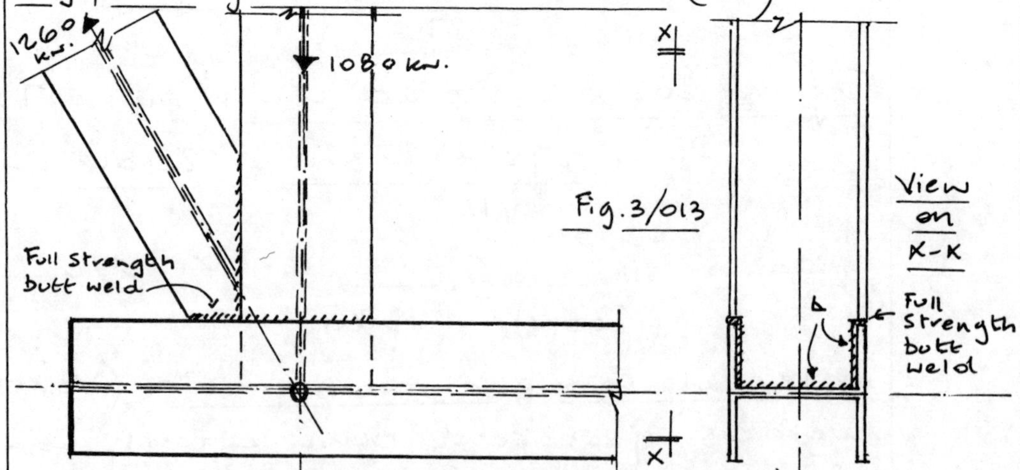

Fig. 3/013

Check diagonal (tension member)

Weld length = 2(75 + 250) = 650 mm.

Ref 3/1 Use 8mm F.S.B.W. : Capacity = $\underline{2.2}$ kN/mm

$\therefore$ Resistance = 2.2 × 650 = $\underline{1430\ kN}$

$\therefore$ Resistance > force

(1430) (1260)

Ref.	
	Primary truss deflection check:

Using same approximate method as secondary truss (see Fig 3/005) :-

Area of top boom = 11000 mm^2

Area of bottom boom = 7580 mm^2.

Average area = 9290 mm^2.

Service U.D.L. on secondary trusses = 8.4 kN/m : total = $8.4 \times 30 = 252$ kN.

Reaction on primary truss from two secondary trusses = $2 \times \frac{252}{2} = 252$ kN.

Spread point loads as a U.D.L. :-

Total = $5 \times 252 = 1260$ kN.

$$I_{approx.} = \frac{9290 \times 5000^2}{2} = 1.16 \times 10^{11} \text{ mm}^4$$

$$\therefore \Delta = \frac{5}{384} \times \frac{1260 \times 10^3 \times (30000)^3}{205 \times 10^3 \times 1.16 \times 10^{11}} = 18.6 \text{ mm}$$

$$\text{Allowable} = \frac{30000}{360} = 83.3 \text{ mm} \quad \checkmark \text{ o.k.}$$

Step 3 - Supporting columns.

A) Main central column - braced CHS pylon.

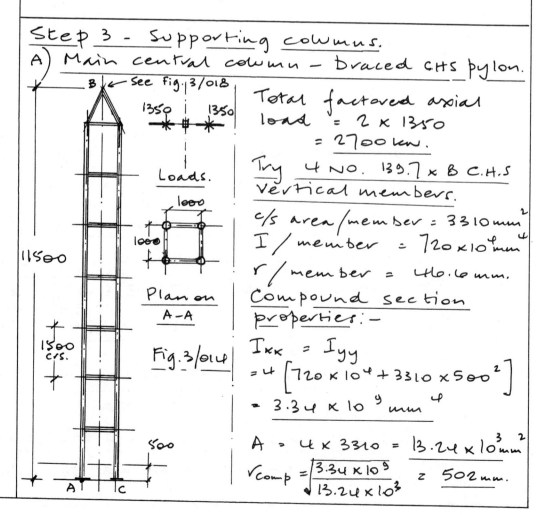

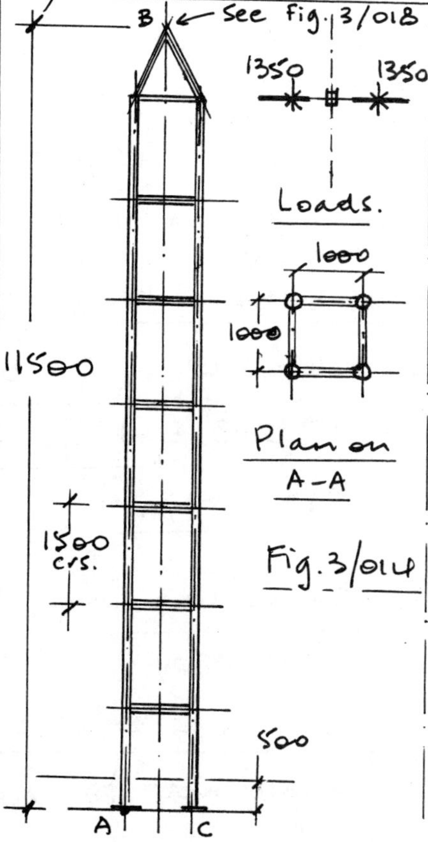

Fig. 3/014

Total factored axial load = 2×1350 = 2700 kN.

Try 4 No. 139.7 × 8 C.H.S vertical members.

c/s area/member = 3310 mm^2

I / member = $720 \times 10^4 \text{ mm}^4$

r / member = 46.6 mm.

Compound section properties:-

$I_{xx} = I_{yy}$

$= 4 \left[720 \times 10^4 + 3310 \times 500^2\right]$

$= 3.34 \times 10^9 \text{ mm}^4$

$A = 4 \times 3310 = 13.24 \times 10^3 \text{ mm}^2$

$$r_{comp} = \sqrt{\frac{3.34 \times 10^9}{13.24 \times 10^3}} = 502 \text{ mm}.$$

Ref.	
BS5950 Fig 20	Pylon is a propped cantilever, $L_E = 0.85L$
	$\therefore L_E = 0.85 \times 11500 = 9775$ mm.
	$(l_e/r) = \frac{9775}{502} = 19.5$
Table 27(c)	From Table 25, for a compound rolled section column, take Table 27(c).
	For $p_y = 275$ N/mm², $p_c = 271$ N/mm²
	$\therefore P_c = \frac{271 \times 13.24 \times 10^3}{10^3} = 3588$ kN.
	$\therefore P_c > F_c$ ✓ o.k.
	(3588) (2700)
	Individual leg buckling:
BS5950 Fig. 20	$l_e = 1500$ mm (see Fig. 3/014)
	$\therefore (l_e/r) = \frac{1500}{46.6} = 32.2$
Table 27(a)	For $p_y = 275$ N/mm², $p_c = 266$ N/mm²
	$\therefore P_c = \frac{266 \times 3310}{10^3} = 880$ kN.
	Force/leg = 2700/4 = 675 kN
	$\therefore P_c > F_c$ ✓ o.k.
	(880) (675)
	Foundation: Safe bearing = 250 kN/m²
	Service axial load $\simeq \frac{2700}{1.5} = 1800$ kN.
	Try a 3 m. sq. base × 0.8 m thick.
	Self weight = 3 × 3 × 0.8 × 24 = 172.8 kN.
	Total load = 1800 + 172.8 = 1973 kN.
	Pressure = $\frac{1973}{3^2}$ = 219 kN/m²
	Pressure < Safe bearing ✓ o.k.
	(219) (250)
	$\therefore$ Make base 3 m. sq. × 0.8 m. deep.
	(Note! Reinforcement design not provided)

Ref.	B) Column supporting secondary trusses (Grids 1 & 3 – Fig 3/001)
	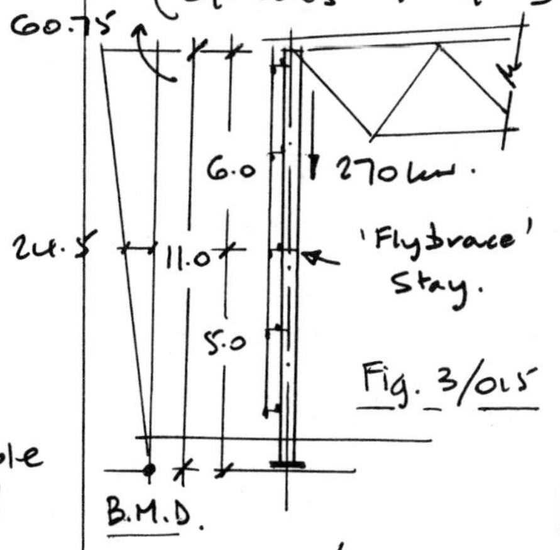
	Fig. 3/015
	Try a 254 x 254 x 73 U.C
	Moment at top of column = 270 x (0.125 + 0.1) = <u>60.75 kN.m.</u>
	Axial load = <u>270 kN.</u>
	<u>Check classification</u> :–
Table 7	(p_y = 275 N/mm² : ε = 1.0)
	b/T = <u>8.94</u> < 15 ε
	For web: 'squash' load = c/s area x p_y = $\frac{9290 \times 275}{10^3}$ = <u>2555 kN.</u>
	$R = \frac{270}{2555}$ = <u>0.1</u> (& positive)
Cl. 3.5.4	∴ Limiting $d/t = \frac{120\,\varepsilon}{1+1.5R} = \frac{120}{1+0.15}$ = <u>104.4</u>
	d/t = <u>23.3</u> < 104.4
	∴ <u>Section 'not slender'.</u>
Cl. 4.8.3.3.1	Check the o/all buckling, top length on:– $\frac{F}{A_g . p_c} + \frac{m . M_x}{M_b} < 1.0$
	<u>For calculation of p_c</u> :–
	For x–x axis, L_E = 1.5 L = 1.5 x 11.0 = <u>16.5 m.</u>
	For y–y axis, L_E = <u>6.0 m.</u>
Table 27(b)	$\lambda_x = \frac{16500}{111}$ = <u>149</u> : p_{cx} = 75 N/mm² } p_c = <u>75 N/mm²</u>
Table 27(c)	$\lambda_y = \frac{6000}{64.6}$ = <u>93</u> : p_{cy} = 137 N/mm² }
	∴ $P_c = \frac{75 \times 9290}{10^3}$ = <u>697 kN.</u>
	<u>For calculation of M_b</u> :–
	L_E is greater of either <u>0.85 x 5.0</u> or <u>6.0 m.</u>
	∴ <u>L_E = 6.0 m.</u>
Table 14.	$\lambda = \frac{6000}{64.6}$ = <u>93</u> : $\frac{\lambda}{x} = \frac{93}{17.3}$ = 5.4 : <u>V = 0.79</u>
Cl. 4.3.7.5	∴ λ_{LT} = 1.0 x 0.849 x 0.79 x 93 = 62 : <u>p_b = 207 N/mm²</u>
Table 11	∴ M_b = 207 x 989 / 10³ = <u>205 kN.m.</u>
	<u>For calculation of 'm'</u>
	From Fig. 3/015, $\beta = + \frac{24.5}{60.75}$ = <u>0.40</u>

Ref.	
	$\therefore m = 0.74$
Table 18	& $\bar{M} = 0.74 \times 60.75 = \underline{45}$ kN.m.
Cl. 4.8.3.3.1	$\therefore \dfrac{270}{697} + \dfrac{45}{205} = \underline{0.61} < 1.0$ ✓ o.k.
	Note! over-designed, but 254 mm wide flange required for detail (see Fig. 3/018)
	$\therefore$ Use $\underline{254 \times 254 \times 73 \text{ U.C.}}$

c) Column supporting primary trusses

(columns A2 & L2 – see Fig. 3/001)

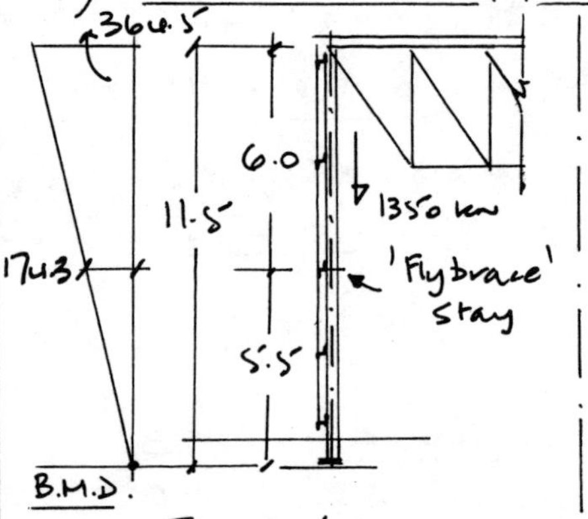

Fig 3/016

Try a 305 x 305 x 198 U.C.

$(p_y = 265 \text{ N/mm}^2)$

Moment at top of column $= 1350 (0.17 + 0.1)$

$= \underline{364.5}$ kN.m.

Axial load $= \underline{1350 \text{ kN}}$.

Ref.	
Table 7	Check Classification: $\varepsilon = \sqrt{\dfrac{275}{265}} = \underline{1.02}$
	Flanges: $b/T = \underline{5} < 15\varepsilon \ (= 15.3)$
	Web: 'squash load' = c/s area x p_y
	$= \dfrac{265 \times 25200}{10^3} = \underline{6675 \text{ kN}}$.
Cl. 3.5.4	$\therefore R = \dfrac{1350}{6675} = \underline{0.2}$ (& positive)
	$\therefore$ Limiting $d/t = \dfrac{120\,\varepsilon}{1 + 1.5R} = \dfrac{120 \times 1.02}{1 + 1.5 \times 0.2} = \underline{94}$
	$d/t = \underline{12.8} < 94$
	$\therefore$ Section 'not slender'
Cl. 4.8.3.3.1	Check the o/all buckling, top length on:–
	$\dfrac{F}{A_g . p_c} + \dfrac{m . M_x}{M_b} \leq 1.0$
	For calculation of p_c:–
	For x–x axis, $L_E = 1.5L = 1.5 \times 11.5 = \underline{17.25 \text{ m}}$.
	For y–y axis, $L_E = \underline{6.0 \text{ m}}$.
	$\lambda_x = \dfrac{17250}{142} = \underline{121}$: $p_{cx} = 104.5 \text{ N/mm}^2$; $\lambda_y = \dfrac{6000}{80} = \underline{75}$: $p_{cy} = 167 \text{ N/mm}^2$ } $p_c = \underline{104.5 \text{ N/mm}^2}$

Ref.	
	∴ $P_c = \frac{104.5 \times 25200}{10^3} = 2633$ kN.
	For calculation of M_b :-
	L_E is greater of either 0.85×5.5 or 6.0 m
	∴ $L_E = 6.0$ m.
Table 14	$\lambda = \frac{6000}{80} = 75$: $\frac{\lambda}{x} = \frac{75}{10.2} = 7.3$: $v = 0.72$
Cl. 4.3.7.5	∴ $\lambda_{LT} = 1.0 \times 0.854 \times 0.72 \times 75 = 46$
Table 11	$p_b = 241$ N/mm^2
	∴ $M_b = \frac{241 \times 3440}{10^3} = 829$ kN.m.
	For calculation of 'm' :-
Table 18	From Fig. 3/016, $\beta = +\frac{174.3}{364.5} = 0.48$: $m = 0.75$
	∴ $\bar{M} = 0.75 \times 364.5 = 273.3$ kN.m.
Cl. 4.8.3.3.1	∴ $\frac{1350}{2633} + \frac{273.3}{829} = 0.842 < 1.0$ ✓ O.K.
	∴ use 305 × 305 × 198 U.C.
	Step 4 : Wind bracing: wind on elevations on grids 'A' & 'L' (see Fig. 3/001) only considered. On grids '1' & '3', wind moment assumed taken by portal action on the columns. In Fig. 3/001, one pair of horizontal trusses each 'end' assumed, with one set of side bracing per 'end'.
	For calculation of W_k :-
	From brief, $V = 46$ m/sec.
	$S2 = 0.70$ (outskirts of a large town)
CP3 ChV Pt.2	∴ $V_s = 46 \times 0.7 = 32.2$ m/sec.
	$q = \frac{0.613 \times 32.2^2}{10^3} = 0.635$ kN/m^2
Table 10	Building is square : $\frac{l}{w} = \frac{b}{d} = 1.0$
	Height / breadth $= 11.5/60 = 0.2 < 1/2$
	∴ $C_f = 0.9$.
Table 2	∴ $1.4\, W_k = 1.4 \times 0.9 \times 0.635 = 0.8$ kN/m^2

Ref.	
	Assuming gable posts @ 7.5m. crs (see Fig 3/017) with a height of post of 11m., reaction at roof level = 0.8 × 7.5 × 11/2 = 33 kN.

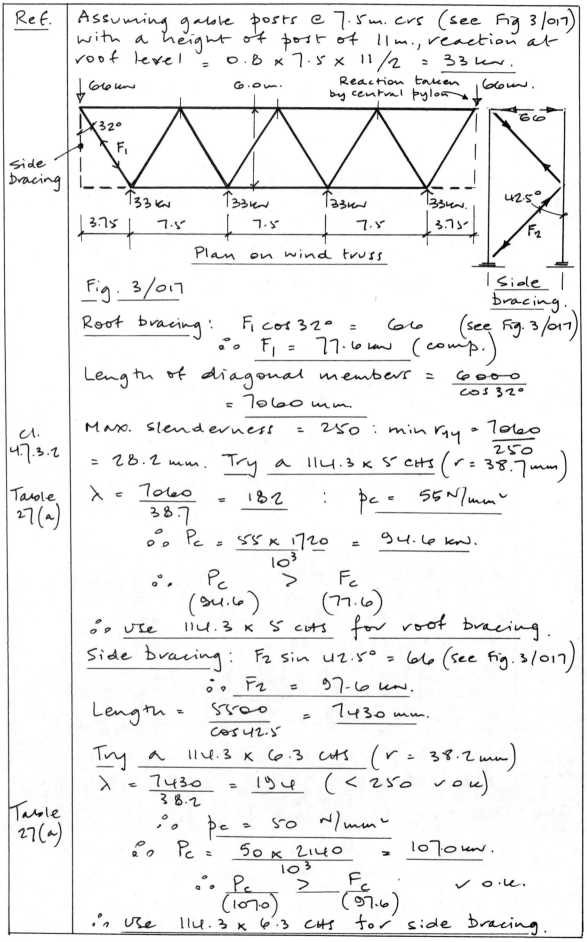

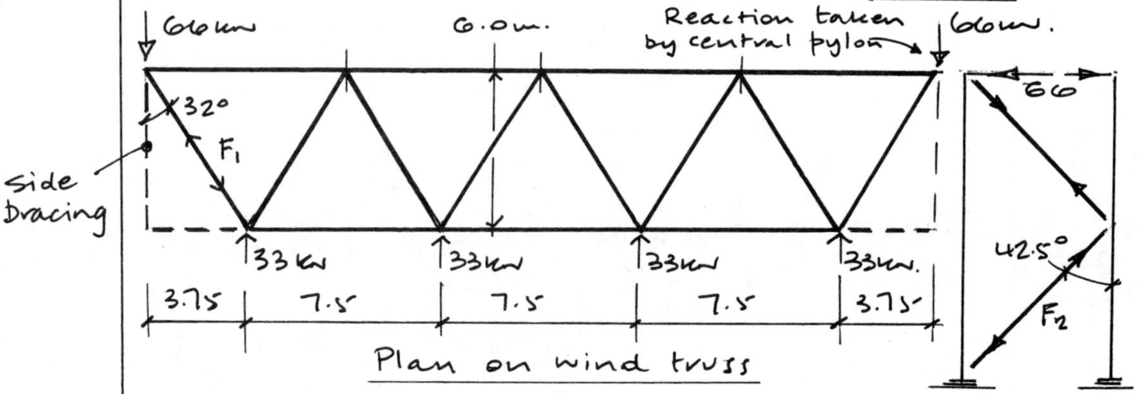

Plan on wind truss

Side bracing.

Fig. 3/017

Ref.	
	Root bracing: $F_1 \cos 32° = 66$ (see Fig. 3/017)
	∴ $F_1 = 77.6$ kN (comp.)
	Length of diagonal members $= \frac{6000}{\cos 32°}$
	= 7060 mm
Cl. 4.7.3.2	Max. slenderness = 250 : min $r_{yy} = \frac{7060}{250}$
	= 28.2 mm. Try a 114.3 × 5 CHS (r = 38.7 mm)
Table 27(a)	$\lambda = \frac{7060}{38.7} = 182$: p_c = 55 N/mm²
	∴ $P_c = \frac{55 \times 1720}{10^3} = 94.6$ kN.
	∴ P_c (94.6) > F_c (77.6)
	∴ Use 114.3 × 5 CHS for roof bracing.
	Side bracing: $F_2 \sin 42.5° = 66$ (see Fig. 3/017)
	∴ $F_2 = 97.6$ kN.
	Length $= \frac{5500}{\cos 42.5} = 7430$ mm.
	Try a 114.3 × 6.3 CHS (r = 38.2 mm)
	$\lambda = \frac{7430}{38.2} = 194$ (< 250 ✓ ok)
Table 27(a)	∴ p_c = 50 N/mm²
	∴ $P_c = \frac{50 \times 2140}{10^3} = 107.0$ kN.
	∴ P_c (107.0) > F_c (97.6) ✓ o.k.
	∴ Use 114.3 × 6.3 CHS for side bracing.

Ref.	
	Check central pylon for the wind truss reaction ($1.4W_k + 1.0G_k$).

In Fig. 3/014, wind truss reaction (2 No. trusses) is a horizontal force at B.

Force = 2 × 66 = 132 kN.

Overturning moment at base

= 132 × 11.5 = 1518 kN.m.

Taking moments about A, reaction at point C = R_c.

∴ Resistance moment = $R_c \times 1$

1518 = $R_c \times 1$: R_c = 1518 kN.

∴ Reaction /leg = $\frac{1518}{2}$ = 759 kN.

$1.0G_k$: secondary truss reaction

= $0.5 \times 6.0 \times \frac{30}{2}$ = 45 kN

2 No. reactions / point load = 2×45 = 90 kN.

Primary truss reaction = $\frac{5 \times 90}{2}$ = 225 kN.

∴ $1.0G_k$ = 2 × 225 = 450 kN.

Load /leg = $\frac{450}{4}$ = 112.5 kN.

∴ $1.4W_k + 1.0G_k$ = 759 + 112.5 = 871.5 kN.

From pylon calcs, P_c for one leg

= 880 kN.

∴ $P_c > F_c$ ✓ o.k.

(880) (871.5)

Notes!

1) No check provided for deflection of pylon due to wind.

2) Normally: large trusses subject to stress reversal due to wind uplift. In this case, longitudinal bracing to bottom booms required.

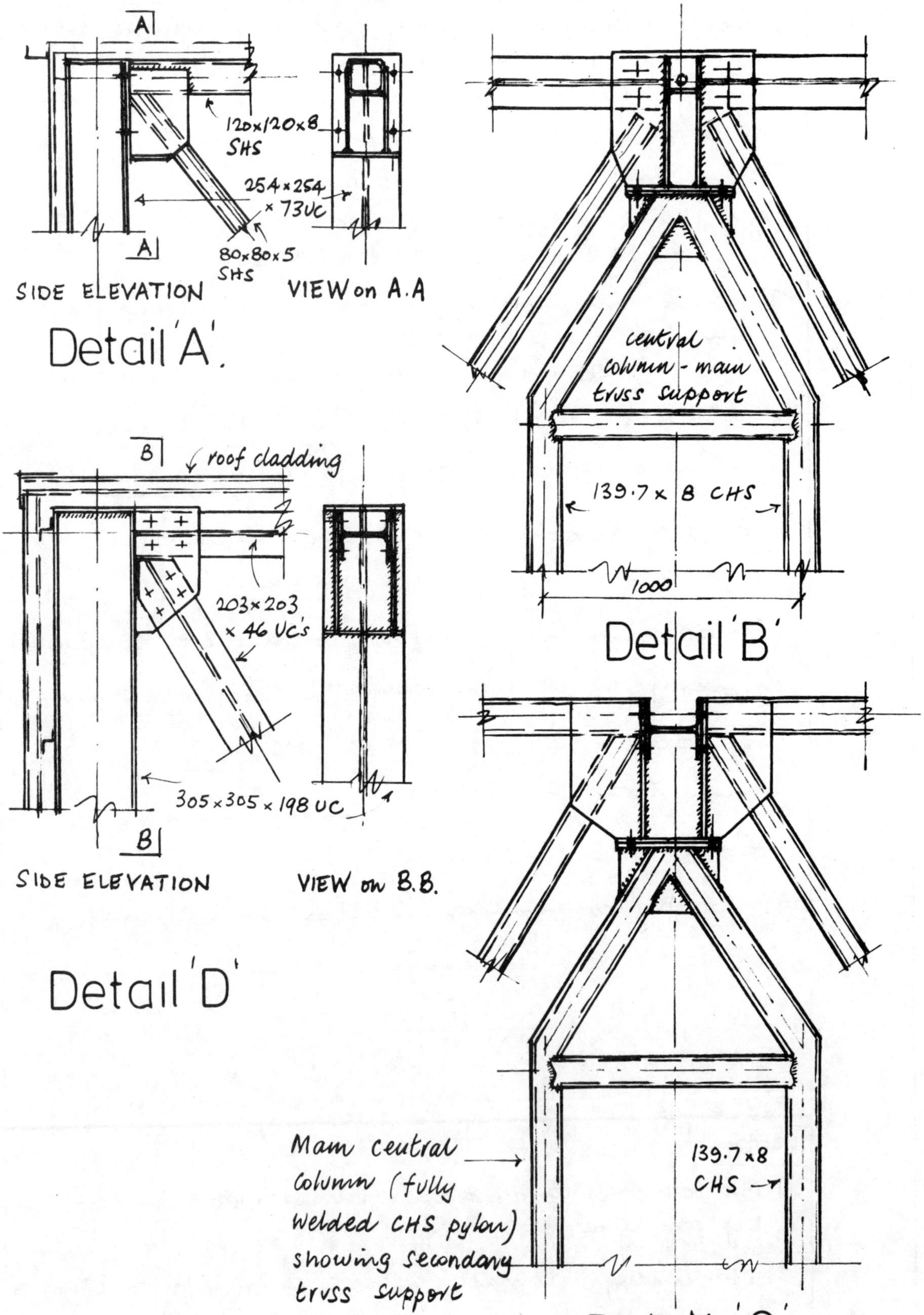
A
120×120×8 SHS
254×254 ×73UC
80×80×5 SHS
SIDE ELEVATION
VIEW on A.A
Detail 'A'.
central column - main truss support
139.7 × 8 CHS
1000
Detail 'B'
B
roof cladding
203×203 × 46 UC's
305×305×198 UC
SIDE ELEVATION
VIEW on B.B.
Detail 'D'
Main central column (fully welded CHS pylon) showing secondary truss support
139.7×8 CHS
Detail 'C'.

Fig. 3/018

Chapter 4 Dockside potash store facility

A Port Authority requires a potash store facility, having a central discharge conveyor and a portal scraper, as shown in Fig. 4/001. The building is to be triangular in section, with gable ends covered in steel cladding, with one door each end, 6.0 m. wide x 5.0 m. high. The building is to be free of internal columns, and the roof is to be of lightweight steel cladding (no natural light required) and insulation boards. The angle of repose of the stored material is estimated to be 25 degs., as shown in Fig. 4/001. The design chosen is a steel portal frame, and this provides the basis for design in this example.

Loading – qk

Characteristic roof loading = 1.0 kN/m^2 (snow + services).
Conveyor, allow 3 kN/m. run for **qk**.

Loading – gk

Roof central conveyor zone = 6.0 kN/m^2.
Roof covering = 0.4 kN/m^2.
Portal scraper unit self-supporting.

Loading – wk

Basic wind speed 50 m/sec. (city centre site, adjacent the sea)

Site conditions

High water table, with stiff clay at 1.75 m. down. Pilecaps on friction piles (some raking) chosen.

Design Code of Practice

The building is to be designed in accordance with BS 5950, Structural Steelwork.

Materials

All structural steelwork, including rolled steel purlins, to be EN10025 S275 steel. Potassium chloride, or potash, is closely related to common salt, and is thus highly corrosive when damp. The steelwork must thus be galvanised to 140 microns to give a life to first maintenance of 10 – 20 years in accordance with BS 5463.
Concrete in the bases to be of fcu = 35 N/mm^2 characteristic strength.

Frame Analysis

In this example, a hand method has been used to calculate bending moments due to:-

1.6 Qk + 1.4 Gk
1.4 Wk(uplift) + 1.0 Gk

Coefficients for the rigid frame were taken from 'Kleinlogel' charts (Ref. 4/1)

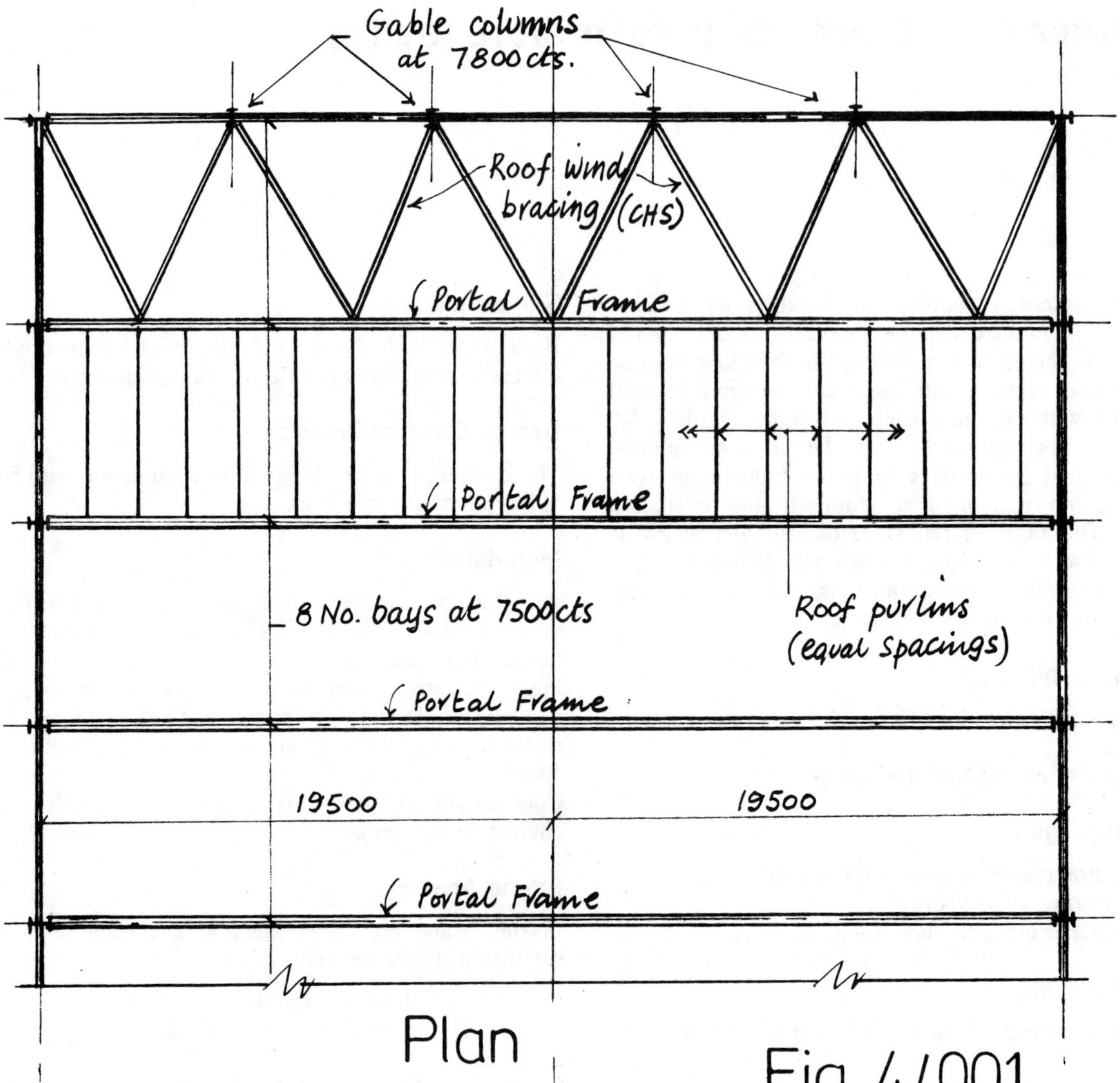
Gable columns at 7800 cts.
Roof wind bracing (CHS)
Portal Frame
Portal Frame
8 No. bays at 7500 cts
Roof purlins (equal spacings)
Portal Frame
19500
19500
Portal Frame
Plan

Fig. 4/001

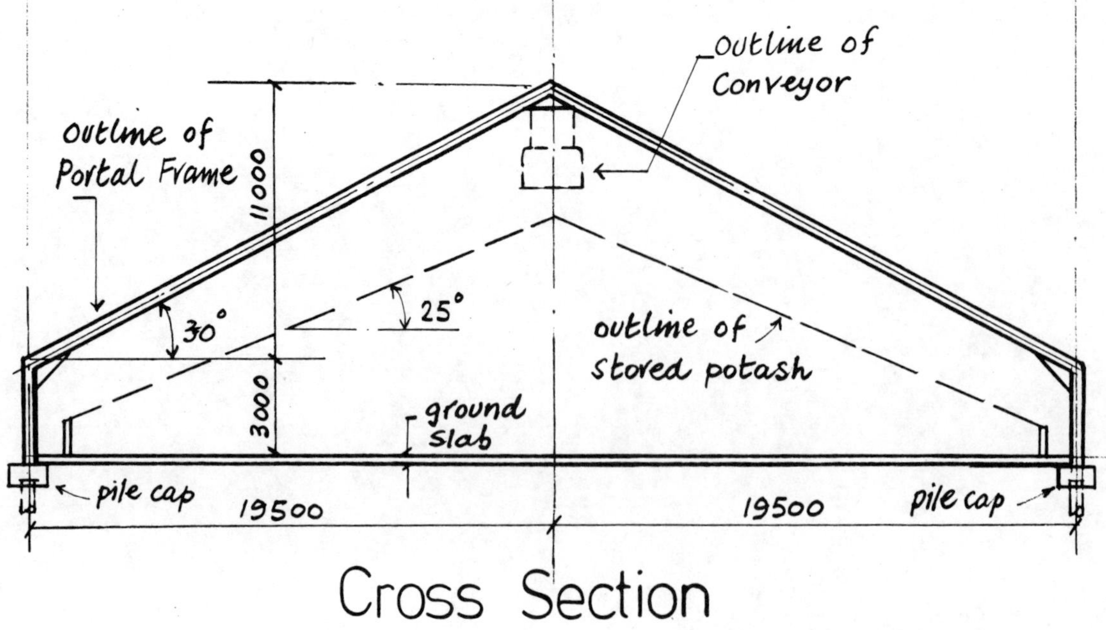
outline of Conveyor
outline of Portal Frame
11000
30°
25°
outline of stored potash
3000
ground slab
pile cap
19500
19500
pile cap
Cross Section

Fig. 4/002

Ref. (BS 5950)

Design loadings on portal frames (7.5 m. crs.)

L.C.I – 1.6 Qk + 1.4 Gk.

Roof loading = 1.6 × 1.0 + 1.4 × 0.4
= 2.16 kN/m² : U.D.L = 7.5 × 2.16 = 16.2 kN/m.

Conveyor loading (take as ridge point load)

q_k = 6.0 kN/m² : 2 m. wide

∴ U.D.L = 12.0 kN/m.

g_k = 3.0 kN/m (see brief)

∴ Design loading = 1.6 × 12.0 + 1.4 × 3.0
= 23.4 kN/m.

Point load/frame = 23.4 × 7.5 = 175.5 kN.

L.C.II – 1.4 Wk (uplift) + 1.0 Gk.

Wind loading : V = 50 m/sec. (brief)

$V_s = V \times S_2 = 50 \times 0.95 = 47.5$ m/sec.

∴ $q = \dfrac{0.613 \times 47.5^2}{10^3} = 1.38$ kN/m²

Cp coeffts: $h/w = 14/40 \simeq 0.3 < 1/2$

Fig. 4/003

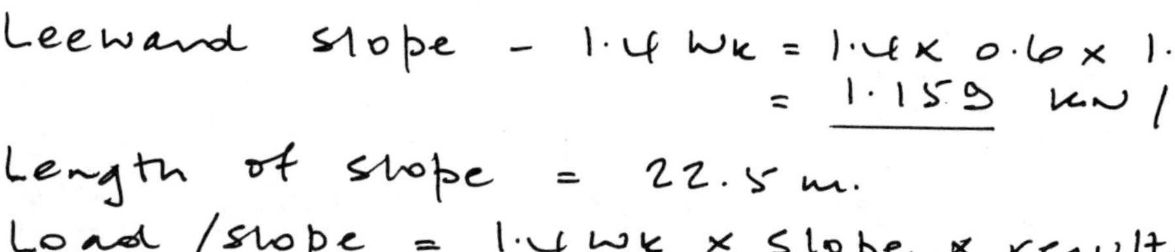

Windward slope – 1.4 Wk = 1.4 × 0.2 × 1.38
= 0.386 kN/m².

Leeward slope – 1.4 Wk = 1.4 × 0.6 × 1.38
= 1.159 kN/m²

Length of slope = 22.5 m.

Load/slope = 1.4 Wk × slope × resultant × frame centres.

Fig. 4/004
1.4 Wk.

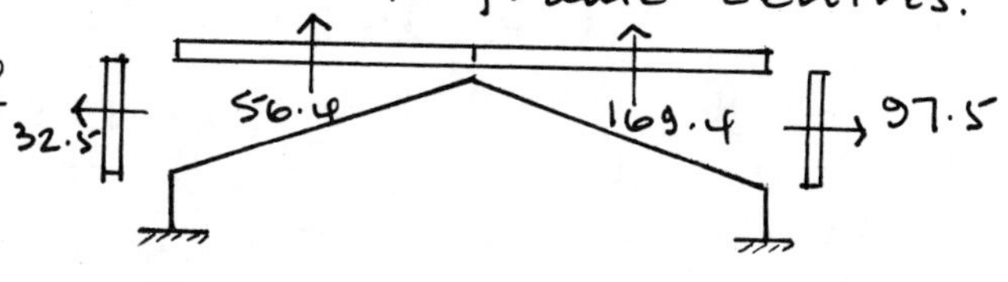

Ref.

Now, 1.0Gk = 0.4 kN/m² (see brief).

Fig 4/005

1.0Gk.

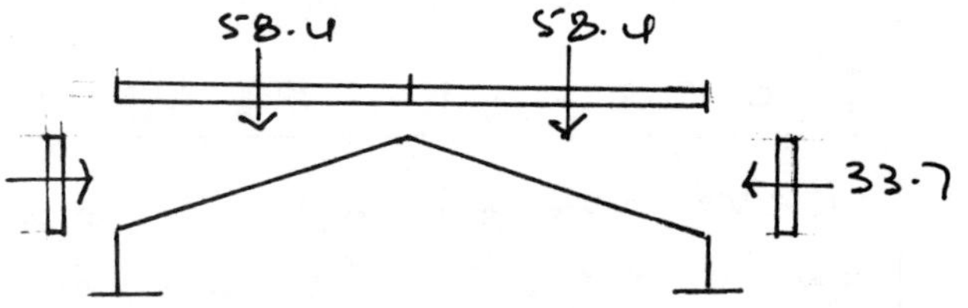

1.4Wk + 1.0Gk By inspection of Figs. 4/004 & 4/005, windward forces just about balance: therefore load case will be analysed for leeward forces only :-

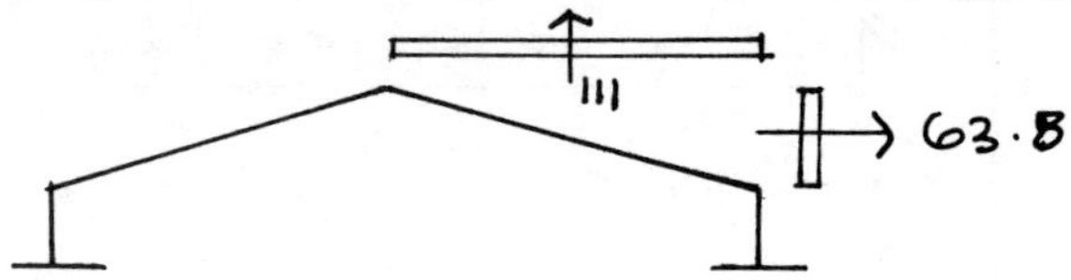

Fig. 4/006

1.4Wk + 1.0Gk.

Ref 4/1

'Kleinlogel' rigid frame coefficients

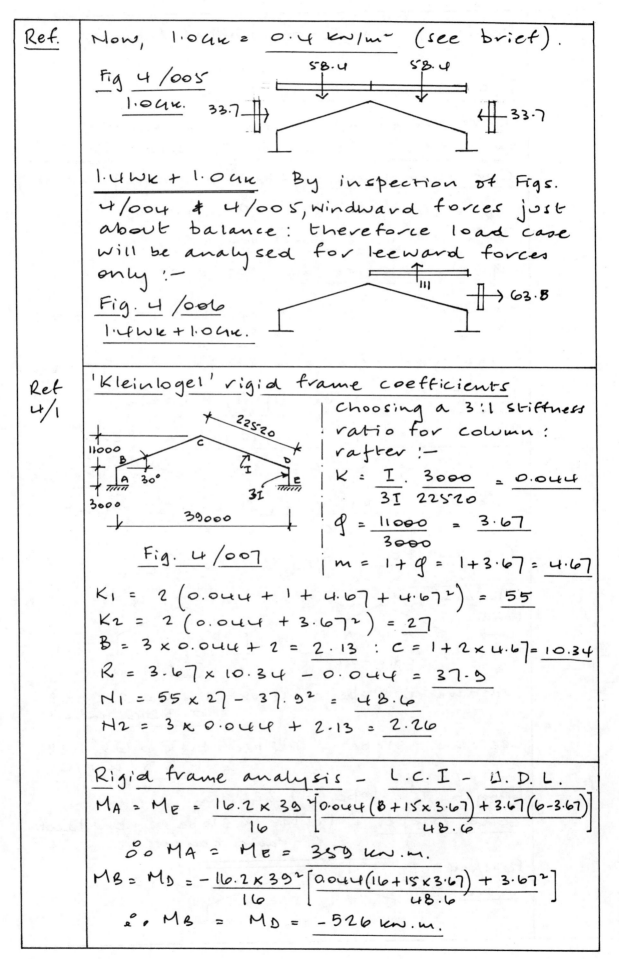

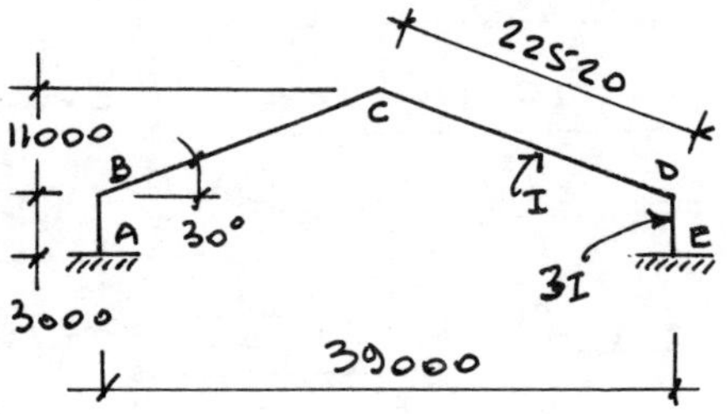

Fig. 4/007

Choosing a 3:1 stiffness ratio for column : rafter :-

$K = \frac{I}{3I} \cdot \frac{3000}{22520} = \underline{0.044}$

$\phi = \frac{11000}{3000} = \underline{3.67}$

$m = 1 + \phi = 1 + 3.67 = \underline{4.67}$

$K_1 = 2(0.044 + 1 + 4.67 + 4.67^2) = \underline{55}$

$K_2 = 2(0.044 + 3.67^2) = \underline{27}$

$B = 3 \times 0.044 + 2 = \underline{2.13}$: $C = 1 + 2 \times 4.67 = \underline{10.34}$

$R = 3.67 \times 10.34 - 0.044 = \underline{37.9}$

$N_1 = 55 \times 27 - 37.9^2 = \underline{48.6}$

$N_2 = 3 \times 0.044 + 2.13 = \underline{2.26}$

Rigid frame analysis - L.C.I - U.D.L.

$M_A = M_E = \frac{16.2 \times 39^2}{16}\left[\frac{0.044(8 + 15 \times 3.67) + 3.67(6 - 3.67)}{48.6}\right]$

$\therefore M_A = M_E = \underline{359 \text{ kN.m.}}$

$M_B = M_D = -\frac{16.2 \times 39^2}{16}\left[\frac{0.044(16 + 15 \times 3.67) + 3.67^2}{48.6}\right]$

$\therefore M_B = M_D = \underline{-526 \text{ kN.m.}}$

Ref.	
	$M_C = \frac{16.2 \times 39^2}{8} - 3.67 \times 359 + 4.67 \times 526$

$\therefore$ $M_C = -694$ kN.m. (N.B. not positive)

Reactions:

$V_A = V_E = \frac{16.2 \times 39}{2} = 315.9$ kN.

$H_A = H_E = \frac{359 + 526}{3} = 295$ kN.

L.C.I – point load (at c)

$M_A = M_E = \frac{3 \times 175.5 \times 39(0.044 + 2 \times .044 \times 3.67 + 3.67)}{4 \times 48.6}$

$\therefore$ $M_A = M_E = 426$ kN.m.

$M_B = M_D = \frac{-3 \times 175.5 \times 39 \times 0.044 \times 4.67}{2 \times 48.6}$

$\therefore$ $M_B = M_D = -43.4$ kN.m

$M_C = \frac{175.5 \times 39}{4} - 3.67 \times 426 - 4.67 \times 43.4$

$\therefore$ $M_C = -55$ kN.m. (N.B. not positive)

Reactions:

$V_A = V_E = 175.5/2 = 87.75$ kN.

$H_A = H_E = \frac{426 + 43.4}{3} = 156.5$ kN.

Total: L.C. I

Fig. 4/008
(Moments)

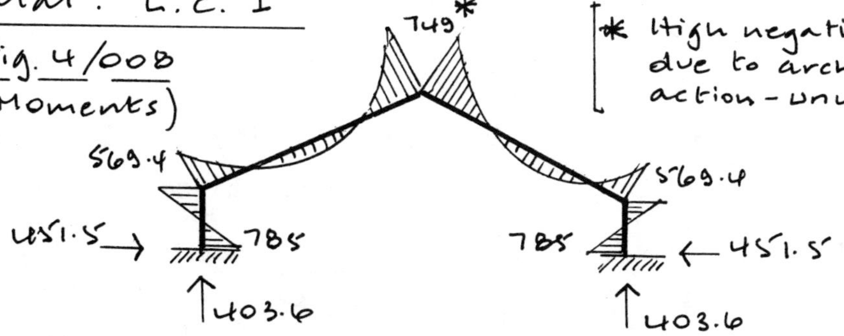

[* High negative mt. due to arching action – unusual]

L.C.II Loading in Fig. 4/006 input into planeframe program; results below:-

Fig. 4/009
(Moments)

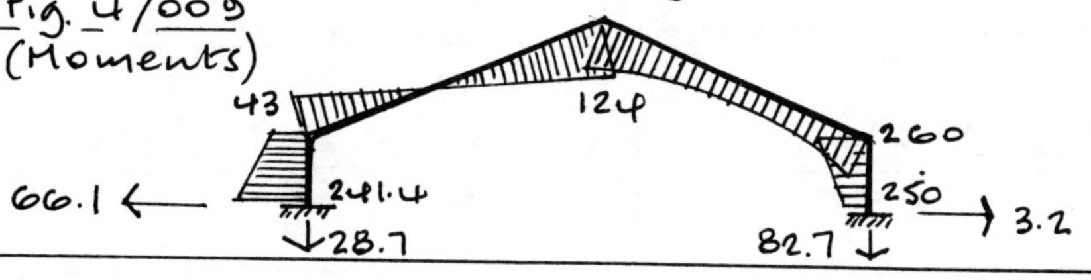

Ref.	Member designs.

A: Column: Try a 686 x 254 x 170 U.B.

Ref 4/2

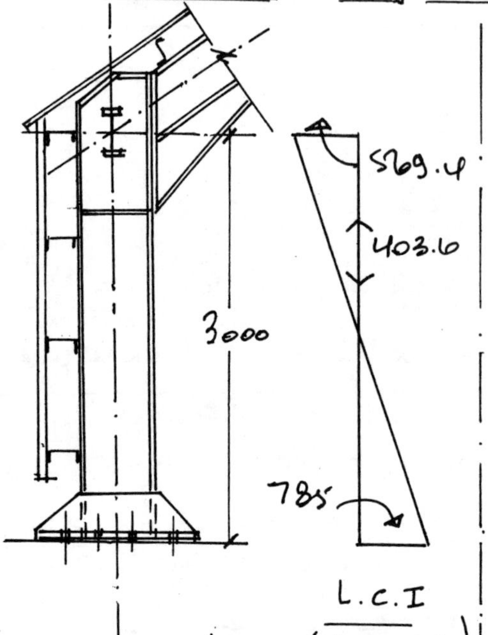

Fig. 4/010

L.C.I (Moments)

$T = 23.7$ mm. : $p_y = 265\ N/mm^2$

Classification:

$\varepsilon = \sqrt{\frac{275}{265}} = 1.02$

Ref 4/2

Flange: $b/T = 5.40$

Plastic limit $= 8.5\varepsilon = 8.7$

∴ Flange 'plastic'

Ref 4/2

Web: $d/t = 42.4$

Ignoring small axial load; $d/t \leq 79\varepsilon = 81$

∴ Web 'plastic'

Section 'plastic'

Ref 4/2

Section properties: $u = 0.872$; $x = 31.8$; $r_{yy} = 55.3$ mm.; $A = 21700\ mm^2$; $S_{xx} = 5620\ cm^3$

Cl. 4.8.3.2

a) Local capacity: $\left(\frac{M_x}{M_{rx}}\right)^2 < 1.0$

∴ $M_x < M_{rx}$.

M_{rx}: $A_g \cdot p_y = \frac{21700 \times 265}{10^3} = 5750$ kN.

∴ $n = \frac{403.6}{5750} = 0.07 < 0.431$

∴ $K_1 = 5620$; $K_2 = 8100$

∴ $S_{rx} = 5620 - 8100 \times (0.07)^2 = 5580\ cm^3$

∴ $M_{rx} = \frac{5580 \times 265}{10^3} = 1479$ kN.m

∴ $M_x < M_{rx}$: local capacity o.k.
(785) (1479)

Cl. 4.8.3.3.1

b) O/all buckling: $\frac{F}{A_g \cdot p_c} + \frac{m M_x}{M_b} < 1.0$

Table 14

$\lambda_y = \frac{3000}{55.3} = 54$: $\frac{\lambda}{x} = \frac{54}{31.8} = 1.7$: $v = 0.97$

Ref.	
Table 11	$\therefore \lambda_{LT} = 1.0 \times 0.872 \times 0.97 \times 54 = \underline{46}$
	$\underline{p_b = 241\ N/mm^2}$
	$\therefore M_b = \dfrac{241 \times 5620}{10^3} = \underline{1354\ kN.m.}$
	$\underline{\underline{m}}$: $\beta = -\dfrac{569.4}{785} = \underline{-0.72}$: $\underline{m = 0.43}$
Table 27(b)	$\underline{\underline{P_c}}$: $\lambda_y = 54$: $\underline{p_{cy} = 215\ N/mm^2}$
	$\therefore P_{cy} = \dfrac{215 \times 21700}{10^3} = \underline{4665\ kN.}$
	$\therefore \dfrac{403.6}{4665} + \dfrac{0.43 \times 785}{1354} = \underline{0.34 < 1.0}$
	$\therefore$ <u>O/all buckling</u> <u>o.k.</u>
	<u>Note!</u> Although appearing to be over-designed; the column section adds to the stiffness of the frame by reducing the eaves deflection $(1.0 g_k + 1.0 q_k)$ to $\underline{\underline{2.7\ mm}}$ (plane frame result). Limit is $h/1000 = 3000/1000 = \underline{3\ mm.}$
	B: <u>Rafter</u> : <u>try</u> <u>a 533 × 210 × 109 U.B.</u>
	(stanchion : rafter inertias 3:1 assumed).
	Fig 4/011 constructed from statics:-
	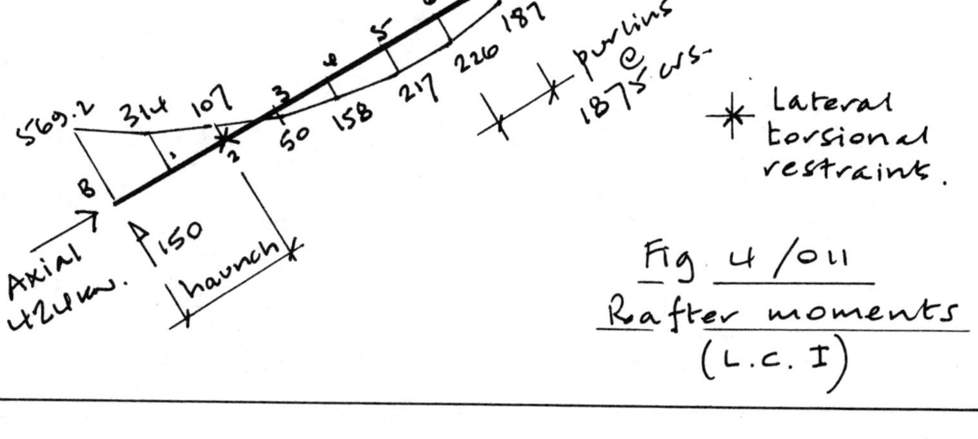
	<u>Fig 4/011</u>
	<u>Rafter moments</u>
	(L.C. I)

Ref.	

a) Check rafter between B9 (L.T.R.) & B11 (L.T.R & end of ridge haunch)

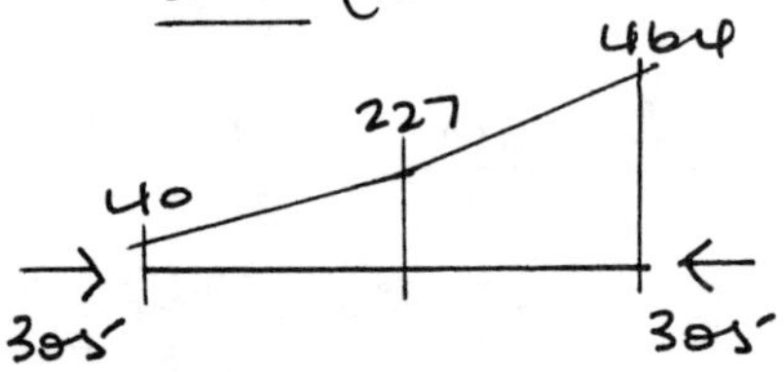

Fig. 4/012
B9 → B11

Axial load: resolved U.D.L down slope
$= 16.2 \times 22.5 \sin 30°$
$= 158$ kN

Axial load at B9
$= 424 - \frac{16.875 \times 158}{22.5}$
$= 305$ kN.

Section properties for a 533 × 210 × 109 U.B

Ref 4/1 — 'Plastic' section: T = 18.8 mm., $p_y = 265$ N/mm²

$U = 0.875$: $x = 30.9$ mm : $r_{yy} = 46$ mm :
$A = 13900$ mm² : $S_{xx} = 2820$ cm³

L_e, between restraints, $= 2 \times 1875 = 3750$ mm

$\therefore \lambda = \frac{3750}{46} = 81.5$: $\frac{\lambda}{x} = \frac{81.5}{30.9} = 2.64$

$\therefore v = 0.93$ (for N = 0.5)

Conservatively, take $n = 1.0$;

$\therefore \lambda_{LT} = 1.0 \times 0.875 \times 0.93 \times 81.5 = 66$

Table 11 — $p_b = 195$ N/mm²

$\therefore M_b = 195 \times 2820 / 10^3 = 550$ kN.m.

Table 27(b) — P_{cy} : $\lambda = 81.5$: $p_{cy} = 173$ N/mm²

$\therefore P_{cy} = \frac{173 \times 13900}{10^3} = 2405$ kN.

Cl. 4.8.3.3.1 — $\therefore$ Combined axial & bending =

$\frac{305}{2405} + \frac{464}{550} = 0.97 < 1.0$ ✓ ok.

b) Check rafter between B5 & B6 (restraint from purlins on comp. flange)

Fig. 4/013 B5 → B6

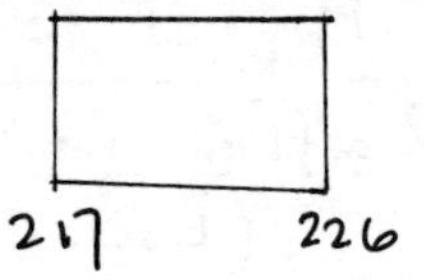

$L_e = 1875$ mm

$\therefore \lambda = \frac{1875}{46} = 41$

$\frac{\lambda}{x} = \frac{41}{30.9} = 1.32$: $v = 0.98$

Ref.	
Table 11	∴ $\lambda_{LT} = 1.0 \times 0.875 \times 0.98 \times 41 = \underline{35}$ $p_b = 265\ N/mm^2$ ∴ $M_b = 265 \times 2820 / 10^3 = \underline{747\ kN.m}$
Table 27(b)	P_{cy}: $\lambda = 41$: $p_{cy} = 240\ N/mm^2$ ∴ $P_{cy} = \dfrac{240 \times 13900}{10^3} = \underline{3336\ kN}$.
	Axial load, F $= 424 - \dfrac{0.375}{22.5} \times 158 = \underline{358\ kN}$.
Cl. 4.8.3.3.1	∴ Combined axial & bending = $\dfrac{358}{3336} + \dfrac{226}{747} = \underline{0.41} < 1.0$ ✓ o.k.
	∴ For L.C.I, 686 × 254 × 170 U.B. OK. for stanchions; 533 × 210 × 109 U.B. O.K. for rafters. From Fig. 4/009, an inspection of the bending moments indicates that this load case is <u>not</u> critical.
Refs. 4/3, 4/4, & 4/5	<u>Connections</u>: Haunch (eaves) connections dealt with in other publications (Ref 4/4 for example) along with ridge connection. However, a fixed-base stanchion bolt design follows, method as Ref 4/3, details as ref. 4/5 Using 8.8 Grade H/D bolts, $p_t = 450\ N/mm^2$. p_b = allowable bearing pressure on concrete $p_b = 0.4\ f_{cu}$: if $f_{cu} = 35\ N/mm^2$ (see brief) ∴ $p_b = 14\ N/mm^2$ For all dimensions, see Fig 4/014. $x = 15.p_b.d/(15p_b + p_t) = \dfrac{15 \times 14 \times 940}{(15 \times 14 + 450)}$ $x = \underline{299\ mm}$. $M' = M + W.a$ = mt. about centre of bolts. $M' = 785 + 403.7 \times 0.44 = \underline{963\ kN.m}$

Ref.	
	C = force in compression = $M'/(d - x/3)$
	$\therefore$ $C = \dfrac{963 \times 10^6 \times 10^{-3}}{(940 - 299/3)} = \underline{1146\ kN.}$
	p_{max} = max. pressure = $2C/bx$
	$p_{max} = \dfrac{2 \times 1146 \times 10^3}{600 \times 2} = \underline{12.8\ N/mm^2} < 14$ ✓ O.K.
	T = tension in bolts = 1146 − 403.6 = $\underline{742\ kN}$
	Capacity, M24 bolt = $\dfrac{353 \times 450}{10^3} = \underline{159\ kN.}$
	5 No. bolts = $\underline{795\ kN}$ ✓
	$\therefore$ use 5 No. M24 H/D bolts/side = $\underline{10\ No.\ total.}$
	Fig. 4/014
	(see figure below)
	Longitudinal stability: The 2 bays of wind trusses (see Fig. 4/002) for wind on the gable, plus the gable steel will have to be designed for wind pressure.

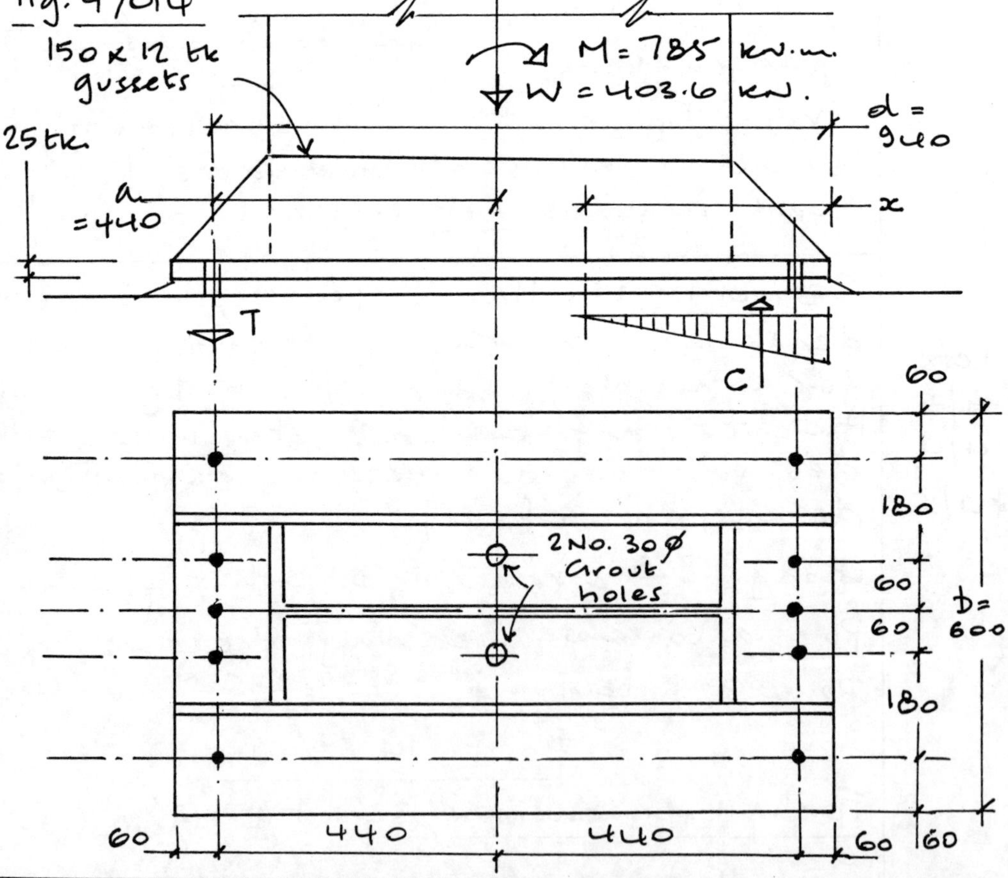

Chapter 5 Library/exhibition hall building

A Local Authority requires a prestigious public building, which combines a new Library with a large Exhibition Hall. The proposed scheme is shown in section and plan in Fig. 5/001. The Library floor, which provides the roof of the Exhibition Hall, is hung from a large roof truss, because the structural zone of the floor (600 mm.) is too shallow to span a Universal Beam or Column section, or even a built-up beam. The external elevations are to be clad in lightweight glass curtain walling fixed to floors and columns, and the floors themselves are to be of screeded pre-cast units on ledger angles. The Exhibition Hall is to have two roller-shutter doors in each gable of the building, central to the Hall centreline, 6.0 m. x 6.0 m. in area. Although the building has stiff cores for lifts, services, and staircases, it cannot be assumed that these will be adequate to deal with lateral wind forces. In this solution, only the main structural elements, bracing and foundations are designed.

Loading – live qk (For grid references, see Fig. 5/001)

Library floor between grids A and D	– 10.0 kN/m^2
Plant Room floor between grids B and C	– 7.5 kN/m^2
Office floors (including partitions)	– 5.0 kN/m^2
Roof (snow)	– 0.75 kN/m^2
Roof (plant suspended from roof)	– 0.5 kN/m^2

Loading – gk

Roof sheeting and insulation	– 1.0 kN/m^2

Loading – wk

Basic wind speed 44 m/sec (city centre site)

Site conditions

All point loads to be taken through pilecaps, 0.5 m. below ground level, on end-bearing piles on Grade III chalk 5.0 m. below ground level. Between ground level and the chalk is a band of loose gravel suitable for supporting ground-bearing slabs only.

Design Code of Practice

The building is to be designed in accordance with BS 5950, Structural Steelwork.

Other Design Aids

For reasons of space and copyright, the pre-cast floor units have been taken from the manufacturer's tables without publication in this book.

Pre-cast floor units (library floor) – from the Hollow Core Floor and Roof Units table, 200 mm. thick units with non-structural screed will carry an unfactored live load of 10 kN/m^2 over a span of 5.0 m. simply-supported.

Pre-cast floor units (plant room) – as above, 150 mm. thick units will carry an unfactored live load of 7.5 kN/m^2 over a span of 5.0 m. simply-supported.

Pre-cast floor units (office floors) – as above, 110 mm. thick units will carry an unfactored live load of 5 kN/m^2 over a span of 4.5 m. simply-supported.
Note! In the following calculations, **gk** = 5.0 kN/m^2, for units + screed has been taken for all 3 cases.

Materials

All structural steelwork to be EN 10025 S275

Fig.5/001

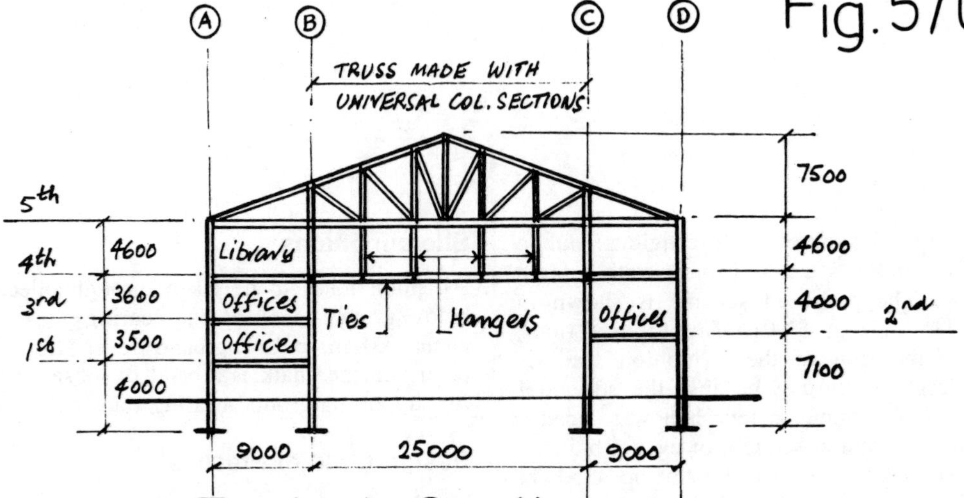

Typical Section

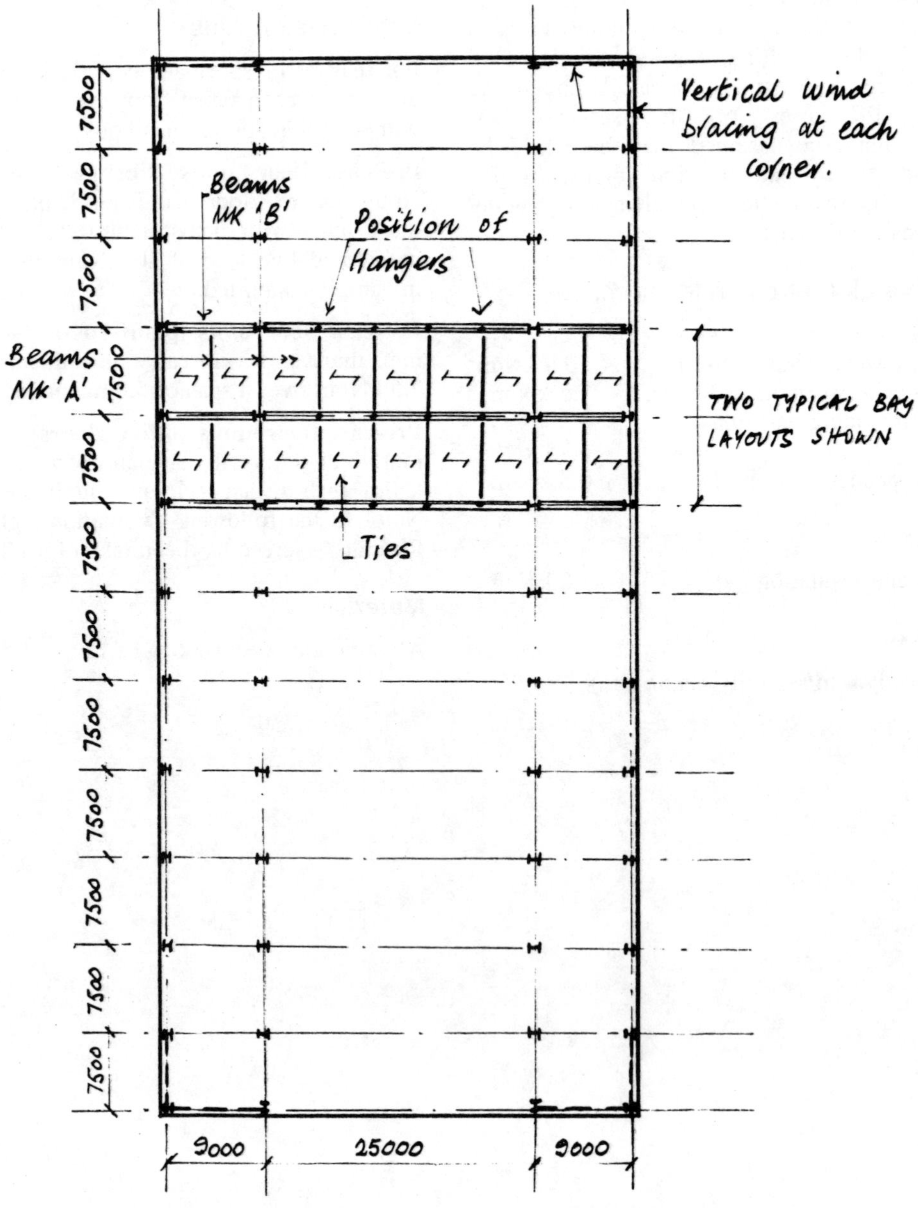

Plan at Library Floor Level

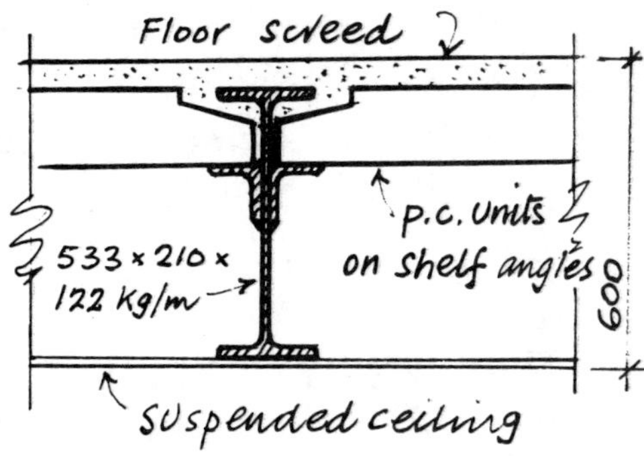

Beam / Slab
Detail

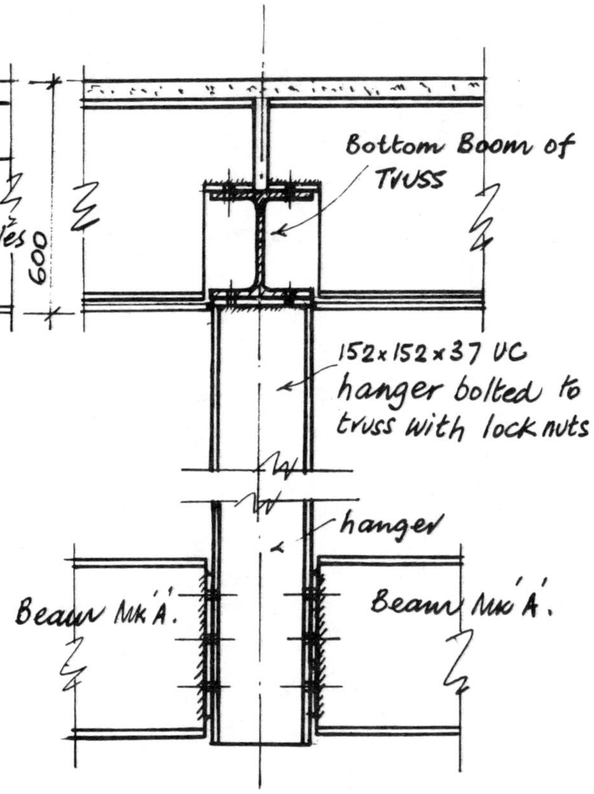

Hanger Details

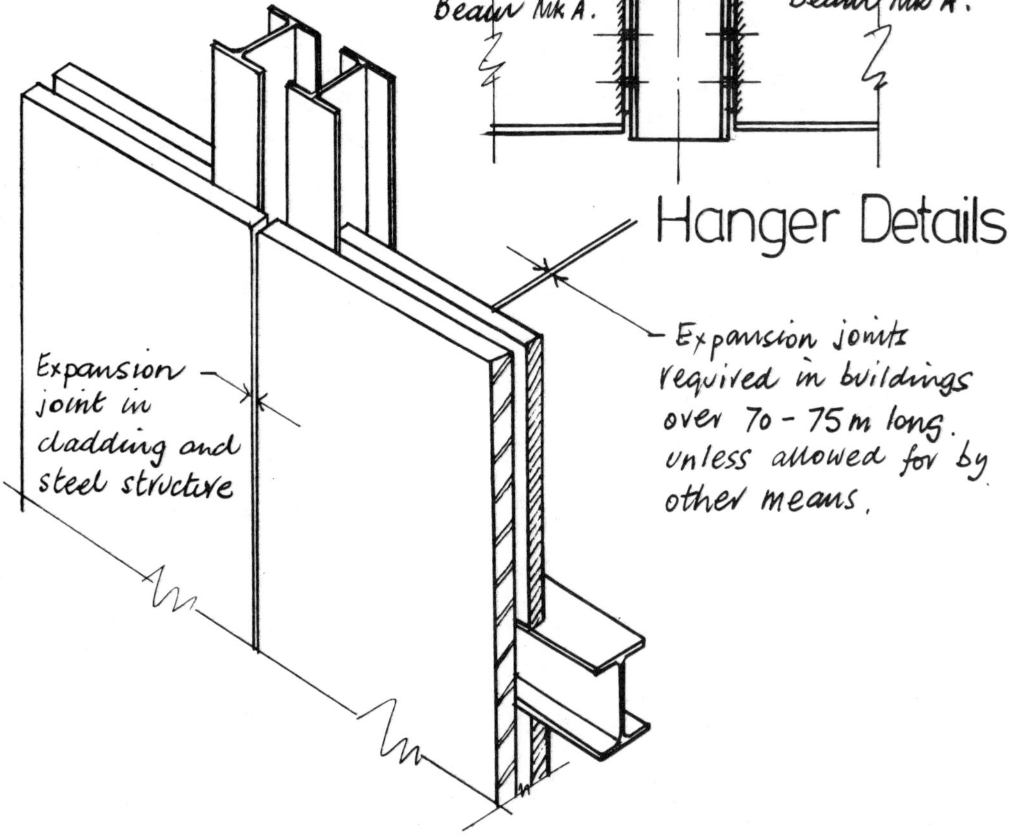

Typical Detail at Expansion Joint

Fig.5/002

Ref	
(B.S 5950)	**Design Floor loadings:-**
	Allow $g_k = 5.0\ kN/m^2$ for pre-cast slab + finishes (see brief)
Table 2	Library floor $= 1.6 \times 10 + 1.4 \times 5$
	$= 23.0\ kN/m^2$
	Plant room floor $= 1.6 \times 7.5 + 1.4 \times 5$
	$= 19.0\ kN/m^2$ (grids B → C)
	Office floors $= 1.6 \times 5 + 1.4 \times 5$
	$= 15.0\ kN/m^2$
	Design roof loading $= 1.6(0.75 + 0.5) + 1.4 \times 1.0$
	$= 3.4\ kN/m^2$
	Beams Mk. 'A' [Fig 5/001]
	Span 7.5 m. – fully restrained.
	Beams at 5.0 m spacing (Fig 5/001)
	∴ U.D.L. $= 23 \times 5 = 115\ kN/m.$
	∴ $M_{MAX} = \frac{115 \times 7.5^2}{8} = 808\ kN.m.$
	F_v = shear $= \frac{115 \times 7.5}{2} = 431\ kN.$
	Try a 533 × 210 × 122 kg/m (must be less than 600 mm deep)
Table 6	Check classification: T = 21.3 mm : $p_y = 265\ N/mm^2$
Table 7	∴ $\varepsilon = \sqrt{\frac{275}{265}} = 1.02$
Ref 5/1	Flange: $b/T = 4.97 < 8.5\varepsilon = 8.7$ – 'plastic'
	Web: $d/t = 37.2 < 79\varepsilon = 80.6$ – 'plastic'
	∴ section 'plastic'
	Check shear: $F_v = 431\ kN$
Cl. 4.2.3	$P_v = 0.6 \times 265 \times 12.8 \times 544.6 / 10^3$
	$P_v = 1108\ kN$
	∴ $P_v > F_v$ (1108) (431) ✓ o.k.
Cl. 4.2.4	Check bending: for 'low shear',
	$0.6 P_v = 0.6 \times 1108 = 665\ kN$
	∴ $0.6 P_v > F_v$ – 'low shear' (665) (431)

Ref.	Calculation
Ref. (B.S. 5950)	∴ Use $M_{cx} = p_y . S_{xx}$
	$M_{cx} = 265 \times 3203 \times 10^3/10^6 = 849$ kN.m
Cl. 4.2.5	∴ $M_{cx} > M_{MAX}$ ✓ o.k.
	(849) (808)
Cl. 2.5.1	**Check Deflection:** $q_k = 10 \times 5 = 50$ kN/m
	$\Delta = \frac{5}{384} \times \frac{50 \times (7500)^4}{205 \times 10^3 \times 76207 \times 10^4} = 13.2$ mm
Table 5	Allowable $= \frac{7500}{360} = 20.8$ mm ✓ o.k.
	∴ Mk. 'A': 533 × 210 × 122 kg/m U.B.
	Beams Mk. 'B' [Fig. 5/001]
	Span 9.0 m. — restrained at supports & at centre point.
	Between gridlines A & B and C & D, beams 'A' are at 4.5 m. crs.
	∴ Reaction from beams 'A' $= \frac{4.5 \times 431 \times 2}{5}$
	= 776 kN.
	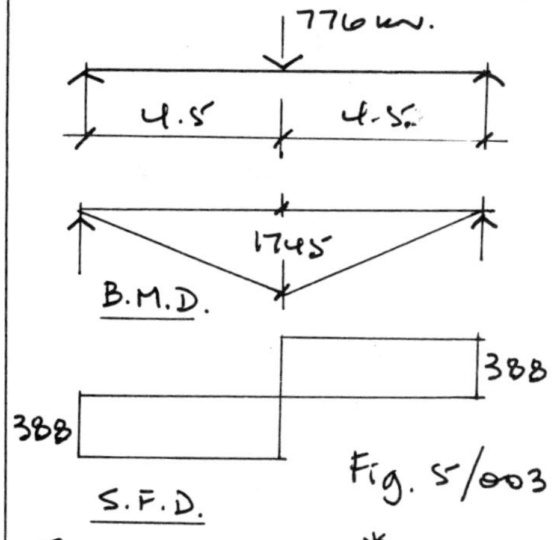 Fig. 5/003
	$M_{MAX} = \frac{WL}{4} = \frac{776 \times 9}{4}$
	= 1745 kN.
	$F_v = \frac{776}{2} = 388$ kN.
	Unrestrained length = 4.5 m. (see Fig 5/003)
	∴ $L_e = 4500$ mm
	Try a 762* × 267 × 197 kg/m U.B.
Ref. 5/1 & Table 6	T = 25.4 mm, $p_y = 265$ N/mm²
	($r_{yy} = 57.1$ mm, $u = 0.869$, $x = 33.2$, $S_{xx} = 7170$ cm³)
	For a beam not loaded between restraints, 'n' = 1.0 & 'm' varies. (see Table 13)
	$\lambda = \frac{4500}{57.1} = 78.8$: $\frac{\lambda}{x} = \frac{78.8}{33.2} = 2.4$
Table 14	∴ 'v' = 0.94
	∴ $\lambda_{LT} = 1.0 \times 0.869 \times 0.94 \times 78.8 = 55.3$
Table 11	For $p_y = 265$ N/mm², $p_b = 219.0$ N/mm²

Ref.	
	$\therefore M_b = \frac{219 \times 7170}{10^3} = \underline{1570 \text{ kN.m.}}$
(B.S. 5950) Table 18	<u>Calculate $\bar{M}$</u>: $\beta = 0$ & therefore $m = 0.57$
	$\bar{M} = m.M_{MAX} = 0.57 \times 1745 = \underline{995 \text{ kN.m}}$
Cl. 4.3.7.1	$\therefore \underline{M_b > \bar{M}}$ ✓ O.K.
	(1570) (995)
Cl. 4.2.5	Also beam must satisfy Clause 4.2.5 for M_{MAX}.
Cl. 4.2. .	$P_v = \frac{0.6 \times 265 \times 770 \times 15.6}{10^3} = \underline{1910 \text{ kN.}}$
	$0.6P_v = 0.6 \times 1910 = \underline{1146 \text{ kN.}}$
	Shear at M_{MAX} position $= \underline{388 \text{ kN.}}$ [Fig 5/003]
	$\underline{F_v < 0.6P_v}$: 'low shear' situation.
	$\therefore M_{cx} = \frac{265 \times 7170}{10^3} = \underline{1900 \text{ kN.m.}}$
	$\therefore \underline{M_{cx} > M_{MAX}}$ ✓ O.K.
	(1900) (1745)
	<u>Check shear</u>: $F_v = 388$ kN
Cl. 4.2.3	$P_v = 1910$ kN (see above)
	$\therefore \underline{P_v > F_v}$ ✓ O.K.
	(1910) (388)
	<u>Check Deflection</u>:
Cl. 2.5.1	q_k from beams 'A' $= 50 \times \frac{4.5}{5} \times 7.5$
	$= \underline{337.5 \text{ kN.}}$
	$\therefore \Delta = \frac{1}{48} \times \frac{337.5 \times 10^3 \times 9000^3}{205 \times 10^3 \times 240000 \times 10^4} = \underline{10.4 \text{ mm.}}$
Table 5	Allowable $= \frac{9000}{360} = \underline{25 \text{ mm.}}$ ✓ O.K.
	$\therefore$ <u>Mk. 'B' : 762* × 267 × 197 kg/m U.B.</u>
	(* Assuming that floor depth not a problem above offices)
	<u>Note</u>! Design of plant room beams similar.

Ref. (BS 5950)

Design of hangers (supported by bottom boom of truss) - see Figs. 5/001 & 5/002

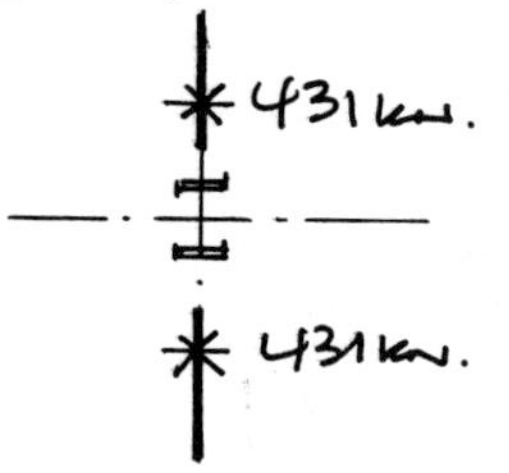

Fig. 5/004

Load Case 1 (Fig. 5/004)
(Full live load each side of hanger) - Beams 'A'

Total tensile load = 862kN

Assuming $p_y = 275\ N/mm^2$;

Area required $= \dfrac{862 \times 10^3}{275} = 3130\,mm^2$

Ref 5/1

Try a 152 x 152 x 37 U.C ($A = 4740\ mm^2$)

$T = 11.5\ mm$: $p_y = 275\ N/mm^2$ ✓ o.k.

$\therefore P_t = \dfrac{275 \times 4740}{10^3} = 1303\ kN.$

$\therefore \quad P_t > F_t$ ✓ o.k.
$(1303) \quad (862)$

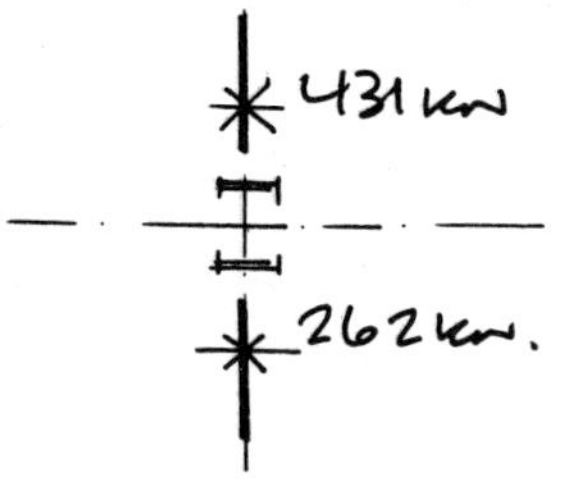

Fig. 5/005

Load Case 2 (Fig. 5/005)
(Full live load one side only) - Beams 'A'

Cl. 2.2.1

Total tensile load = 693kN.

Eccentricity of loads

$= \dfrac{D}{2} + 100 = \dfrac{150}{2} + 100 = 175\,mm$

$\therefore M_x = (431 - 262) \times 0.175 = 30\ kN.m$

From Cl. 4.8.2: $\dfrac{F_t}{P_t} + \dfrac{M_x}{M_{cx}} \le 1.0$

Cl. 3.5.1

Check Classification: ($\varepsilon = 1.0$)
No need, as member in tension.

$\therefore M_{cx} = \dfrac{275 \times 310}{10^3} = 85\ kN.m$

$\therefore \dfrac{693}{1303} + \dfrac{30}{85} = 0.88 < 1.0$ ✓ o.k.

$\therefore$ For hangers use 152 x 152 x 37 U.C.'s

Ref.	
(B.S. 5950)	Main truss design: span 25.0 m, trusses at 7.5 m. centres.

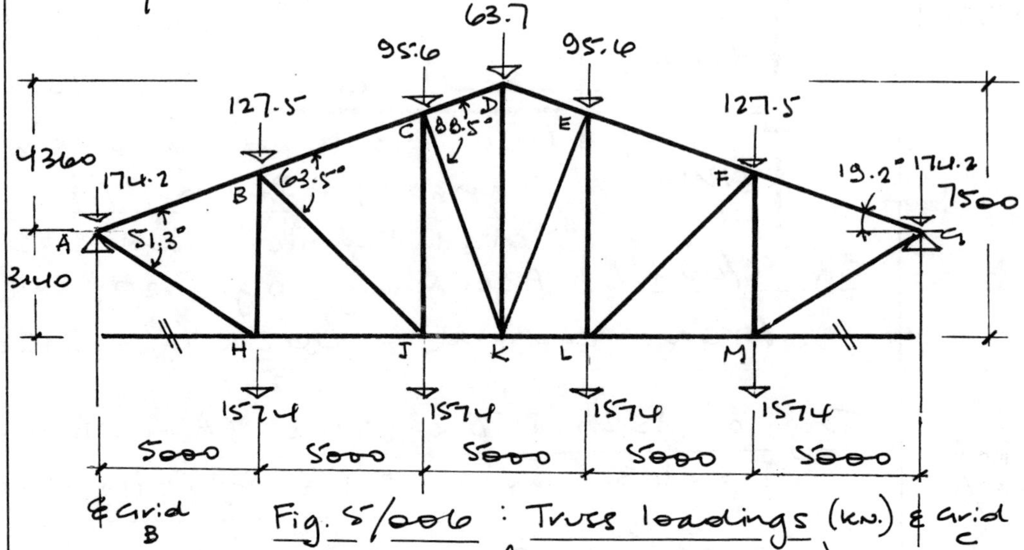

Fig. 5/006: Truss loadings (kN.) (see below for calculations)

Node loadings (see 'floor & roof design loadings' – first page of calculations)

a) **Top boom loadings**:

At B & F: Load = 3.4 × 7.5 × 5 = **127.5 kN.**

At A & G: Load = ½ panel load + reaction from sloping roof beam between grids A & B and C & D (see Fig. 5/002)

Load $= 3.4 \times 7 \times \frac{5}{2} + \frac{1}{2}\left[3.4 \times 9 \times 7.5\right] =$ **174.2 kN.**

At D: Load = 3.4 × 7.5 × 2.5 = **63.7 kN.**

At C & E: Load $= 3.4 \times 7.5 \times \left(\frac{5 + 2.5}{2}\right) =$ **95.6 kN.**

b) **Bottom boom loadings**: (H, J, L & M)

Load from hangers (max.) = **862 kN.**

Reactions from plant room floor beams:

Load $= 2 \times \frac{1}{2}\left[19.0 \times 5 \times 7.5\right] =$ **712.5 kN.**

Total = **1574 kN.**

Reactions at R_A & R_G:

$$R_A = R_G = \frac{1}{2}\left[2 \times 174.2 + 2 \times 127.5 + 2 \times 95.6 + 63.7 + 4 \times 1574\right]$$

= **3577 kN.**

Ref.
(BS 5950)

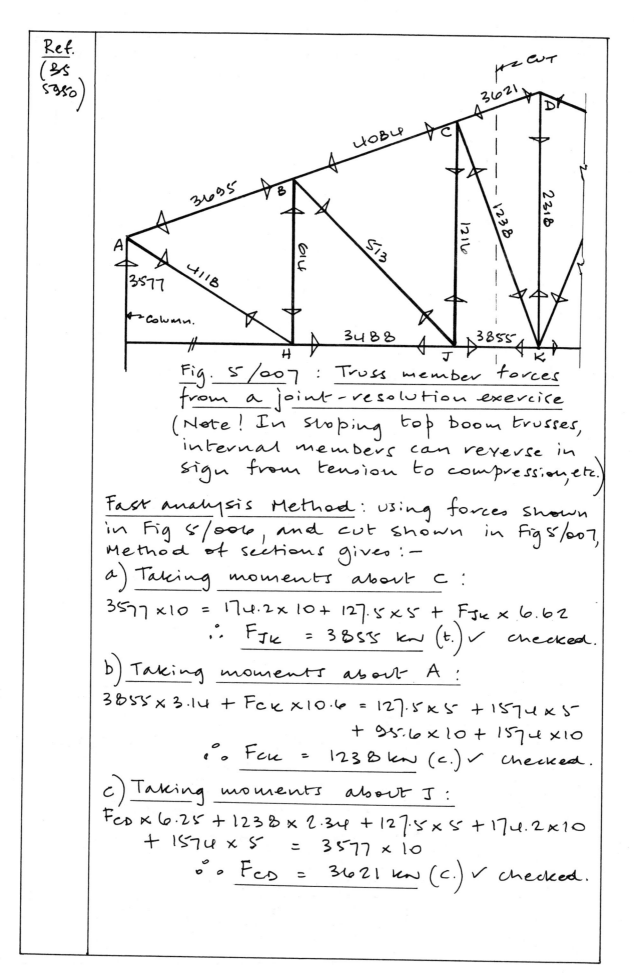

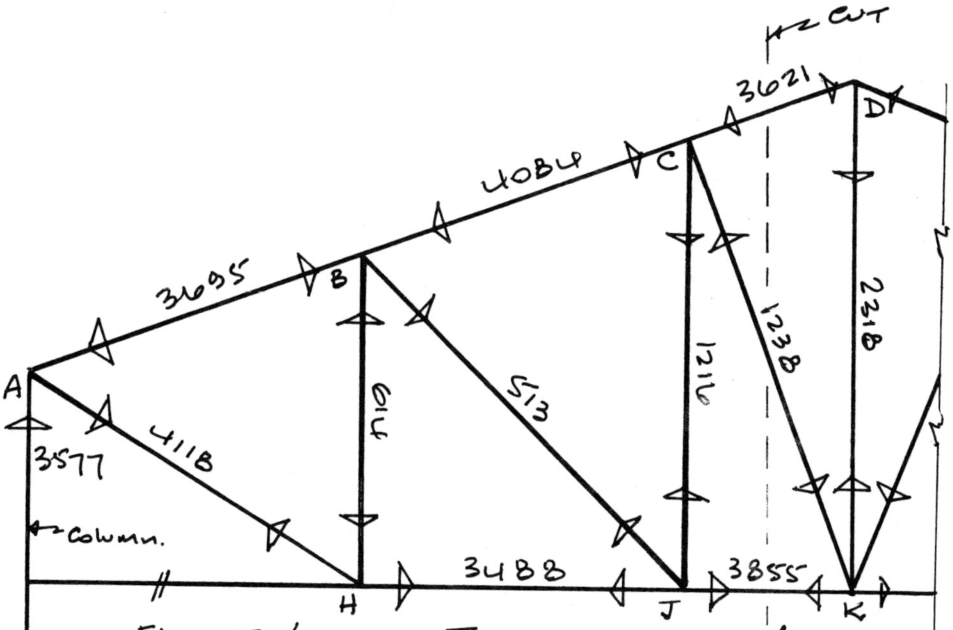

Fig. 5/007 : Truss member forces from a joint-resolution exercise
(Note! In sloping top boom trusses, internal members can reverse in sign from tension to compression, etc.)

Fast analysis Method: using forces shown in Fig 5/006, and cut shown in Fig 5/007, Method of sections gives:-

a) Taking moments about C:

$$3577 \times 10 = 174.2 \times 10 + 127.5 \times 5 + F_{JK} \times 6.62$$

$\therefore$ $F_{JK} = 3855$ kN (t.) ✓ checked.

b) Taking moments about A:

$$3855 \times 3.14 + F_{CK} \times 10.6 = 127.5 \times 5 + 1574 \times 5 + 95.6 \times 10 + 1574 \times 10$$

$\therefore$ $F_{CK} = 1238$ kN (c.) ✓ checked.

c) Taking moments about J:

$$F_{CD} \times 6.25 + 1238 \times 2.34 + 127.5 \times 5 + 174.2 \times 10 + 1574 \times 5 = 3577 \times 10$$

$\therefore$ $F_{CD} = 3621$ kN (c.) ✓ checked.

Ref (B.S. 5950)	
	Truss member design: use Universal column sections, with webs in plane of truss.
	a) **Top boom**: critical length = BC : $F_c = 4084$ kN., $Le = 5300$ mm.
Ref 5/1 Table 6	**Try a 305 x 305 x 198 UC.**
	$T = 31.4$ mm ; $py = 265$ N/mm² $(T > 16$ mm$)$
	Check section 'not slender'
Table 7	$\varepsilon = \sqrt{\frac{275}{265}} = \underline{1.03}$
Ref 5/1 Table 7	$b/T = \underline{5} < 15 \times 1.03 = \underline{15.45}$
	∴ flange 'not slender'
Ref 5/1 Table 7	$d/t = \underline{12.8} < 39 \times 1.03 = \underline{40.2}$
	∴ web 'not slender'
	∴ section 'not slender'
Cl. 4.7.4	∴ $P_c = A_g . p_c$.
Ref 5/1	$r_{yy} = 80.2$ mm., $A_g = 25200$ mm²
Cl. 4.7.3.2	$\lambda = \frac{5300}{80.2} = \underline{66} < \underline{180}$ ✓ o.k.
Table 27(c)	From Table 27(c), $p_c = 184$ N/mm²
	∴ $P_c = \frac{184 \times 25200}{10^3} = \underline{4637 \text{ kN}}$.
	∴ $P_c > F_c$ ✓ o.k.
	(4637) (4084)
	b) **Bottom boom**: critical length = JK :
Cl. 4.6.1	$F_t = 3855$ kN. Assume $py = 265$ N/mm².
	Area required $= \frac{3855 \times 10^3}{265} = \underline{14547 \text{ mm}^2}$
Ref 5/1 Table 6	**Try a 305 x 305 x 118 U.C.** $T = 18.7$ mm,
	$py = 265$ N/mm² : $A = 15000$ mm².
	$P_t = \frac{265 \times 15000}{10^3} = \underline{3975 \text{ kN}}$.
Cl. 4.6.1	∴ $P_t > F_t$ ✓ o.k.
	(3975) (3855)

Ref (B.S. 5950)	
	c) Diagonals. Check AH; Force = 4118 kN (tens.)
	Assume $p_y = 265\ N/mm^2$
Cl. 4.6.1	Area required $= \frac{4118 \times 10^3}{265} = 15540\ mm^2$
Ref s/1	Try a 305 × 305 × 137 U.C. T = 21.7 mm, $p_y = 265\ N/mm^2$: A = 17500 mm²
	$P_t = \frac{265 \times 17500}{10^3} = 4637\ kN.$
Cl. 4.6.1	$\therefore \underset{(4637)}{P_t} > \underset{(4118)}{F_t}$ ✓ o.k.
	For same section, check CK – compressive force = 1238 kN.
Ref s/1	$L_e = L = 7150\ mm$, $r_{yy} = 78.2\ mm.$
Cl 4.7.3.2	$\therefore \lambda = \frac{7150}{78.2} = 91.4 < 180$ ✓ o.k.
	Check section 'not slender'
Table 7	$\varepsilon = \sqrt{\frac{275}{265}} = 1.03$
Ref s/1	$b/T = 7.11 < 15 \times 1.03 = 15.45$
Table 7	∴ flange 'not slender'
Ref s/1	$d/t = 17.9 < 39 \times 1.03 = 40.2$
Table 7	∴ web 'not slender'
	∴ Section 'not slender'
Cl. 4.7.4.	$\therefore P_c = A_g \cdot p_c.$
Table 27(c)	Table 27(c), $p_c = 137\ N/mm^2.$
	$\therefore P_c = \frac{137 \times 17500}{10^3} = 2397\ kN.$
	$\therefore \underset{(2397)}{P_c} > \underset{(1238)}{F_c}$ ✓ o.k.
	d) Verticals. Check DK; Force = 2318 kN (tens.)
	Assume $p_y = 275\ N/mm^2$
Cl. 4.6.1	Area required $= \frac{2318 \times 10^3}{275} = 8429\ mm^2$
Ref s/1	Try a 254 × 254 × 73 U.C. T = 15.4 mm, $p_y = 275\ N/mm^2$ (✓ o.k.) : A = 9290 mm²

Ref. (BS 5950)	
	$P_t = \frac{275 \times 9290}{10^3} = \underline{2555\ kN}.$
	$\therefore \underline{P_t > F_t}$ ✓ o.k.
	(2555) (2318)
	For same section, check BH - compressive force = 614 kN.
	Check section 'not slender'
	$\varepsilon = \sqrt{\frac{275}{275}} = \underline{1.0}$
	$b/T = \underline{8.94} < 15\varepsilon = \underline{15.0}$
	$\therefore$ flange 'not slender'
	$d/t = \underline{23.9} < 39\varepsilon = \underline{39.0}$
	$\therefore$ web 'not slender'
	$\therefore$ section 'not slender'
	$\therefore P_c = A_g.p_c.$
Ref 5/1	$l_e = L = 4900$ mm : $r_{yy} = 64.6$ mm
	$\lambda = \frac{4900}{64.6} = \underline{76} < 180$ ✓ o.k.
Table 27(c)	From Table 27(c) : $p_c = 169\ N/mm^2$
	$\therefore P_c = \frac{169 \times 9290}{10^3} = \underline{1570\ kN}.$
	$\therefore \underline{P_c > F_c}$ ✓ o.k.
	(1570) (614)
	Sections:-
	Top boom: 305 x 305 x 198 U.C. Bottom boom: 305 x 305 x 118 U.C. Diagonals: 305 x 305 x 137 U.C.'s Verticals: 254 x 254 x 73 U.C.'s
	Connections: see Fig. 3/013, Chapter 3, for typical connection for a heavy truss, with U.C. members connecting. One typical connection for the hanger / truss junction is given overleaf.

Ref. (BS 5950)

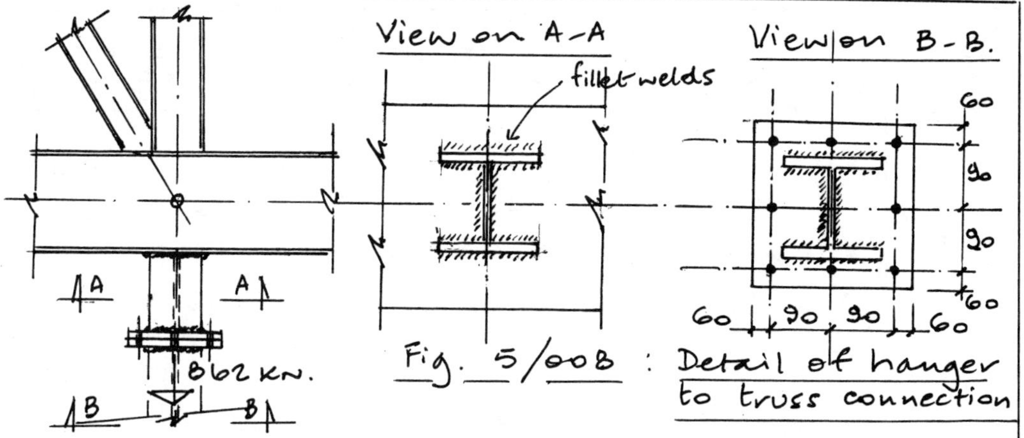

Fig. 5/008 : Detail of hanger to truss connection

Factored tension on hanger = 862 kN.

Check weld strength – see Fig. 5/008

Length of weld = 2 × 150 + 4 × 65 + 2 × 125
= 810 mm.

Ref 5/1

Strength of an 8 mm fillet weld = 1.2 kN/mm

∴ Capacity of weld = 1.2 × 810 = 972 kN.

∴ Capacity > tension.
(972 kN) (862 kN.)

Check bolts between stub end plate & hanger end plate. (see 'View on B-B')

Ref 5/1

Try 8 No. M20 8.8 bolts : P_t = 110 kN

Capacity = 8 × 110 = 880 kN.

∴ Capacity > tension
(880 kN) (862 kN.)

2 No. 20 tk end plates should be checked for 'prying forces'.

Design of a supporting column, grid line 'C'

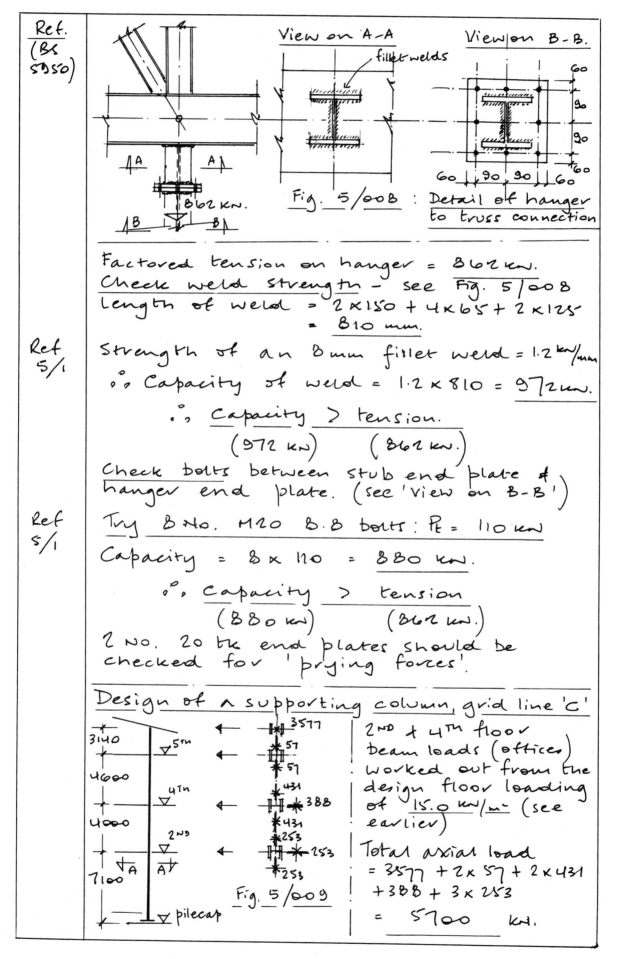

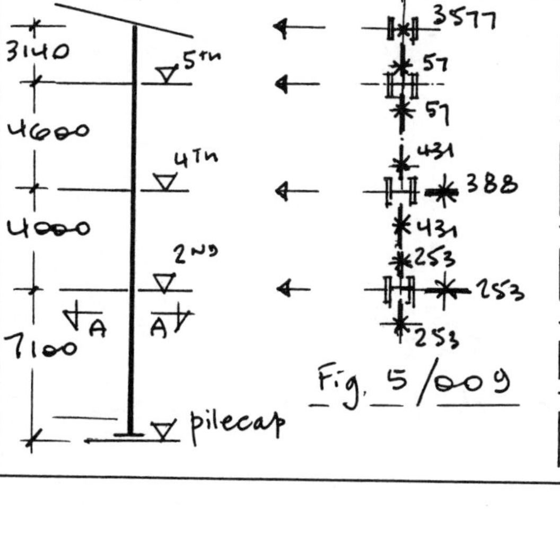

Fig. 5/009

2ND & 4TH floor beam loads (offices) worked out from the design floor loading of 15.0 kN/m² (see earlier)

Total axial load
= 3577 + 2 × 57 + 2 × 431
+ 388 + 3 × 253
= 5700 kN.

Ref. (BS 5950)	Check pile-cap to 2ND floor, section A–A (Fig 5/009): axial load = 5700 kN.
	Effective length: 2ND floor, restrained in direction & position, pilecap,
Table 24.	restrained in direction: $L_E = 0.85L$
	Try a 356 x 406 x 287 kg/m U.C.
Cl. 4.7.7	Ecc^y on x–x axis = 0.1 + 0.15 = 0.25 m.
	Eccentric load = 253 kN
	$\therefore M_{xx}$ = 253 x 0.25 = 63.2 kN.m.
	I/L (lower) = 1/7.1 = 0.141
	I/L (upper) = 1/4 = 0.250
	Σ = 0.391
Cl. 4.7.7	$\therefore M_{xx}$ lower (at section A–A)
	$= 63.2 \times \frac{0.141}{0.391} = 22.8$ kN.m.
	$L_E = 0.85 \times 7.1 = 6.0$ m.
Ref 5/1	r_{yy} = 103 mm.; A = 36600 mm^2; S_x = 5820
	$\therefore$ For calculation of P_c : T = 36.5; p_y = 265
Cl. 4.7.3.2	$\lambda = \frac{6000}{103} = 58 < 180$
	From Table 27(c), p_c = 199 N/mm^2
	$\therefore P_c = \frac{199 \times 36600}{10^3} = 7283$ kN.
Cl. 4.7.7 & Table 11	For calculation of M_{bs} :
	$\lambda_{LT} = \frac{0.5 \times 7100}{103} = 35$: p_b = 265 N/mm^2
	$\therefore M_{bs} = \frac{265 \times 5820}{10^3} = 1542$ kN.m.
Cl. 4.7.7 & Cl. 4.8.3.3.1	Interaction formula:
	$\frac{F}{P_{cy}} + \frac{M_x}{M_{bs}} \leq 1.0$
	$\therefore \frac{5700}{7283} + \frac{22.8}{1542} = 0.80 < 1.0$ ✓ o.k.
	$\therefore$ For cols on grid 'C' : 356 x 406 x 287 U.C o.k.

Ref (BS 5950)

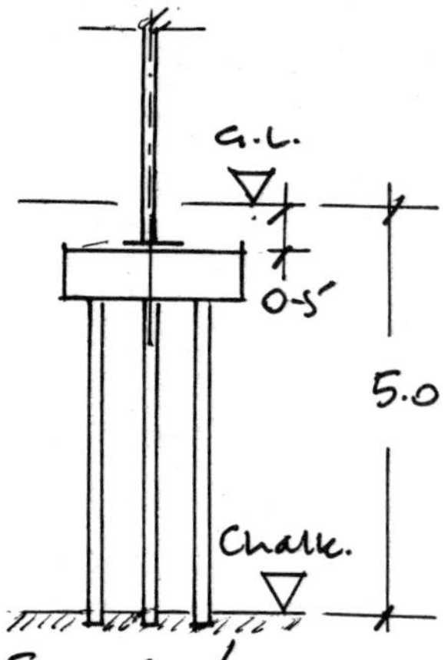

Fig. 5/010 Typical Foundation.

U.L.S. axial load on pilecap = 5700 kN.

∴ Service load ≏ $\frac{5700}{1.5}$ = 3800 kN.

Take pilecap as 3.0 m. sq × 0.6 m deep (see Chapter 8 for full design of pilecap)

∴ S.W.T. = $3.0^2 \times 0.6 \times 24$
= 130 kN.

Total load = 3930 kN.

Ref 5/2 Table 4.3

Use bored-and-cast-in-place piles, end bearing on the chalk. From Ref 5/2 the safe bearing for a 500 mm ø pile on Grade III chalk is 4 MN/m² = 4000 kN/m².

∴ End bearing/pile = $\frac{\pi \times 0.5^2}{4} \times 4000$

= 785 kN.

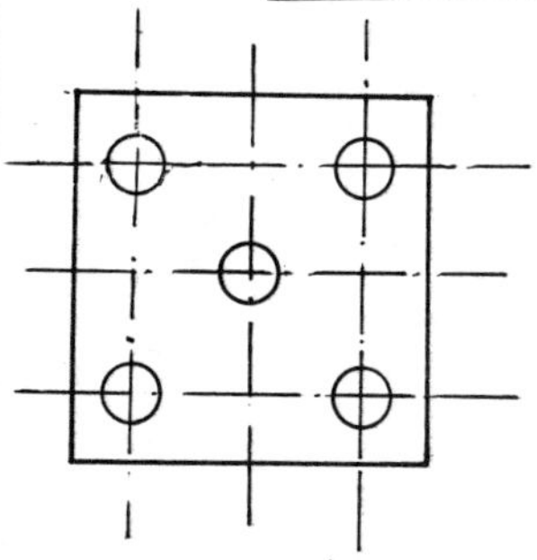

Fig. 5/011

Try 5 No. piles (see Fig /011)

Working load = 5 × 785
= 3925 kN.

∴ Working load ≏ Service load

Ref 5/2 Table 2.12

N.B.! Piles to be checked in compression between chalk & pilecap. Min. 5 No. T16 longitudinal bars, 40 mm cover.

Overall stability of building: 2 sets of wind bracing on each side & end. Wind on side designed for, here; 90° direction similar.

CP3 Ch V Pt 2.

Calculation of q.

V = 44 m/sec.

S2 = 0.78 (h = 15.5 m)

∴ V_s = 0.78 × 44 = 34.3 m/sec.

∴ q = $\frac{0.613 \times 34.3^2}{10^3}$ = 0.72 kN/m²

Ref. (BS 5950) CP3. Chv. Pt 2 Table 7

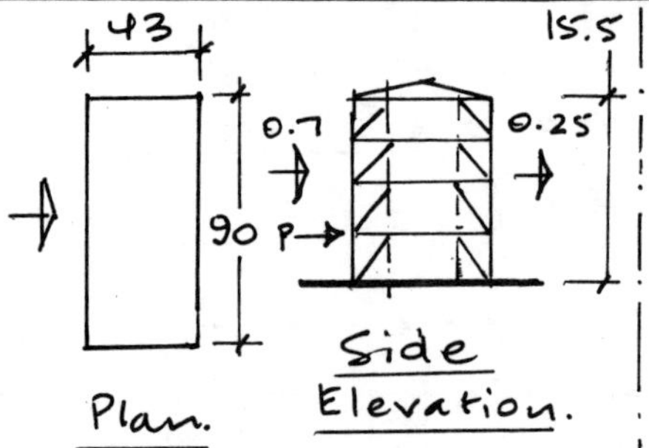

Fig. 5/012

Cpe coefficients (see Fig. 5/012)

$\frac{h}{w} = \frac{15.5}{43} = 0.36 \left(\leq \frac{1}{2}\right)$

$\frac{l}{w} = \frac{90}{43} = 2.1 \left(\frac{3}{2} < \frac{l}{w} < 4\right)$

Windward side, Cpe = +0.7

Leeward side, Cpe = −0.25

Total = 0.95 ⟶

Calculation of 1.4 Wk.

Wind load on complete side of building

= 1.4 × 0.95 × 0.72 × 90 × 15.5 = 1336 kN.

One set of bracing each end:

Force per set $= \frac{1336}{2} =$ 668 kN.

As a conservative quick design, concentrate this force as a shear, P, at 1st floor level. (see Figs. 5/012 & 013)

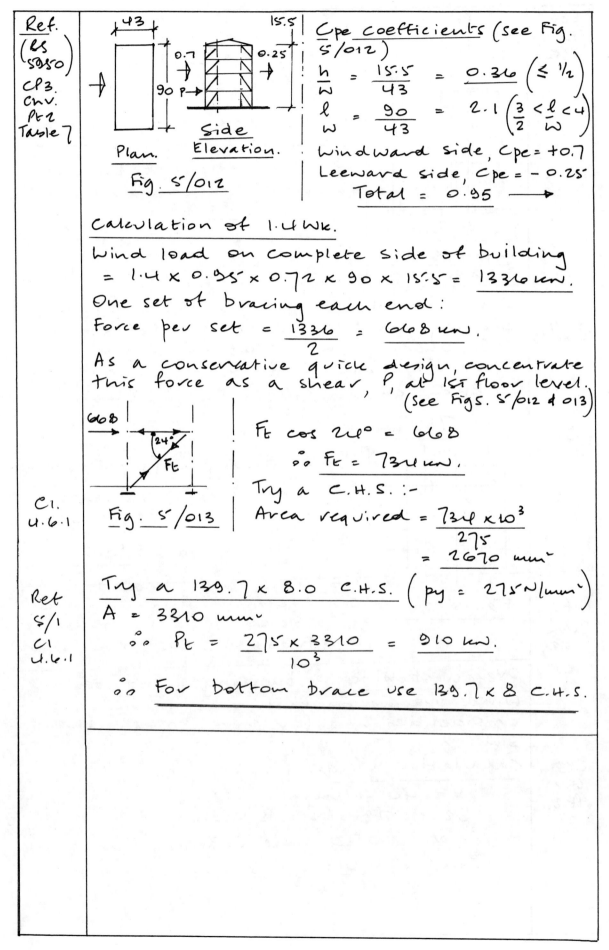

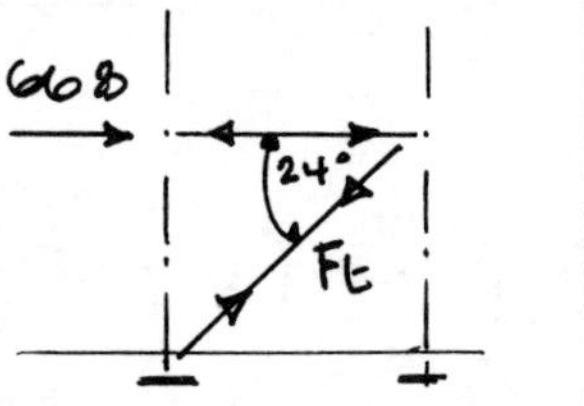

Fig. 5/013

$F_t \cos 24° = 668$

∴ $F_t = 731$ kN.

Try a C.H.S. :-

Cl. 4.6.1 Area required $= \frac{731 \times 10^3}{275}$

= 2670 mm²

Ref 5/1 Try a 139.7 × 8.0 C.H.S. (p_y = 275 N/mm²)

A = 3310 mm²

Cl 4.6.1 ∴ $P_t = \frac{275 \times 3310}{10^3} = 910$ kN.

∴ For bottom brace use 139.7 × 8 C.H.S.

Chapter 6 New grandstand at a football stadium

A prominent Premier League football club require a new stand and Hospitality boxes. The proposal stipulates a requirement for an unimpeded view of the pitch, a stand of 60 m. long, with a roof 24 m. wide and a minimum of 15 m. to the underside. The proposed section is shown in Figure 6/001, whilst the proposed plans at raker beam level, and roof level are shown in Figures 6/004 and 6/007, respectively. Details of the raker beam and precast concrete terracing are shown in Fig. 6/002. The space under the boxes and gangway is to be a circulation area, whilst the space under the rakers is to be a void. The structure is to be as aesthetically pleasing as possible.
The scheme chosen has a rear framework of CH Sections, connected by bosses to the middle and top of the cantilever columns, with SHS trusses connected to the top of the column, and suspended at mid-span by the framework. The trusses have a system of longitudinal bracing to the bottom booms, as shown in Fig. 6/007. With regard to the wind uplift load case on the truss, it must be noted that a simplification has been made, whereby the wind uplift is taken as vertical, and *not* at right-angles to the sheeting. Foundations and precast concrete terracing are not designed in this example. It is also assumed that the shape of the roof trusses can be cambered to cater for deflection.

Loading – qk

Roof – snow loading	= 0.60 kN/m^2
Roof – services	= 0.25 kN/m^2
Terracing, boxes and gangway	= 5.00 kN/m^2

Loading – gk

Roof – decking + insulation + steel = 0.56 kN/m^2
Terracing – to be calculated

Loading – wk

The football ground is in the middle of the town: v = 46 m/sec.

Site Conditions

The soil profile, from a thorough ground investigation, showed 0.75 m. of made ground over 1.5 m. of loose gravel lying on 25 m. of firm chalk. The ground water table was not apparent throughout the investigation.

Codes of Practice

BS 5950 – Structural Steelwork
CP3 Ch5 Pt. 2 – Wind Loading Code

Materials

Steel to BS EN 10025: S275 for the superstructure

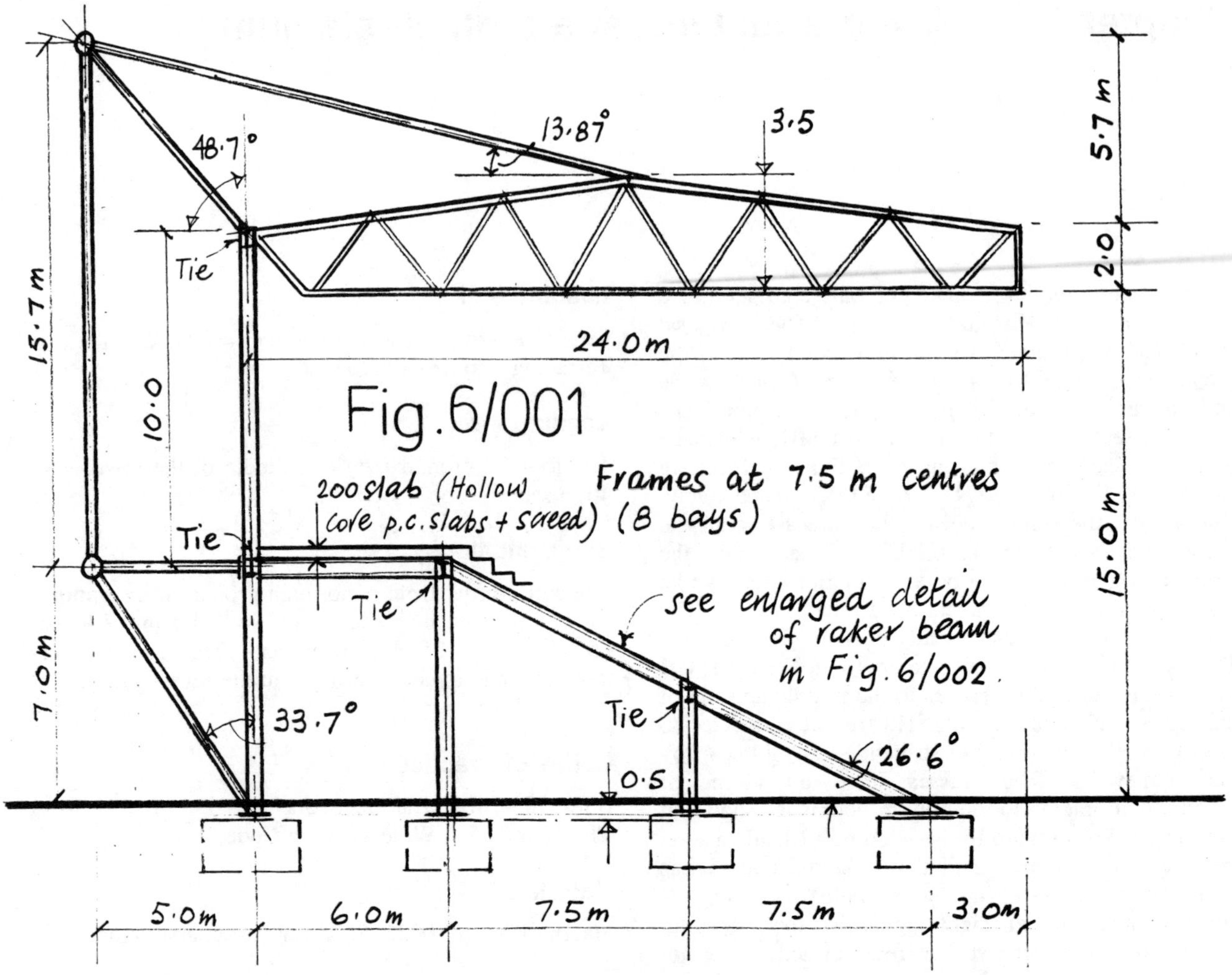

Fig.6/001

TYPICAL SECTION showing STEEL FRAME

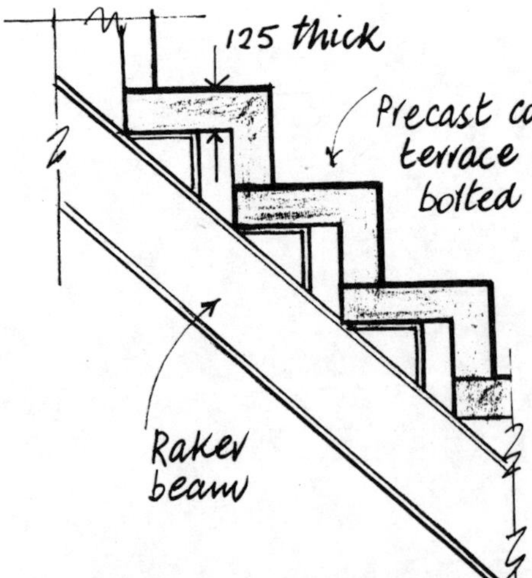

TYPICAL DETAIL of RAKER BEAM

Fig.6/002

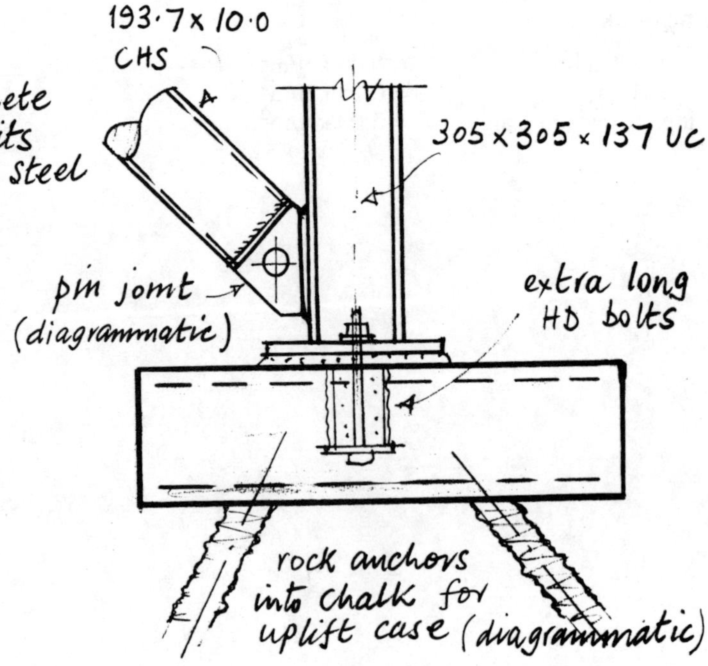

BASE DETAIL

Fig.6/003

Ref.	
(BS 5950)	**Design loadings:** (see brief)
	Roof – Load case I – $1.6q_k + 1.4g_k$
	$q_k = 0.60$ (snow) $+ 0.25$ (services)
	$q_k = 0.85\ kN/m^2$
	∴ Loading $= 1.6 \times 0.85 + 1.4 \times 0.56$
	$= 2.14\ kN/m^2$
	Roof – Load case II – $1.4w_k$ (uplift) $+ 1.0g_k$
CP3 Ch V Pt. 2 Table 3(3)	Calculation of w_k :
	$V = 46$ m/sec (see brief)
	$S2 = 0.81$ ($h = 17m$, class 'c')
	∴ $V_s = 0.81 \times 46 = 37.3$ m/sec.
	∴ $q_k = w_k = \frac{0.613 \times 37.3^2}{10^3} = 0.85\ kN/m^2$
	C_{pe} (uplift on canopy) $= 1.2$
	∴ ↑ $1.4w_k = 1.2 \times 1.4 \times 0.85 = 1.43\ kN/m^2$
	↓ $1.0g_k = 0.56\ kN/m^2$
	↑ Net $1.4w_k + 1.0g_k = 1.43 - 0.56$
	$= 0.87\ kN/m^2$
	Raker Beams: (see Fig. 6/002)
	g_k : For every metre of plan length, there is $445 + 1000 = 1445$ mm., say 1500 mm. precast steps, 125 mm thick.
	Volume/m $= 0.125 \times 1.5 = 0.188\ m^3$
	S.w.t. $= 24 \times 0.188 = 4.5\ kN/m$
	∴ Design load $= 1.6 \times 5 + 1.4 \times 4.5$
	$= 14.3\ kN/m^2$
	Gangway: (see Fig. 6/001)
	For 200 thk precast slabs, $g_k = 4.8\ kN/m^2$ (Hollowcore + screed)
	∴ Design load $= 1.6 \times 5 + 1.4 \times 4.8$
	$= 14.2\ kN/m^2$

Ref.
(BS 5950)

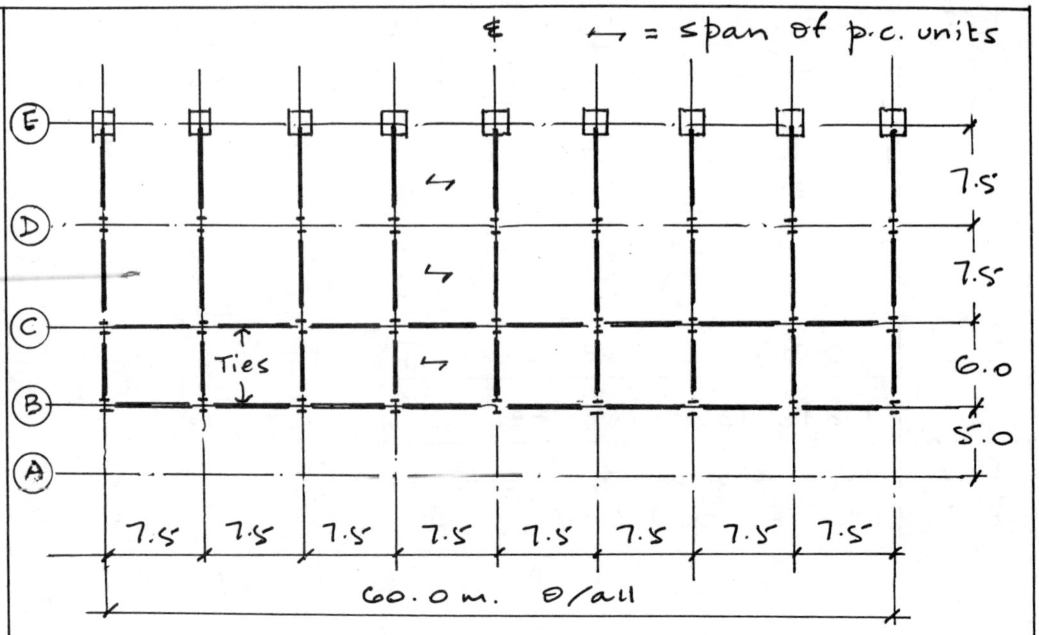

Fig. 6/004 : Plan on rakers.

Design of raker beams

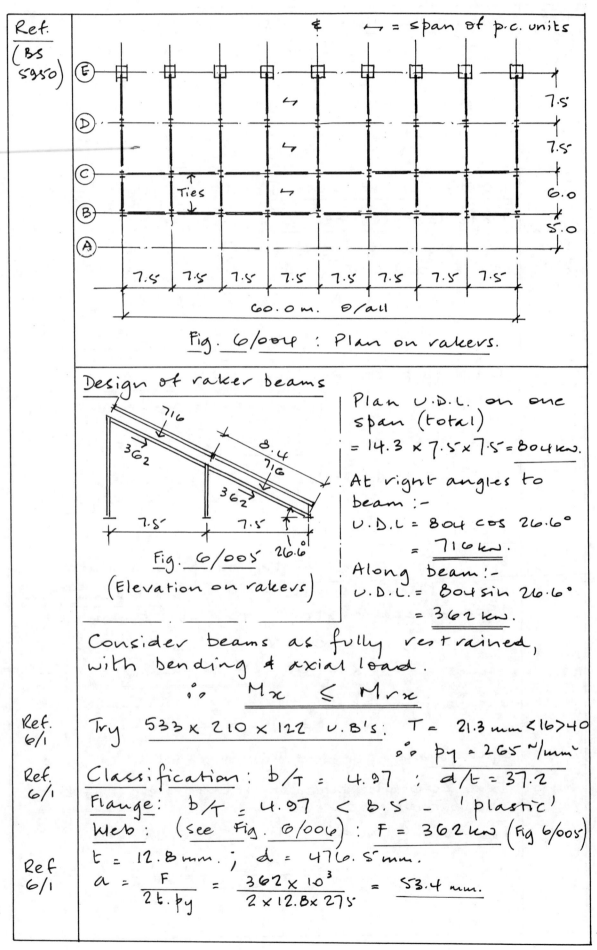

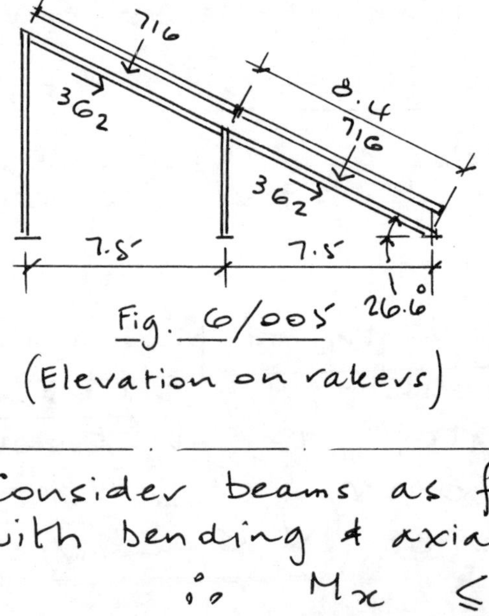

Fig. 6/005
(Elevation on rakers)

Plan U.D.L. on one span (total)
= 14.3 × 7.5 × 7.5 = 804 kN.

At right angles to beam :-
U.D.L = 804 cos 26.6°
= 716 kN.

Along beam :-
U.D.L. = 804 sin 26.6°
= 362 kN.

Consider beams as fully restrained, with bending & axial load.

$$\therefore \quad M_x \leq M_{rx}$$

Ref. 6/1 Try 533 × 210 × 122 U.B's : T = 21.3 mm <16>40
∴ py = 265 N/mm²

Ref. 6/1 Classification : b/T = 4.97 ; d/t = 37.2
Flange : b/T = 4.97 < 8.5 – 'plastic'
Web : (see Fig. 6/006) : F = 362 kN (Fig 6/005)
t = 12.8 mm ; d = 476.5 mm.

Ref 6/1

$$a = \frac{F}{2t.p_y} = \frac{362 \times 10^3}{2 \times 12.8 \times 275} = 53.4 \text{ mm.}$$

Ref. (BS 8110)

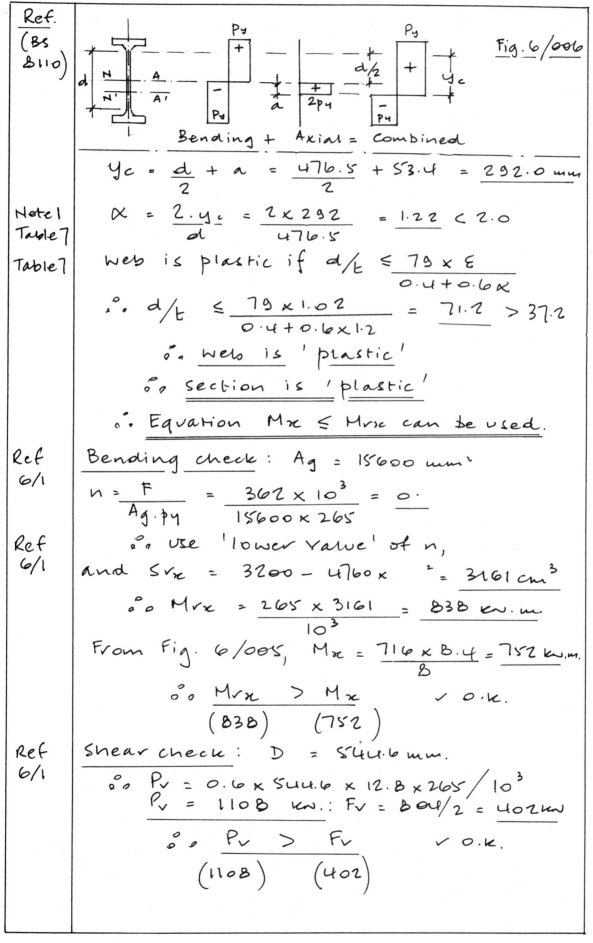

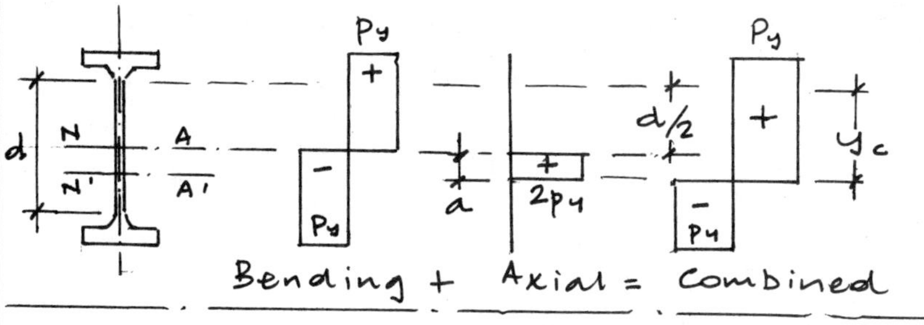

Fig. 6/006

$y_c = \frac{d}{2} + a = \frac{476.5}{2} + 53.4 = \underline{292.0 \text{ mm}}$

Note 1 Table 7 — $\alpha = \frac{2 \cdot y_c}{d} = \frac{2 \times 292}{476.5} = \underline{1.22} < 2.0$

Table 7 — web is plastic if $d/t \le \frac{79 \times \varepsilon}{0.4 + 0.6\alpha}$

$\therefore d/t \le \frac{79 \times 1.02}{0.4 + 0.6 \times 1.2} = \underline{71.2} > 37.2$

$\therefore$ web is 'plastic'

$\therefore$ section is 'plastic'

$\therefore$ Equation $M_x \le M_{rx}$ can be used.

Ref 6/1 — **Bending check**: $A_g = 15600 \text{ mm}^2$

$n = \frac{F}{A_g \cdot p_y} = \frac{362 \times 10^3}{15600 \times 265} = \underline{0\cdot}$

Ref 6/1 — $\therefore$ use 'lower value' of n,

and $S_{rx} = 3200 - 4760 \times {}^2 = \underline{3161 \text{ cm}^3}$

$\therefore M_{rx} = \frac{265 \times 3161}{10^3} = \underline{838 \text{ kN.m}}$

From Fig. 6/005, $M_x = \frac{716 \times 8.4}{8} = \underline{752 \text{ kN.m.}}$

$\therefore$ M_{rx} > M_x ✓ o.k.
(838) (752)

Ref 6/1 — **Shear check**: $D = 544.6$ mm.

$\therefore P_v = 0.6 \times 544.6 \times 12.8 \times 265 / 10^3$

$P_v = 1108$ kN.: $F_v = 804/2 = \underline{402 \text{kN}}$

$\therefore$ P_v > F_v ✓ o.k.
(1108) (402)

Ref	
(BS 5950)	**Deflection check**: $I_{xx} = 55400\ cm^4$
	By proportion of loadings: $q_k = \frac{716 \times 5}{14.3} = 250\ kN$
Ref 6/1	$\therefore \Delta = \frac{5}{384} \times \frac{250 \times 10^3 \times (8400)^3}{205 \times 10^3 \times 76200 \times 10^4} = 12.3\ mm$
	Allowable $= 8400/200 = 42\ mm$ ✓ o.k.
	$\therefore$ **Use 533 × 210 × 122 U.B.'s for rakers.**
	Design of gangway beams (between grids B&C)
	Span 6.0 m.; fully restrained by units.
	Gangway design loading (see previously) $= 14.2\ kN/m^2$
	$\therefore$ U.D.L. $= 14.2 \times 7.5 = 110.4\ kN/m.$
	$\therefore M_{MAX} = 110.4 \times 6^2/8 = 497\ kN.m.$
	S_{xx} required $= \frac{497 \times 10^3}{275} = 1807\ cm^3$
Ref 6/1	Try 457 × 191 × 82 kg/m U.B.'s ($S_{xx} = 1830\ cm^3$)
Ref 6/1	M_{cx}: $T = 16.0\ mm \le 16$: $p_y = 275\ N/mm^2$
	$\therefore M_{cx} = \frac{275 \times 1830}{10^3} = 503\ kN.m.$
	$\therefore M_{cx} > M_{MAX}$ ✓ o.k. (503) (497)
Ref 6/1	**Check shear**: $t = 9.9\ mm$: $D = 460.2\ mm$
	$\therefore P_v = 0.6 \times 275 \times 9.9 \times 460.2/10^3 = 752\ kN.$
	$F_v = 110.4 \times 6/2 = 331.2\ kN.$
	$\therefore P_v > F_v$ ✓ o.k.
	Check deflection: $q_k = 5 \times 7.5 = 37.5\ kN/m.$
	$\therefore \Delta = \frac{5}{384} \times \frac{37.5 \times (6000)^4}{205 \times 10^3 \times 37100 \times 10^4} = 8.32\ mm.$
	Allowable $= 6000/200 = 30\ mm$ ✓ o.k.
	$\therefore$ **Use 457 × 191 × 82 U.B.'s.**

Ref.

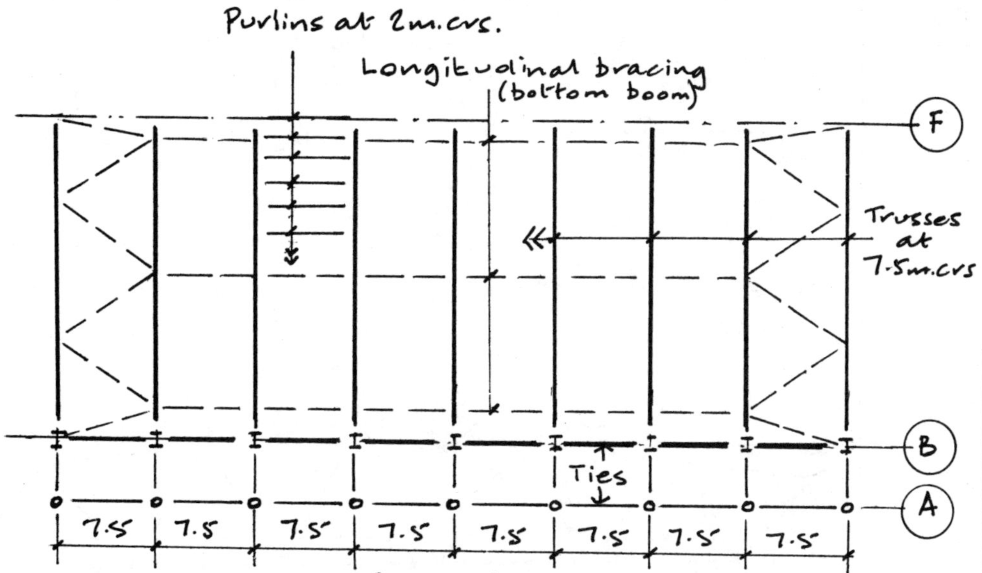

Fig. 6/007
Plan on roof steelwork.

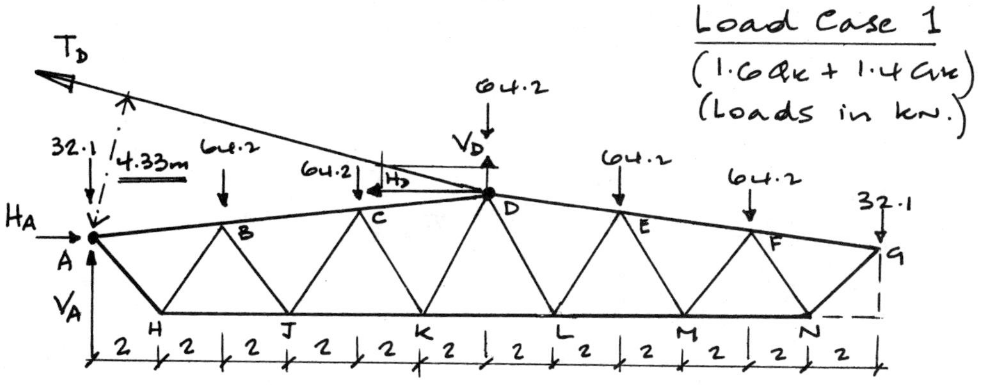

Fig. 6/008 - Truss loading.

Load Case 1 - Design loading = 2.14 kN/m²
(Fig 6/008) (see first page of calcs.)

Purlins are at 2m. crs (see Fig. 6/007)

∴ Purlin point load per truss

$= 2.14 \times 7.5 \times 2 = \underline{32.1\text{ kN}}$.

To obtain truss internal axial loads, concentrate loads at nodes and check later for bending in top boom.

∴ Node loads (B, C, D, E & F) = $2 \times 32.1 = \underline{64.2\text{ kN}}$

Node loads (A & G) = $\underline{32.1\text{ kN}}$

Taking moments about A :-

$$T_D \times 4.33 = 64.2\,[4 + 8 + 12 + 16 + 20] + 32.1 \times 24$$

$$\therefore\ T_D = 1067\text{ kN}.$$

Ref.

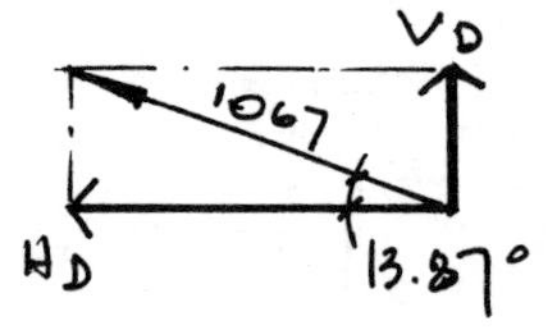

Fig. 6/009

From Fig. 6/009 :-

$V_D = 1067 \sin 13.87° = \underline{256 kN}$

$H_D = 1067 \cos 13.87° = \underline{1036 kN.}$

Putting these reactions, along with the loading in Fig 6/008, in a Planeframe program gave the following results :-

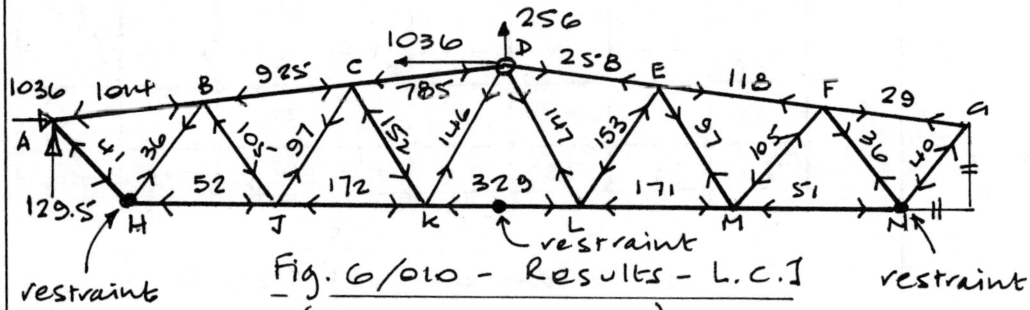

Fig. 6/010 - Results - L.C.I

(Forces in kN's)

Load Case 11

(1.4Wk [uplift] + 1.0Gk)

(Loads in kN.)

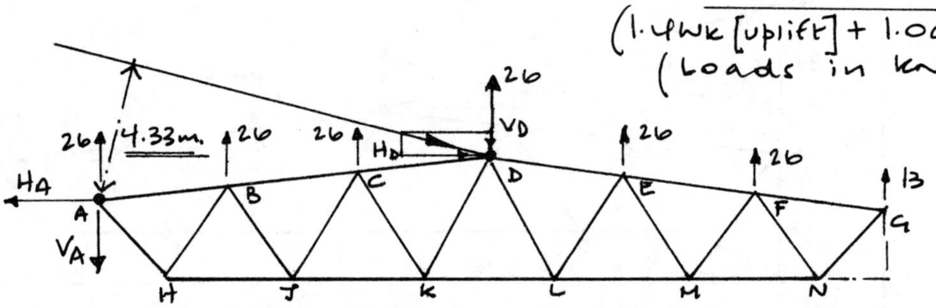

Fig. 6/011 - Truss Loading

Load case 11 - (Fig 6/011)

Design Loading = $\underline{0.87 kN/m^2}$ (uplift) - see first page of calcs.

∴ Purlin point load per truss

$= 0.87 \times 7.5 \times 2 = \underline{13 kN} \uparrow$

Node loads (B, C, D, E & F) = 2 × 13 = $\underline{26 kN} \uparrow$

Node loads (A & G) = $\underline{13 kN.}$

Taking moments about A :-

$C_D \times 4.33 = 26 [4 + 8 + 12 + 16 + 20] + 13 \times 24$

∴ $C_D = 432.3 kN.$

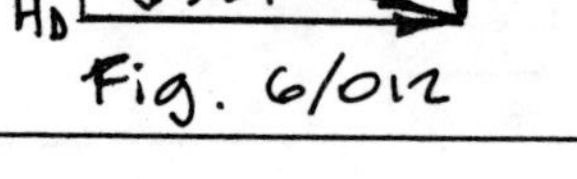

Fig. 6/012

$V_D = 432.3 \sin 13.87 = \underline{103.6 kN.}$

$H_D = 432.3 \cos 13.87 = \underline{419.7 kN.}$

Ref.

Putting these reactions, along with the loading in Fig. 6/011, in a Planeframe program, gave the following results :—

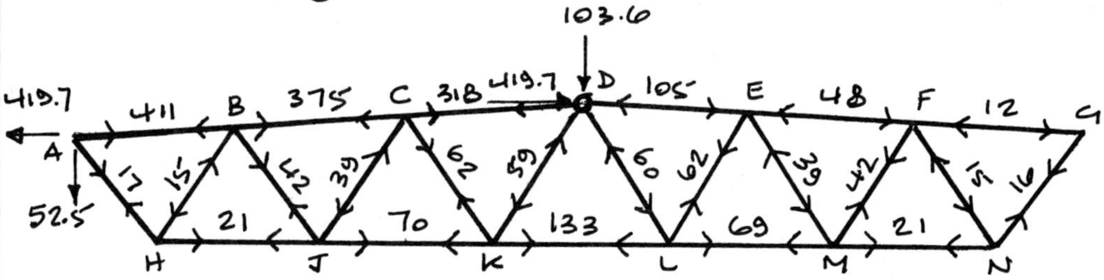

Fig. 6/013 – Results – L.C.11
(Forces in kN.'s)

Truss Member Design.

a) Top boom A → D – panel length on Slope = 4032 mm. : try a 200 x 100 x 12.5 RHS : A = 68.0 cm²; r_{xx} = 6.88 cm; r_{yy} = 3.88 cm; Z_{xx} = 322 cm³; p_y = 275 N/mm²

Ref 6/1

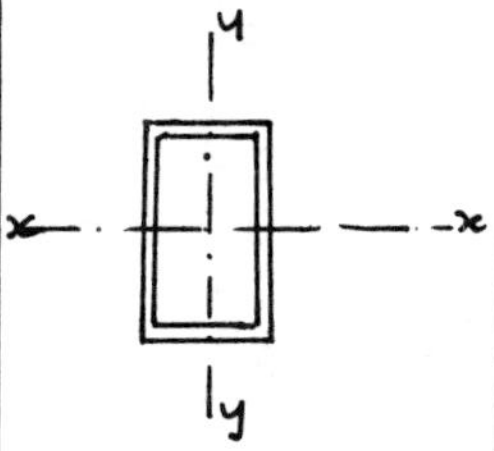

Fig. 6/014

i) L.C. 1 Axial force = 1014 kN (comp.)

Bending:

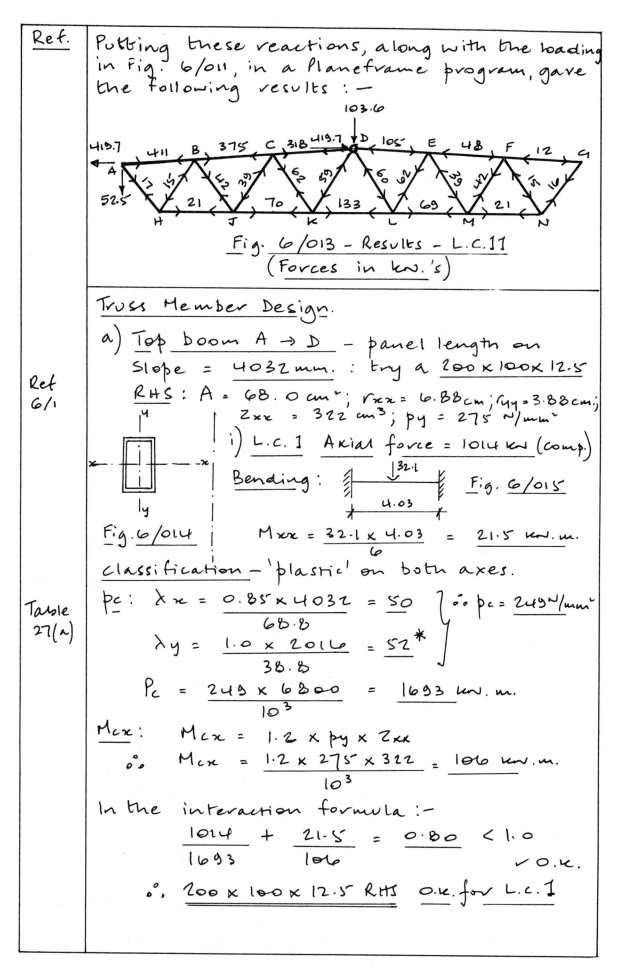

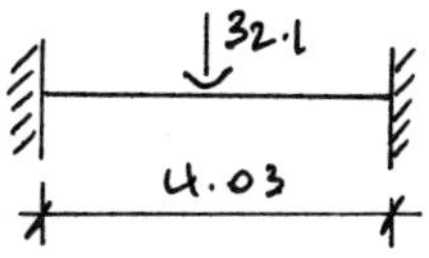

Fig. 6/015

$$M_{xx} = \frac{32.1 \times 4.03}{6} = 21.5 \text{ kN.m.}$$

classification – 'plastic' on both axes.

Table 27(a)

p_c: $\lambda_x = \dfrac{0.85 \times 4032}{68.8} = 50$; $\lambda_y = \dfrac{1.0 \times 2016}{38.8} = 52^*$ } ∴ p_c = 249 N/mm²

$$P_c = \frac{249 \times 6800}{10^3} = 1693 \text{ kN. m.}$$

M_{cx}: $M_{cx} = 1.2 \times p_y \times Z_{xx}$

$$\therefore M_{cx} = \frac{1.2 \times 275 \times 322}{10^3} = 106 \text{ kN.m.}$$

In the interaction formula :–

$$\frac{1014}{1693} + \frac{21.5}{106} = 0.80 < 1.0$$

✓ O.K.

∴ 200 x 100 x 12.5 RHS O.K. for L.C. 1

Ref.

ii) L.C. 11 Axial Force = 411 kN (tension)

Bending: 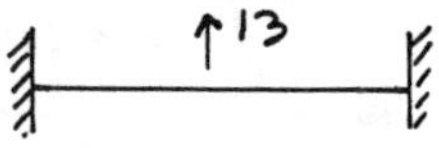 Fig. 6/016

$$M_{xx} = \frac{13 \times 4.03}{6} = 8.73 \text{ kN.m.}$$

P_t: $$P_t = \frac{275 \times 6800}{10^3} = 1870 \text{ kN.}$$

In the interaction formula :–

$$\frac{411}{1870} + \frac{8.73}{106} = 0.30 < 1.0 \checkmark \text{ o.k.}$$

∴ 200 × 100 × 12.5 RHS o.k. for both cases.

b) Top boom D → G – slope length 4032 mm per panel : try a 200 × 100 × 5 RHS – same axes as Fig. 6/014.

Ref. 6/1

$A = 28.9 \text{ cm}^2$; $r_{xx} = 7.23$ cm.; $r_{yy} = 4.2$ cm.

$Z_{xx} = 151 \text{ cm}^3$; $p_y = 275 \text{ N/mm}^2$.

Classification: 'plastic' for bending about x – x axis.

i) L.C. 1 Axial force = 258 kN (tension)

P_t: $$P_t = \frac{275 \times 2890}{10^3} = 795 \text{ kN.}$$

$M_{xx} = 21.5$ kN.m. – as for A → D.

M_{cx}: $$M_{cx} = \frac{1.2 \times 275 \times 151}{10^3} = 49.8 \text{ kN.m.}$$

In the interaction formula :–

$$\frac{258}{795} + \frac{21.5}{49.8} = 0.75 < 1.0 \checkmark \text{ o.k.}$$

∴ 200 × 100 × 5 RHS o.k. for L.C.1

ii) L.C. 11 Axial force = 105 kN (compression)

Table 27(a)

p_c: $$\lambda_x = \frac{0.85 \times 4032}{72.3} = 47.4$$

$$\lambda_y = \frac{1.0 \times 2016}{42} = 48^*$$

$p_c = 253 \text{ N/mm}^2$

∴ $$P_c = \frac{253 \times 2890}{10^3} = 731 \text{ kN.}$$

M_{cx}: $$M_{cx} = \frac{1.2 \times 275 \times 151}{10^3} = 49.8 \text{ kN.m}$$

$M_x = 8.73$ kN.m – as for A → D

Ref.	
	In the interaction formula:-
	$\frac{105}{731} + \frac{8.2}{49.8} = \underline{0.31} < 1.0$ ✓ o.k.
	∴ 200 × 100 × 5 RHS o.k. for both cases
	c) Bottom boom H → N Try a 150×150×10 SHS : A = 55.5 cm^2; $r_{xx} = r_{yy}$ = 5.70 cm.
	i) L.c. 1 Axial force = 329 kN (comp.)
	From Fig. 6/007 & Fig. 6/010, longitudinal restraints are provided at H, N & mid-way between K & L. Therefore, distance between restraints = 10.0 m.
	$L_{ex} = L_{ey}$ = 10000 mm.
	$\lambda = \frac{10000}{57} = 175 < 180$ ✓ o.k.
Table 27(a)	∴ p_c = 60 N/mm^2
	$P_c = \frac{60 \times 5550}{10^3}$ = 330 kN
	∴ $P_c > F_c$ ✓ o.k. (330) (329)
	ii) L.c. II Axial force = 133 kN (tension)
	$P_t = \frac{275 \times 5550}{10^3}$ = 1526 kN.
	∴ $P_t > F_t$ ✓ o.k. (1526) (133)
	∴ 150 × 150 × 10 SHS o.k. for both cases
	d) Internal members (diagonals)
	Try 80 × 80 × 6.3 SHS : A = 18.4 cm^2; $r_{xx} = r_{yy}$ = 3.0 cm.
	i) L.c. 1 { Member LE critical = 153 kN (comp.) ; Member DL critical = 147 kN (tens.) }
	Length = 3605 mm.
Table 27(a)	$\lambda = \frac{0.85 \times 3605}{30} = 102$: p_c = 153 N/mm^2
	∴ $P_c = \frac{153 \times 1840}{10^3}$ = 281 kN.

Ref.	
	∴ $\dfrac{P_c > F_c}{(281) \quad (153)}$ ✓ o.k.
	Check tension: $P_t = \dfrac{275 \times 1840}{10^3} = 506$ kN.
	∴ $\dfrac{P_t > F_t}{(506) \quad (147)}$ ✓ o.k.
	ii) L.C. II { Member DL critical = 60 kN (comp) ; Member LE critical = 62 kN (tens.) }
	By inspection: L.C. II not critical.
	∴ 80 × 80 × 6.3 SHS o.k. for both cases
	Top boom A–D : 200 × 100 × 12.5 RHS Top boom D–G : 200 × 100 × 5.0 RHS Bottom boom : 150 × 150 × 10.0 SHS Internals : 80 × 80 × 6.3 SHS
	Supporting framework : L.C. I
	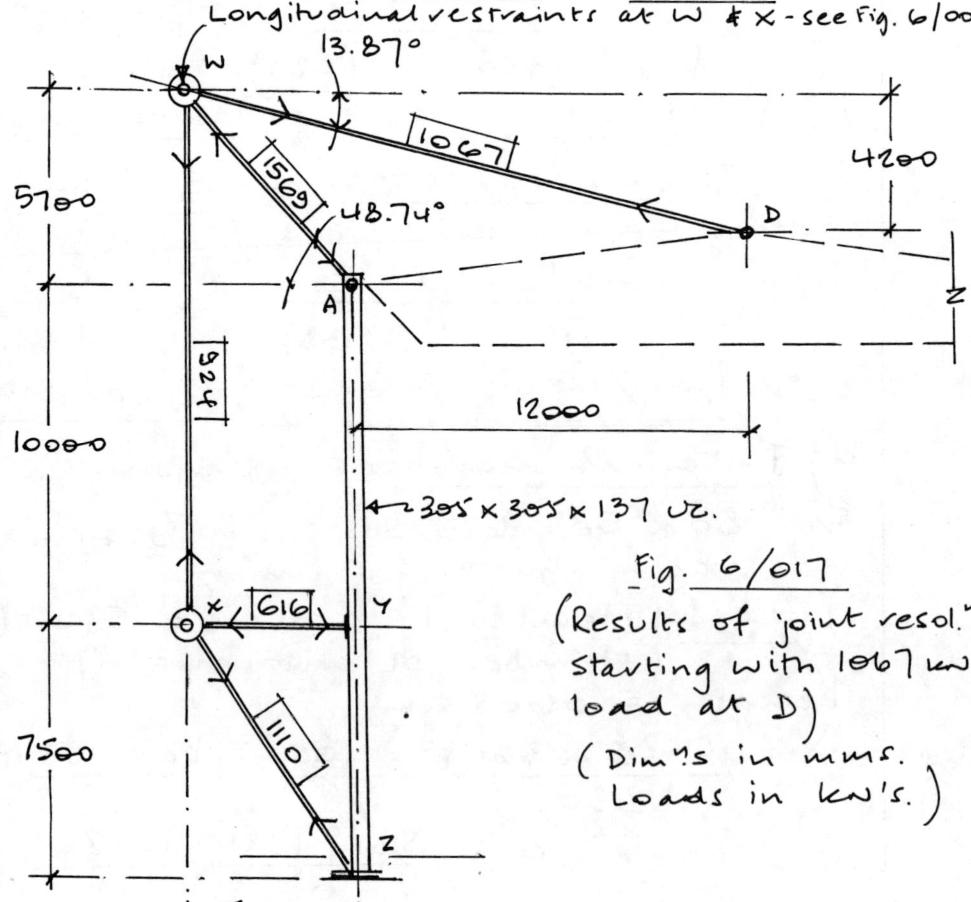

Fig. 6/017
(Results of joint resol.ⁿ, starting with 1067 kN load at D)
(Dimⁿs in mms. Loads in kN's.)

Ref.

Supporting framework : L.C. 11

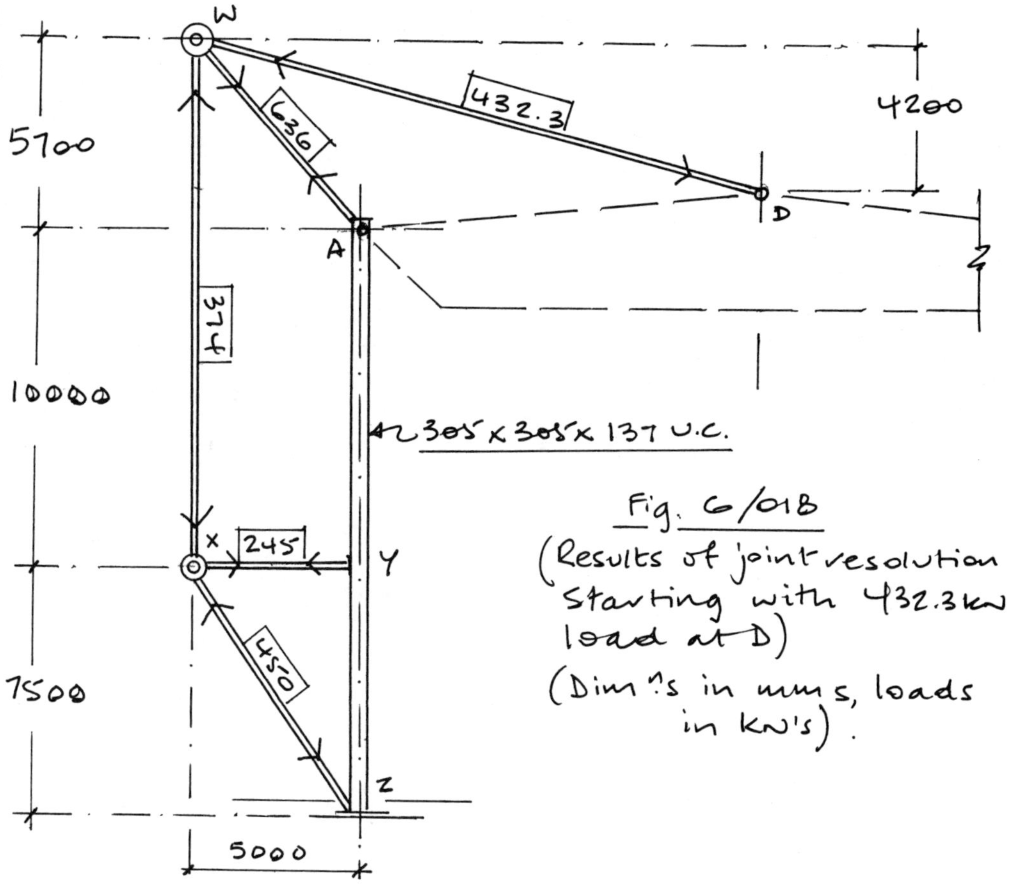

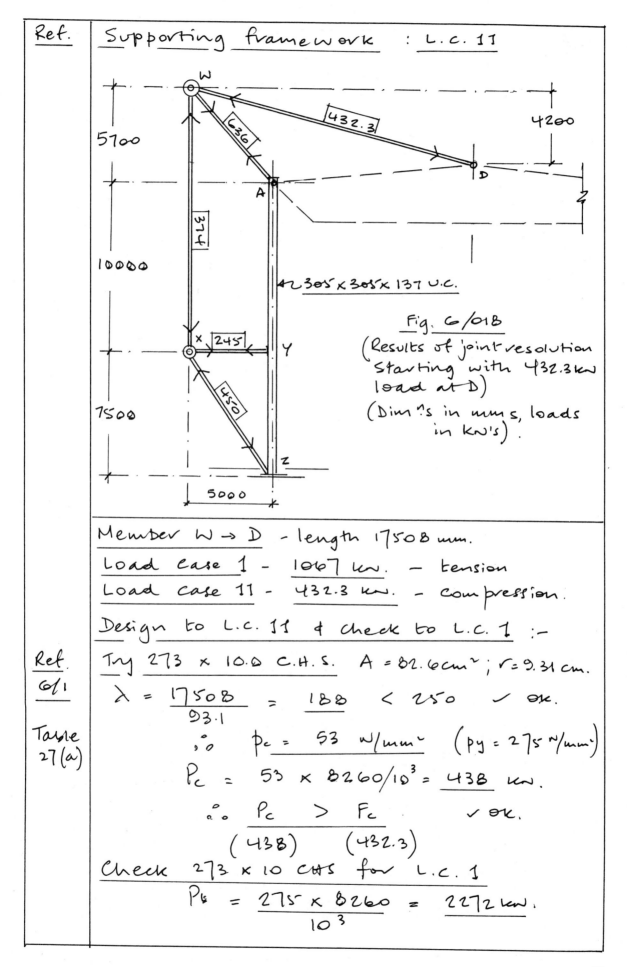

Fig. 6/01B
(Results of joint resolution starting with 432.3 kN load at D)
(Dim^n's in mm's, loads in kN's).

Member W → D - length 17508 mm.

Load case 1 - 1067 kN. - tension

Load case 11 - 432.3 kN. - compression.

Design to L.C. 11 & check to L.C. 1 :-

Ref. 6/1

Try 273 x 10.0 C.H.S. $A = 82.6\text{cm}^2$; $r = 9.31$ cm.

$$\lambda = \frac{17508}{93.1} = 188 < 250 \quad \checkmark \text{ ok.}$$

Table 27(a)

$$\therefore \quad p_c = 53 \text{ N/mm}^2 \quad (p_y = 275 \text{ N/mm}^2)$$

$$P_c = 53 \times 8260/10^3 = 438 \text{ kN.}$$

$$\therefore \quad \underset{(438)}{P_c} > \underset{(432.3)}{F_c} \quad \checkmark \text{ ok.}$$

Check 273 x 10 CHS for L.C. 1

$$P_t = \frac{275 \times 8260}{10^3} = 2272 \text{ kN.}$$

Ref.	
	$\therefore P_t > F_t$ ✓ ok.
	(2272) (1067)
	$\therefore$ 273 × 10.0 CHS ok for W→D
	Member W → A – length 7576 mm.
	Load case I – 1569 kN. – compression.
	Load Case II – 636 kN. – tension.
	Design to L.C. I & check to L.C. II :–
Ref. 6/1	Try 273 × 10.0 CHS : $A = 82.6\ cm^2$; $r = 9.31\ cm$.
	$\lambda = \frac{7576}{93.1} = 81 < 180$ ✓ o.k.
Table 27(a)	$\therefore p_c = 201\ N/mm^2$ ($p_y = 275\ N/mm^2$)
	$\therefore P_c = \frac{201 \times 8260}{10^3} = 1660\ kN.$
	$\therefore P_c > F_c$ ✓ o.k.
	(1660) (1569)
	Check 273 × 10 CHS for L.C. II :–
	$P_t = 2272\ kN.$ (as previous page)
	$\therefore P_t > F_t$ ✓ o.k.
	(2272) (636)
	$\therefore$ 273 × 10.0 CHS ok for W → A
	Member W → X – length 15700 mm.
	Load Case I – 924 kN. – tension
	Load case II – 374 kN. – compression
	Design to L.C. I & check to L.C. II
Ref 6/1	Try 273 × 10.0 C.H.S. : $A = 82.6\ cm^2$, $r = 9.31\ cm$
	$P_t = 2272\ kN$ (as previous page)
	$\therefore P_t > F_t$ ✓ o.k.
	(2272) (924)
	Check 273 × 10 CHS for L.C. II :–
	$\lambda = \frac{15700}{93.1} = 169 < 250$ ✓ o.k.

Ref.	
	$\therefore p_c = 64\ N/mm^2$ ($p_y = 275\ N/mm^2$)
	$\therefore P_c = \frac{64 \times 8260}{10^3} = 529\ kN.$
	$\therefore P_c > F_c$ ✓ ok.
	(529) (374)
	$\therefore$ 273 x 10 CHS o.k. for W → X
	Member X → Y – length 5000 mm.
	Load case I – 616 kN. – compression
	Load case II – 245 kN. – tensile.
	Design to L.C. I & check to L.C. II
Ref 6/1	Try 193.7 x 6.3 CHS : $A = 37.1\ cm^2$; $r = 6.63\ cm$
	$\lambda = \frac{5000}{66.3} = 75 < 180$ o.k. ✓
Table 27(a)	$\therefore p_c = 214\ N/mm^2$ ($p_y = 275\ N/mm^2$)
	$\therefore P_c = \frac{214 \times 3710}{10^3} = 794\ kN.$
	$\therefore P_c > F_c$ ✓ ok.
	(794) (616)
	Check 193.7 x 6.3 CHS for L.C. II
	$P_t = \frac{275 \times 3710}{10^3} = 1020\ kN.$
	$\therefore P_t > F_t$ ✓ ok.
	(1020) (245)
	Member X → Z – length 9014 mm.
	Load case I – 1110 kN – tension.
	Load case II – 450 kN – compression.
Ref. 6/1	Try 193.7 x 10 CHS : $A = 57.7\ cm^2$; $r = 6.50\ cm$.
	Design to L.C. II & check to L.C. I
	$\lambda = \frac{9014}{65} = 139 < 250$ ✓ o.k.
Table 27(a)	$\therefore p_c = 92\ N/mm^2$ ($p_y = 275\ N/mm^2$)
	$\therefore P_c = \frac{92 \times 5770}{10^3} = 531\ kN.$

Ref.

$\therefore$ P_c > F_c ✓ o.k.
(531) (450)

Check 193.7 x 10 CHS for L.C. 1

$$P_t = \frac{275 \times 5770}{10^3} = 1587 \text{ kN.}$$

$\therefore$ P_t > F_t ✓ o.k.
(1587) (1110)

$\therefore$ 193.7 x 10 CHS o.k. for X → Z

Summary:

273 x 10.0 CHS for W → D; W → A & W → X
193.7 x 6.3 CHS for X → Y
193.7 x 10.0 CHS for X → Z

Main Column Design – 305 x 305 x 118 U.C.

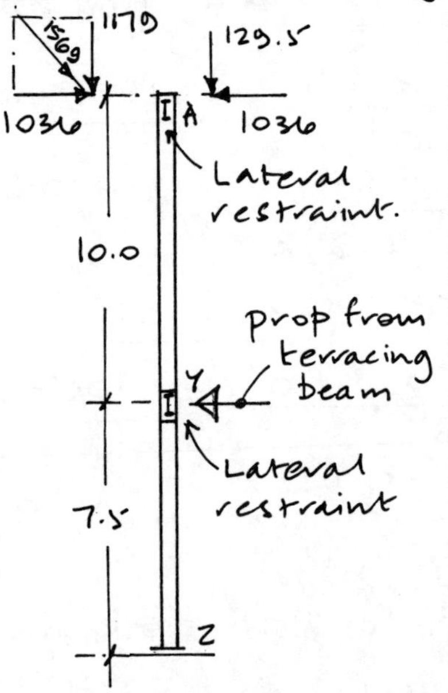

Fig. 6/019

i) Top cantilever length Y → A – L.C. 1 loading shown (L.C. 11 not critical). Truss loads from Fig. 6/010.

Total axial loading from framework & truss
= 1179 + 129.5
= 1308.5 kN.

Effective lengths :–

Table 24. $L_{ex} = 2.0 \times 10 = 20.0$ m.

$L_{ey} = 10.0$ m.

For a 305 x 305 x 137 U.C.; T = 21.7 mm > 16

Ref. 6/1 $\therefore$ $p_y = 265$ N/mm²; $r_{xx} = 13.7$ cm; $r_{yy} = 7.82$ cm.; A = 175 cm²

Table 27(b) $\lambda_x = \frac{20000}{13} = 147$ $p_{cx} = 76$ N/mm²

Table 27(c) $\lambda_y = \frac{10000}{78.2} = 128$ $p_{cy} = 86$ N/mm²

$\therefore$ $p_c = 76$ N/mm²

Ref.

$\therefore P_c = \dfrac{76 \times 17500}{10^3} = 1330 \text{ kN.}$

$\therefore \; P_c \;>\; F_c \qquad$ ✓ o.k.

$\quad (1330) \quad\; (1308.5)$

ii) Length Y–Z :

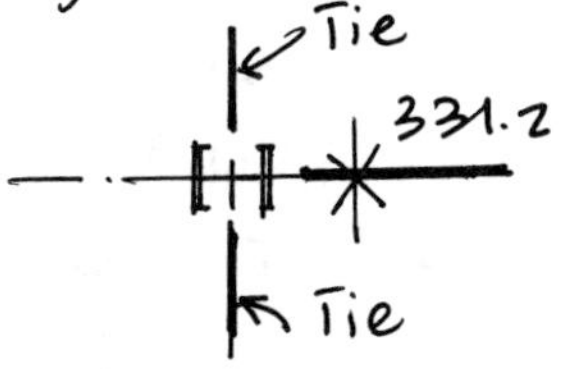

Fig. 6/020
Plan at gangway level

Total axial load on this length $= 331.2 + 1308.5$
$= 1639.7 \text{ kN.}$

Moment, M_{xx}, from gangway beam
$= 331.2\,[0.1 + 0.15]$
$= 82.8 \text{ kN.m.}$

This is split in ratios of stiffnesses, or 50 : 50 :–

$K_{AY} = I_{AY}/10 = 1/10$
$K_{YZ} = I_{YZ}/7.5 = 1/7.5$
ratio $= 10 : 7.5 = 1.33 : 1$

This is $< 1.5 : 1$ – split 50 : 50.

$\therefore M_{xx} = 82.8/2 = 41.4 \text{ kN.m.}$

$\underline{Pc}: \; \lambda = \dfrac{0.85 \times 7500}{78.2} = 82$

Table 27(c) — $p_c = 154 \text{ N/mm}^2 \; (p_y = 265 \text{ N/mm}^2)$

$\therefore P_c = \dfrac{154 \times 17500}{10^3} = 2695 \text{ kN.}$

$\underline{M_b}: \; \lambda_{LT} = \dfrac{0.5 \times 7500}{78.2} = 48.0$

Table 11, Ref 6/1 — $\therefore p_b = 235 \text{ N/mm}^2 \; (p_y = 265 \text{ N/mm}^2)$

$\therefore M_b = \dfrac{235 \times 2300}{10^3} = 540.5 \text{ kN.m.}$

In interaction formula :–

$\dfrac{1639.7}{2695} + \dfrac{41.4}{540.5} = 0.685 < 1.0$ ✓ o.k

$\therefore$ 305 × 305 × 137 U.C. o.k.

Ref.	
	Columns on grid line C (take columns on gridline D as same). See Figs 6/001 & 6/005.

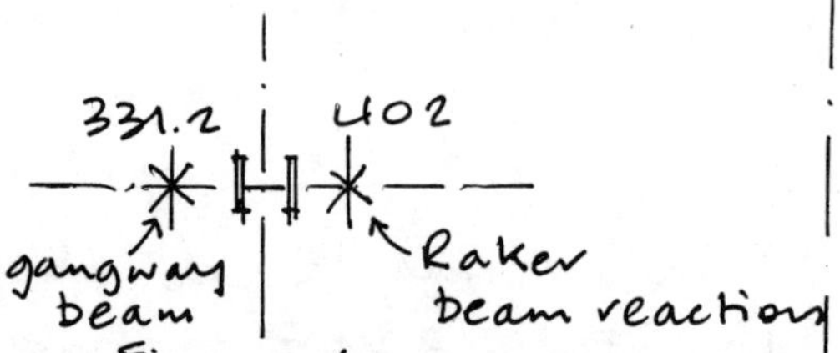

Fig. 6/021
Plan at gangway level

$L_e = 7500$ mm.
Ignore small M_{xx} moment.
Axial load
$= 331.2 + 402$
$= 733.2$ kN.
(ignoring moment)

Ref.	
Ref 6/1	Try a 203 x 203 x 86 U.C. T = 20.5 mm, py = 265
	$r_{yy} = 5.32$ cm ; $A = 110.0$ cm^2
	$\lambda = \frac{7500}{53.2} = 141 < 180$ ✓ o.k.
Table 27(c)	$p_c = 75$ N/mm^2 ($p_y = 265$ N/mm^2)
	$\therefore P_c = \frac{75 \times 11000}{10^3} = 825$ kN.
	$\therefore P_c > F_c$ ✓ ok.
	(825) (733.2)
	$\therefore$ 203 x 203 x 86 U.C's o.k.
	Foundations: a suggested base for the main stanchion is shown in Fig. 6/003.

Chapter 7 Refurbishment of an existing building

A Lancashire textile firm, who own an existing 10-storey building, currently used for storage and offices, wish to create a space at ground floor level, by removing an internal column, marked 'X' on Figure 7/001. The building has a regular grid of columns, 7.3 m. x 3.8 m., and floor to floor dimensions are shown on Figure 7/001. The main beams are 533 x 210 x 92 kg/m. U.B.'s, and the column ties, on the weak axis are 178 x 102 x 19 kg/m. U.B.'s. The connections at each floor level are as shown in Figure 7/001, with the detail also showing that the existing concrete floor is 125 mm. thick, including a 50 mm. structural screed. A further survey showed that the roof is 100 mm. thick, with no screed. The meeting room created by the removal of column 'X' will have a false ceiling, creating 1600 mm. of structural space, as shown in Figure 7/001. The building has several concrete lift and stair cores, considered as being adequate for any lateral loads on the structure. The refurbishment of the building also includes the replacement of blockwork partitions with lightweight movable partitions (see 'loading'). The floors and roof remain the same.

As the Consultant for the project, you have already put forward your scheme, shown on Figure 7/002, which the client has accepted. Calculations to prove the sections chosen are now required by the Local Building Control. The details of the scheme are shown in Figure 7/003 (which indicates that the structural members will fit into a 1600 mm. ceiling space), and the temporary works scheme is shown in Figure 7/004.

Loading – qk

Original loading – roof = 0.75 kN/m^2
– floors = 2.5 kN/m^2 + 4.0 kN/m^2
(block partitions)

New loading – roof = 0.75 kN/m^2
– floors = 3.0 kN/m^2 + 1.0 kN/m^2
(lightweight partitions)

Site conditions

There are no problems of access to the middle of the building, and the Local Building Control have accepted that the existing column foundations around 'X' will be adequate to take any extra loading, as the overall weight of the building has been reduced, due to the difference in partition loading.

Codes of Practice

BS 5950 – Structural Steelwork
BS 6399 – Loading

Materials

Steel : EN 10025 : S355

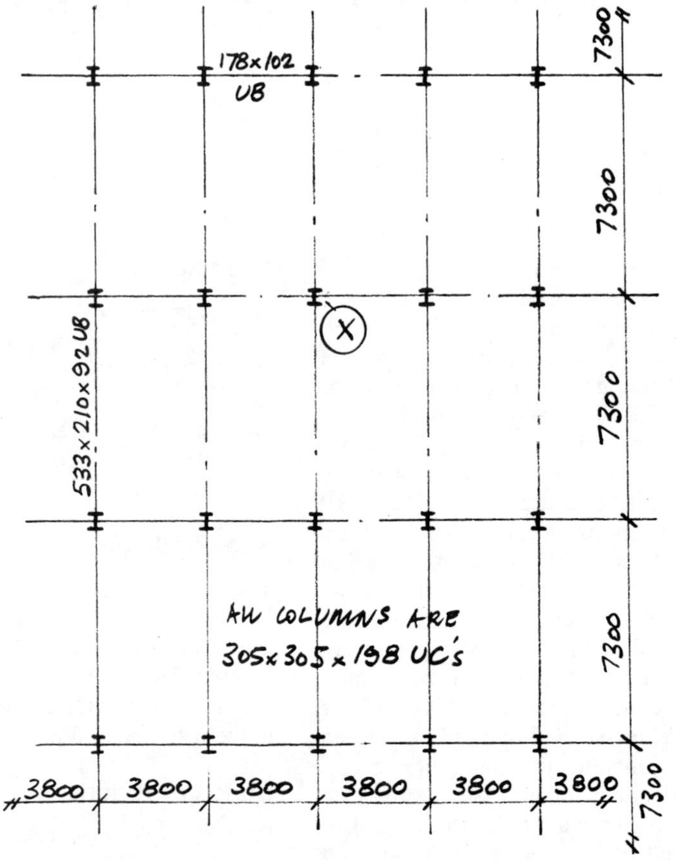

Typical Plan

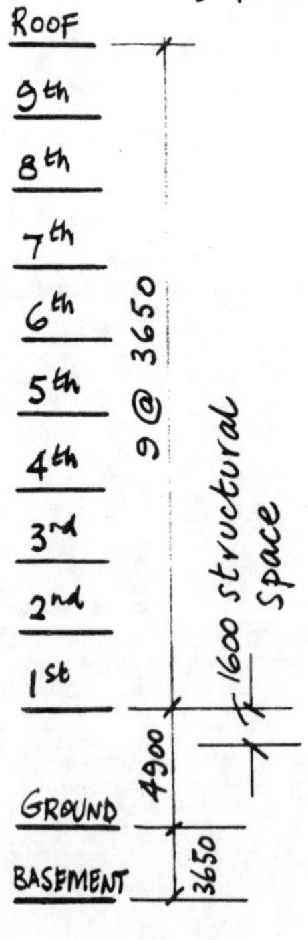

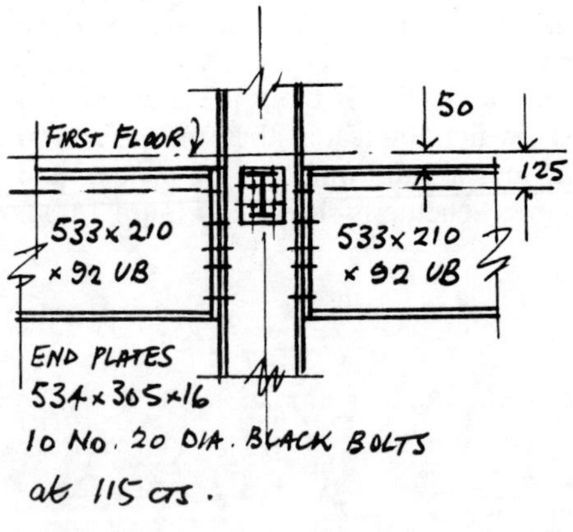

Typical Detail 'X'

Section

Fig. 7/001

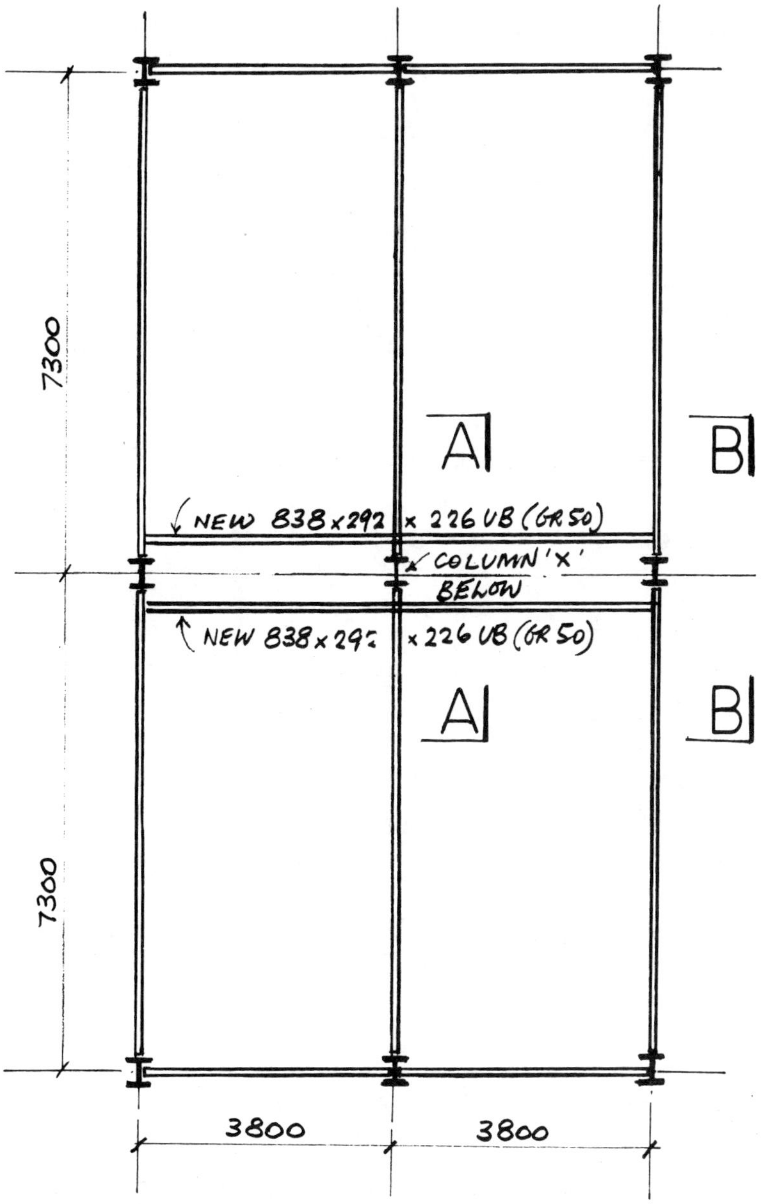

Plan at First Floor Level

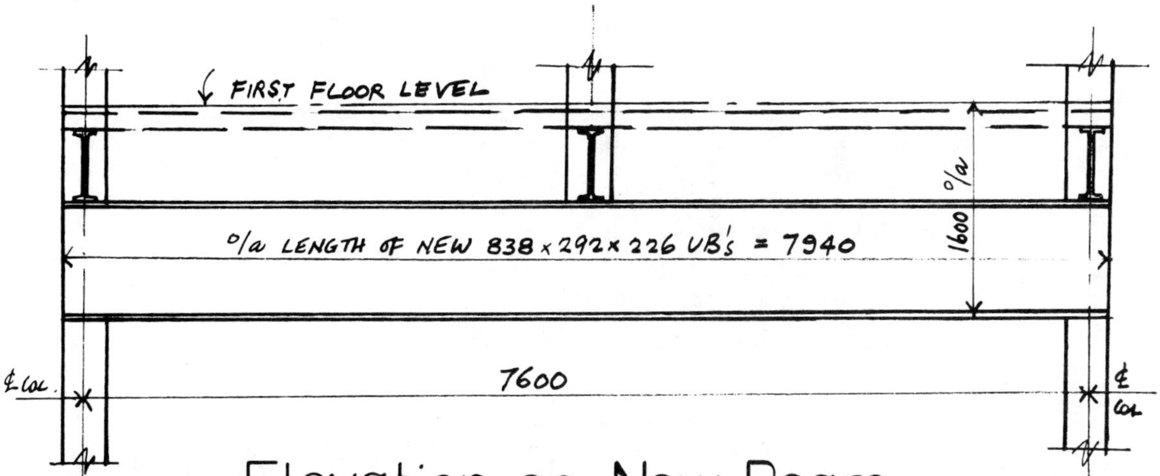

Elevation on New Beam

Fig. 7/002

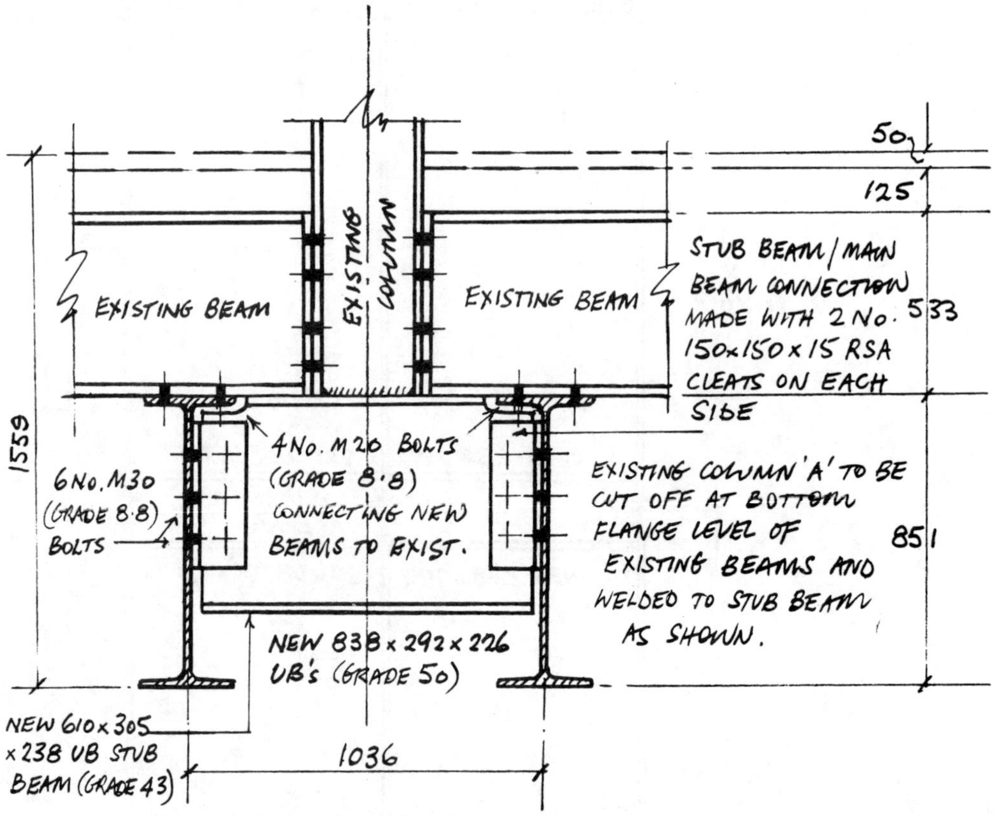

50
125
EXISTING COLUMN
EXISTING BEAM
EXISTING BEAM
STUB BEAM/MAIN BEAM CONNECTION MADE WITH 2 No. 150x150x15 RSA CLEATS ON EACH SIDE
533
1559
4No. M20 BOLTS (GRADE 8·8) CONNECTING NEW BEAMS TO EXIST.
6No. M30 (GRADE 8·8) BOLTS
EXISTING COLUMN 'A' TO BE CUT OFF AT BOTTOM FLANGE LEVEL OF EXISTING BEAMS AND WELDED TO STUB BEAM AS SHOWN.
851
NEW 838x292x226 UB's (GRADE 50)
NEW 610x305 x238 UB STUB BEAM (GRADE 43)
1036

Section A.A.

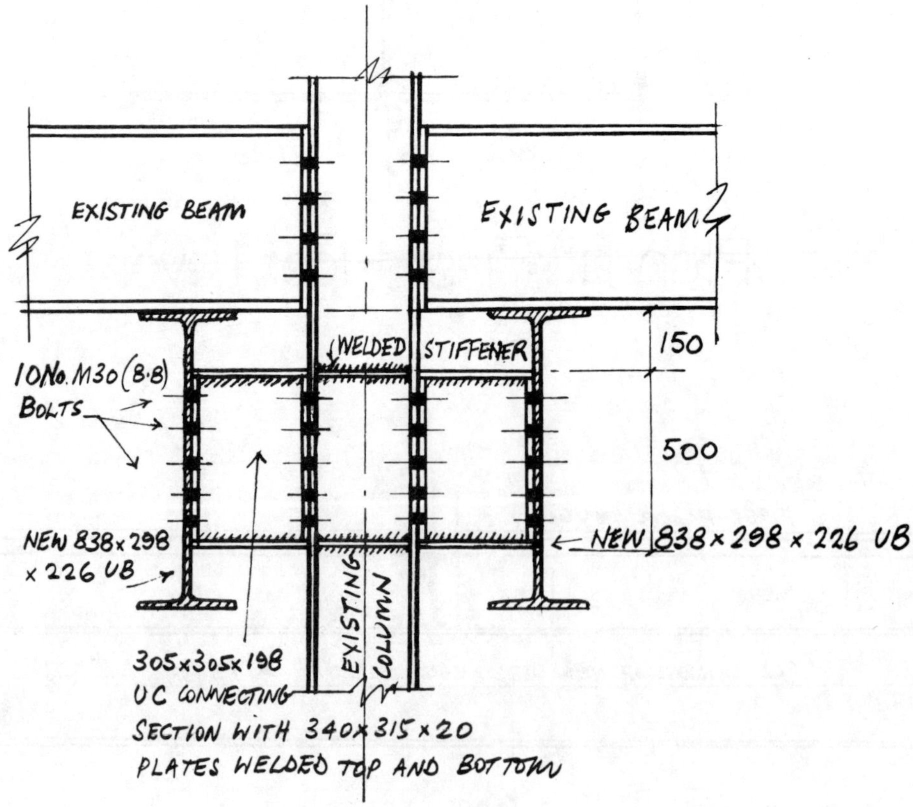

EXISTING BEAM
EXISTING BEAM
WELDED STIFFENER
150
10No. M30 (8·8) BOLTS
500
NEW 838x298 x226 UB
NEW 838x298x226 UB
EXISTING COLUMN
305x305x198 UC CONNECTING SECTION WITH 340x315x20 PLATES WELDED TOP AND BOTTOM

Section B.B.

Fig. 7/003

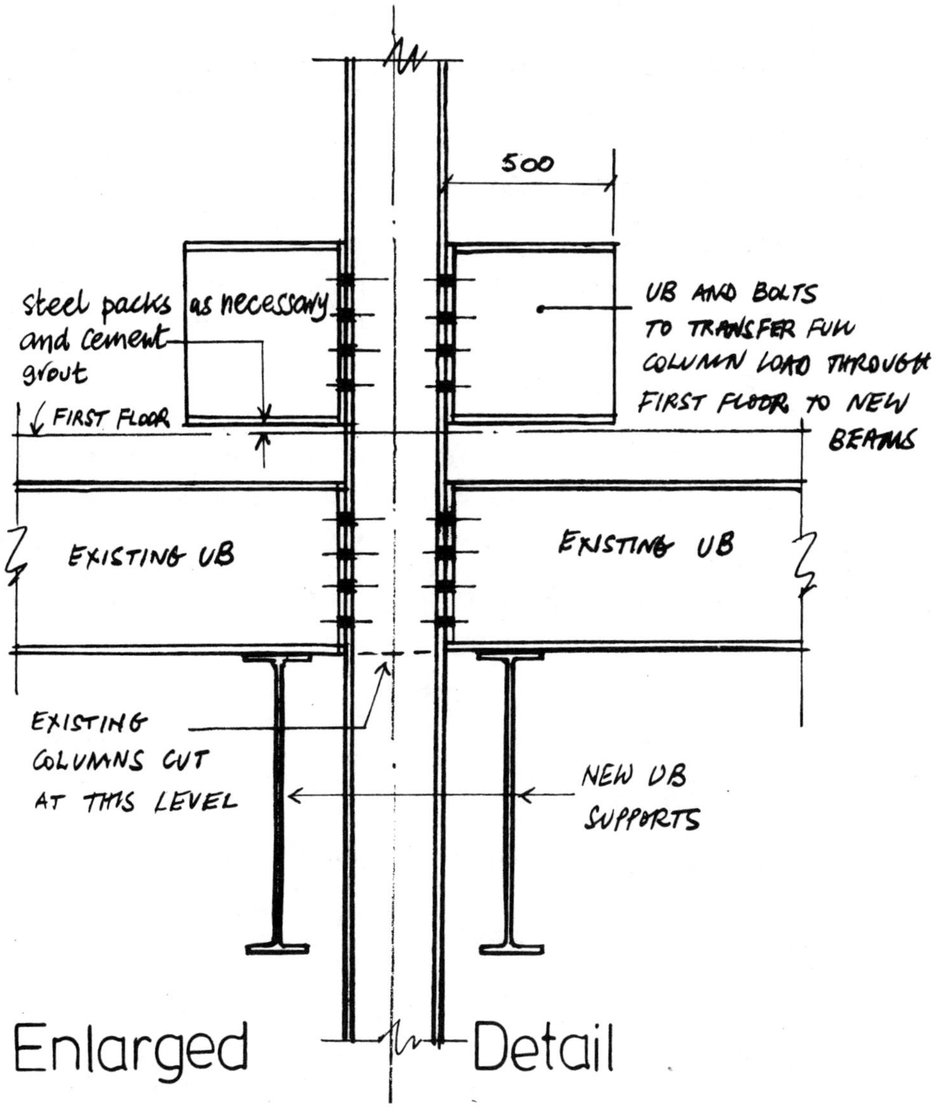

SEQUENCE OF CONSTRUCTION

1. FIX NEW MAIN BEAM SUPPORTS IN POSITION
2. FIX TEMPORARY UB'S TO COLUMN ABOVE FIRST FLOOR AND ENSURE BEAMS AND CONNECTIONS ARE CAPABLE OF FULL COLUMN LOAD INTO NEW BEAMS.
3. CUT COLUMN AT BOTTOM AND TOP AT REQUIRED LOCATIONS
4. FIX NEW STUB BEAM BELOW COLUMN AND WELD
5. REMOVE TEMPORARY UB'S
6. MAKE GOOD AROUND COLUMNS.

Fig.7/004

Ref.	
(BS 5950 U.N.O.)	**Assessment of column loads to be carried by new beams.**
	a) Concrete casings – fire resistance (50mm)
	Main beams = 0.633 × 0.31 × 24 = 4.7 kN/m
	Tie beams = 0.278 × 0.202 × 24 = 1.3 kN/m
	Columns = 0.405 × 0.405 × 24 = 3.9 kN/m
	b) Floor Loads
	Dead: 50mm screed = 1.2 kN/m²
	125 mm R/c slab = 3.0 kN/m²
	4.2 kN/m²
	Live (imposed) – new design loads
	Partitions = 1.0 kN/m²
	Live = 3.0 kN/m²
	4.0 kN/m²
	c) Roof Loads
	Dead: 100 mm R/c slab = 2.4 kN/m²
	Live (imposed): Snow = 0.75 kN/m²
	Dead load carried by column (per floor)
	= (4.2 × 3.8 × 7.3) + (4.7 × 7.3) + (1.3 × 3.8) + (3.9 × 3.65)
	↑ floor ↑ beam ↑ tie ↑ column
	= 170 kN (× 1.4 = 238 kN)
	Dead load carried by column (roof)
	= (2.4 × 3.8 × 7.3) + (4.7 × 7.3) + (1.3 × 3.8) + (3.9 × 3.65)
	↑ roof ↑ beam ↑ tie ↑ column
	= 120 kN (× 1.4 = 168 kN)
	Live load carried by column (per floor)
	= 4.0 × 3.8 × 7.3 = 111 kN (× 1.6 = 178 kN)
BS 6399 Pt.1 Table 2	Reduce total distributed live loads by 40% for 5 → 10 storeys high:–
	= 0.6 × 111 = 67 kN (× 1.6 = 107 kN)
	Live load carried by column (roof)
	= 0.75 × 3.8 × 7.3 = 20.8 kN (× 1.6 = 33.3 kN)
BS 6399 Pt.1 Table 2	No reduction in live load for roof.
	∴ Factored column load just above 1st floor
	= 8 × (238 + 107) + 1 × (168 + 33.3) = 2961 kN.

Ref.	
	From Fig. 7/002 it can be seen that **each** new beam (838 × 292 × 226 UB) carries the reaction from **one** ex. floor beam + 50% of column load.
BS 6399 Pt.1 Table 2	Design floor loading $= 1.4 \times 4.2 + 1.6 \times 4 \times 0.6$ $= 9.72\ kN/m^2$
	U.D.L on ex^g floor beam $= 9.72 \times 3.8 = 36.9\ kN/m$
	Add on casing (4.7 kN/m) – total $= 41.6\ kN/m$
	∴ Reaction, $F_v = \frac{41.6 \times 7.3}{2} = 152\ kN.$
	∴ Central point load on new beam $= \frac{2961}{2} + 152 = 1632.5\ kN.$

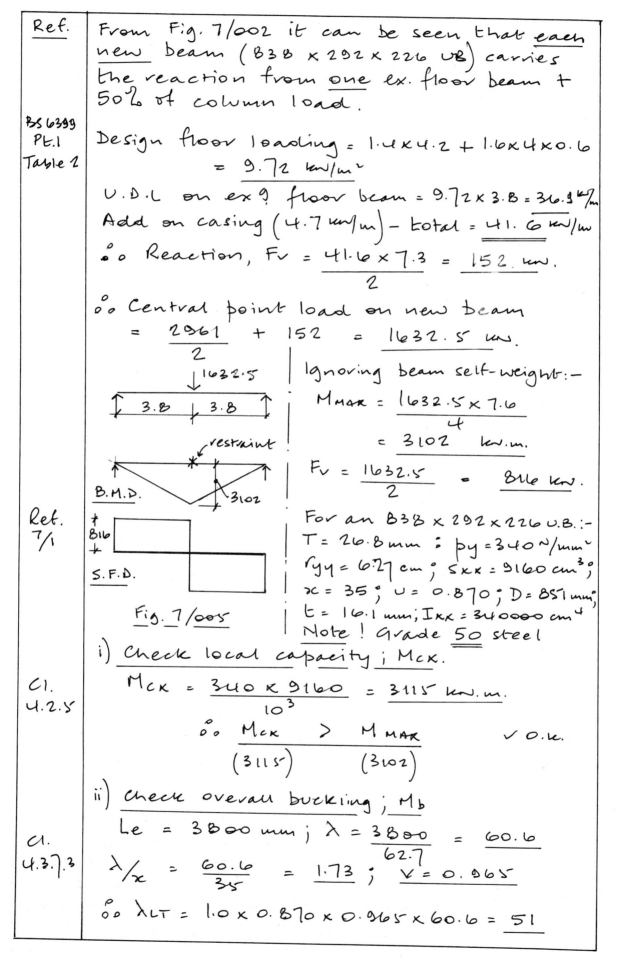

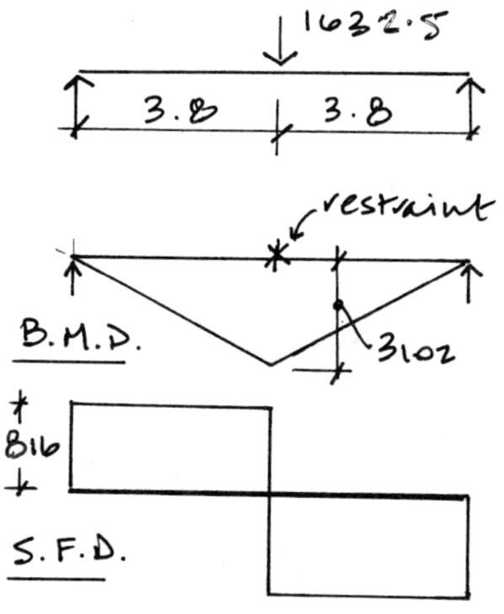

Fig. 7/005

Ignoring beam self-weight:–

$M_{MAX} = \frac{1632.5 \times 7.6}{4} = 3102\ kN.m.$

$F_v = \frac{1632.5}{2} = 816\ kN.$

Ref. 7/1

For an 838 × 292 × 226 U.B.:–
$T = 26.8\ mm$; $p_y = 340\ N/mm^2$
$r_{yy} = 6.27\ cm$; $S_{xx} = 9160\ cm^3$;
$x = 35$; $u = 0.870$; $D = 851\ mm$;
$t = 16.1\ mm$; $I_{xx} = 340000\ cm^4$
Note! Grade 50 steel

i) Check local capacity ; M_{cx}.

Cl. 4.2.5

$M_{cx} = \frac{340 \times 9160}{10^3} = 3115\ kN.m.$

∴ $M_{cx} > M_{MAX}$ (3115) (3102) ✓ o.k.

ii) Check overall buckling ; M_b

Cl. 4.3.7.3

$L_e = 3800\ mm$; $\lambda = \frac{3800}{62.7} = 60.6$

$\lambda / x = \frac{60.6}{35} = 1.73$; $v = 0.965$

∴ $\lambda_{LT} = 1.0 \times 0.870 \times 0.965 \times 60.6 = 51$

Ref.	
	$\therefore$ $p_b = 279\ N/mm^2$ $(p_y = 340\ N/mm^2)$
	$\therefore M_b = \frac{279 \times 9160}{10^3} = 2556\ kN.m.$
Table 18	$\bar{M}$: $\beta = 0$, $m = 0.57$
Cl. 4.3.7.2	$\therefore \bar{M} = m.M_{MAX} = 0.57 \times 3102$
	$= 1768\ kN.m.$
	$\therefore M_b > \bar{M}$ ✓ o.k.
	(2556) (1768)
	iii) Check shear : $F_v = 816\ kN$ (see Fig 7/005)
Cl. 4.2.3	$P_v = 0.6 \times 340 \times 16.1 \times 850.9 / 10^3$
	$= 2795\ kN.$
	$\therefore P_v > F_v$ ✓ o.k.
	(2795) (816)
	iv) Check Deflection :
	Column unfactored live loading
	$= 8 \times 67 + 20.8 = 556.8\ kN.$
	From 1st floor beam,
	U.D.L. $= 4.0 \times 3.8 = 15.2\ kN/m.$
	Reaction $= \frac{15.2 \times 7.3}{2} = 55.5\ kN.$
	$\therefore$ Central point load, per beam
	$= \frac{556.8}{2} + 55.5 = 331\ kN.$
	$\Delta = \frac{1}{48} \times \frac{(331 \times 10^3) \times (7600)^3}{(205 \times 10^3) \times (340000 \times 10^4)} = 4.3\ mm.$
Table 5	Allowable $= 7600/360 = 21.1\ mm$ ✓ o.k.
	$\therefore$ 838 × 292 × 226 kg/m UB's in Grade 50 steel satisfactory
	Check on stub beam (Fig. 7/003) taking load from column on to new beams.
Ref 7/1	Try a 610 × 305 × 238 kg/m UB (Grade 50)
	$T = 31.4\ mm$: $p_y = 340\ N/mm^2$: $S_{xx} = 7460\ cm^3$

Ref.	
	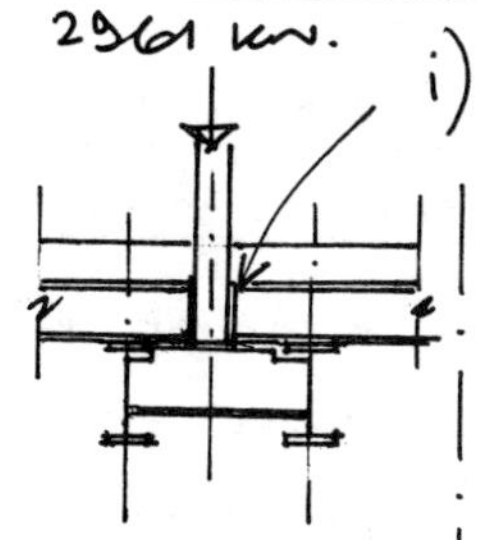 Fig. 7/006 i) Check capacity of existing connection to take column load directly back to new beams:-
Ref 7/1	10 No. M20 4.6 Grade bolts per side in single shear. Capacity = 10 × 39.2 = 392 kN. ∴ Remaining load onto stub beam = 2961 − 2 × 392 = 2177 kN.
	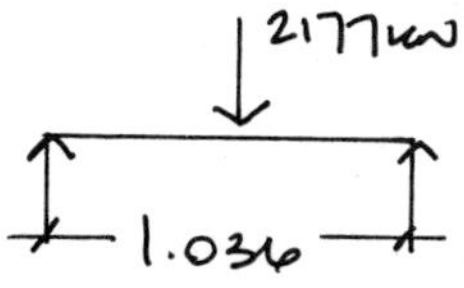 Fig. 7/007 $M_{MAX} = \frac{2177 \times 1.036}{4} = 564$ kN.m.
Cl. 4.2.5	$M_{CX} = \frac{340 \times 7460}{10^3} = 2536$ kN.m. ∴ $M_{CX} > M_{MAX}$ ✓ ok. (2536) (564)
	ii) Check shear: $F_v = \frac{2177}{2} = 1088$ kN.
Ref 7/1	D = 633 mm. ; t = 18.6 mm.
Cl. 4.2.3	∴ $P_v = 0.6 \times 340 \times 633 \times 18.6 / 10^3$ $P_v = 2402$ kN. ∴ $P_v > F_v$ ✓ ok. (2402) (1088) Deflection not a criteria : ∴ 610 × 305 × 238 kg/m UB. O.K. Check connection of stub beam to main beams (Section A-A Fig. 7/003) i) Check bolts in single shear:
Ref 7/1	6 No. M30 8.8 Grade bolts — Capacity = 6 × 210 = 1260 kN. Capacity > Shear ✓ O.K (1260) (1088) ii) Check bolt bearing capacity: Cleats 15 mm thick : beam web 16.1 mm.
Ref 7/1	Capacity = 6 × 247 = 1482 kN. capacity > shear ✓ O.K. (1482) (1088)

Ref.	
	Check on connection of main beam to column – section B–B, Fig. 7/003

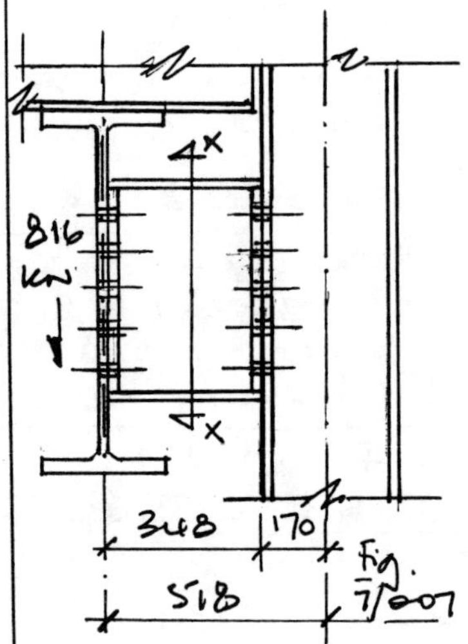

Cantilever moment about face of column:–

$M = 816 \times 0.35$

$= 286$ kN.m.

Section X–X comprises 2 No. 315 × 20 plates welded each side of a 305 × 305 × 198 U.C. cutting.

S_{xx} of welded section:– (taking bending on flanges only):–

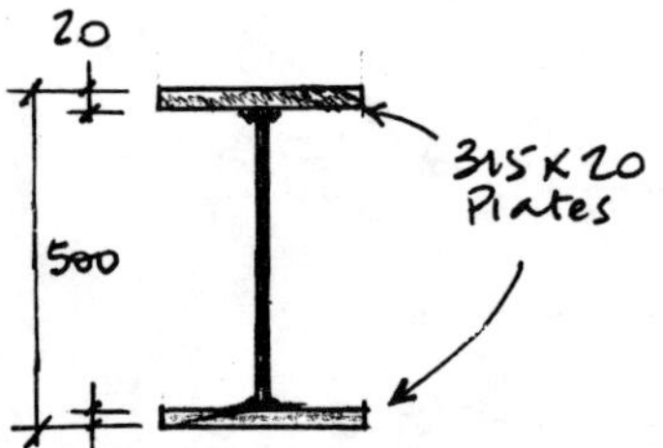

Fig. 7/008
(Section X–X)

$S_{xx} = 2 \times 315 \times 20 \times 240$
$= 3.0 \times 10^6\ \text{mm}^3$

Cl. 4.2.5

$\therefore M_{cx} = \dfrac{340 \times 3.0 \times 10^6}{10^6}$

$M_{cx} = 1020$ kN.m.

$\therefore$ M_{cx} (1020) > M_{MAX} (286) ✓ O.K.

Check shear across the web only:–

Cl. 4.2.3

$P_v = \dfrac{0.6 \times 19.2 \times 460 \times 340}{10^3} = 1802$ kN.

$P_v = 1802$ kN.

$\therefore$ P_v (1802) > F_v (816) ✓ O.K.

Deflection not a criteria.

Check Bolt Group

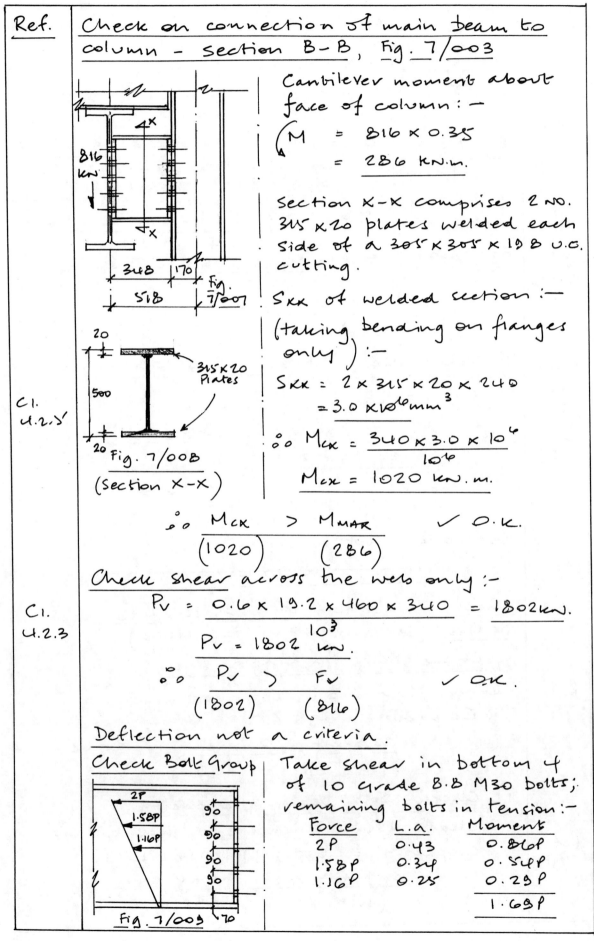

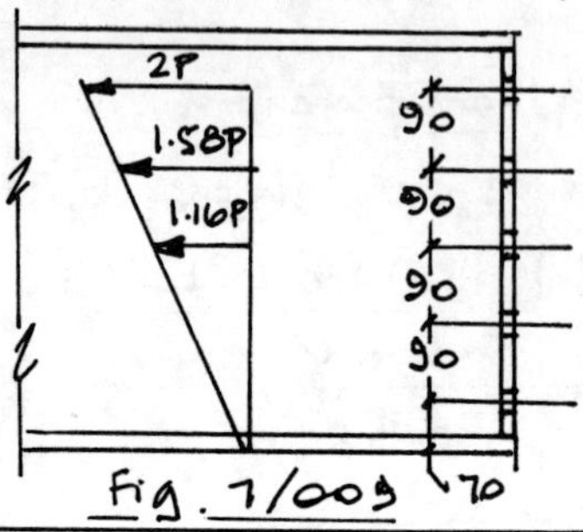

Fig. 7/009

Take shear in bottom 4 of 10 Grade 8.8 M30 bolts; remaining bolts in tension:–

Force	L.a.	Moment
2P	0.43	0.86P
1.58P	0.34	0.54P
1.16P	0.25	0.29P
		1.69P

Ref.	
	Moment at connection = 286 kN.m.
	∴ 286 = 1.69 P
	∴ P = 169 kN.
Ref 6/1	For M30 8.8 bolts : P_t = 252 kN.
	P_t > P ✓ o.k.
	Check bolts in shear :–
	For M30 8.8 bolts : P_v = 210 kN.
	4 × 210 = 840 kN.
	∴ P_v > F_v ✓ o.k.
	(840) (816)

Chapter 8 A new office block in reinforced concrete within retained masonry facades

A large building society requires the renewal of a city centre site for a new headquarters. The Planning Authority insists on the retention of the masonry facades, shown in Section a-a, Fig. 8/001, on the two sides shown in the Plan view. The Planning Authority will allow the construction of a new roof, and also curtain walling on the two new elevations, as shown in Fig. 8/001. In order to line up the floors with the existing windows, and to incorporate air-conditioning, the maximum depth of structural members is to be 425 mm, as shown in Fig. 8/001. This allows for a raised computer floor, and a clear height of 2700, floor-to-ceiling. Further to this, the client requires a minimum column grid of 6 m. centres, and a desire that any columns required should not be placed in front of the windows on the retained facade elevations. The new roof is to be flat, and will mainly support plant in new lightweight enclosures, and absolutely no new loading is to be placed on the retained masonry facades. After demolition of the existing building, with the exception of the two facades, the client requires details of temporary support work to the two facades, which will give resistance to lateral loads, and also to allow new construction to continue (see Figs. 8/017 & 8/018). Finally, the client requires details of how the basement is to be waterproofed (see Fig. 8/019). In the design offered, pattern loading has been ignored, as has progressive collapse. In a full design both of these design criteria would have to be checked for.

Loading – live qk

The roof and floor characteristic loadings are shown in Section a-a, Fig. 8/001, and comprise of 7.5 kN/m^2 on the roof, and 6.0 kN/m^2 (including partitions) on all floors.

Loading – dead gk

An allowance of 1.0 kN/m^2 on floor loadings is required for false floor and ceiling

Loading – wind wk

Basic wind speed 46 m/sec (city centre site)

Soil profile

The results of a comprehensive site investigation are shown in Section a-a, Fig. 8/001, showing that stiff clay will provide the first real resistance to foundation pressures, at a depth of 3.5 m. below existing facade foundation level.

Design Code of Practice

The building is to be designed in accordance with BS 8110, Structural Concrete.

Materials

Concrete – fcu = 40 N/mm^2, cover 20 mm. (superstructure) and 75 mm. (pilecap)
Reinforcement – all high yield steel, fy = 460 N/mm^2.

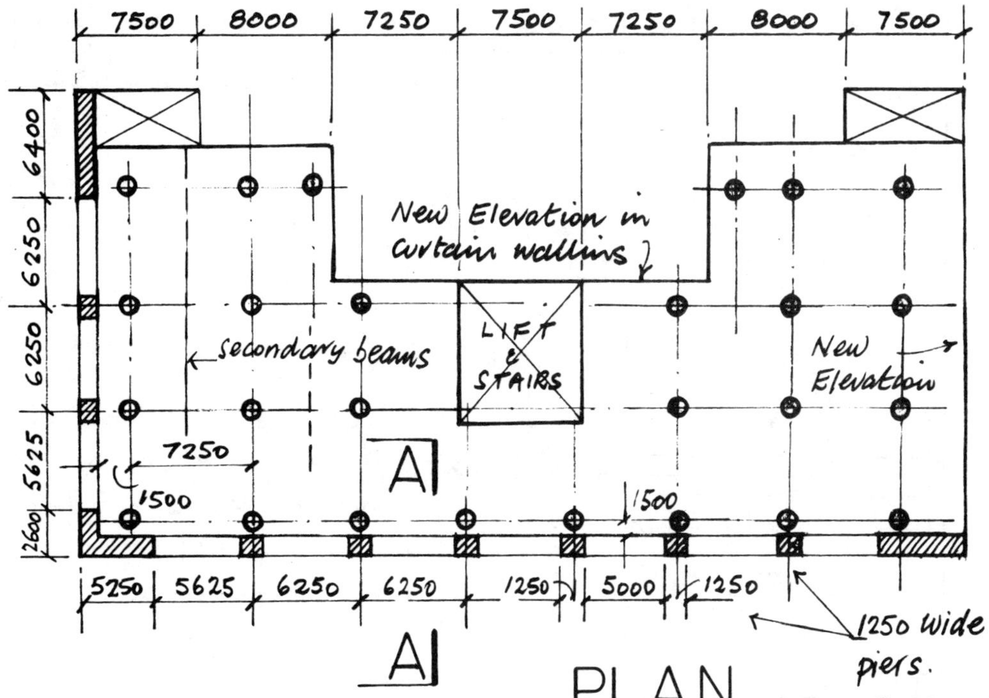

PLAN

Fig.8/001

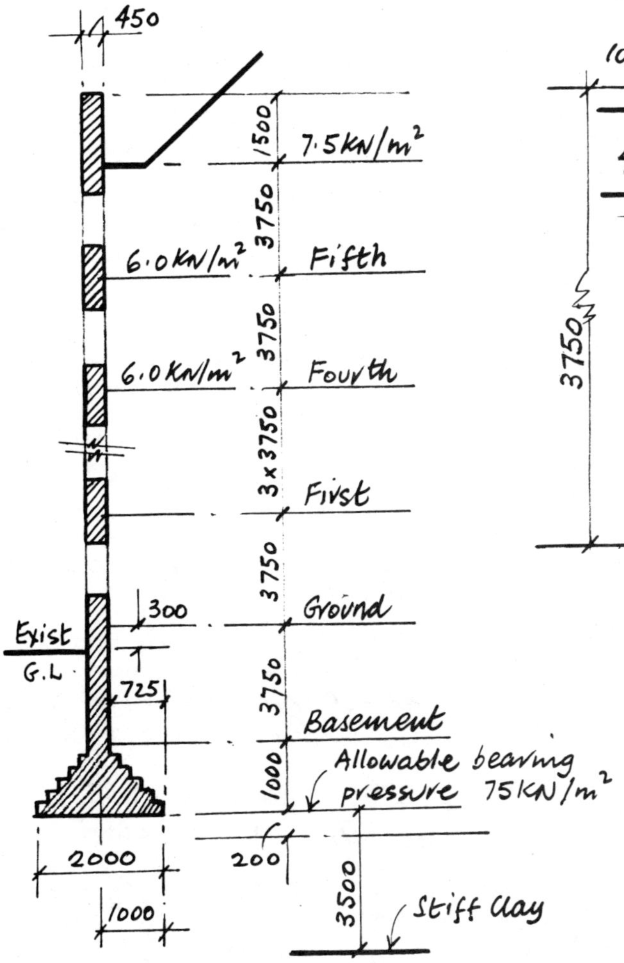

Section A.A.

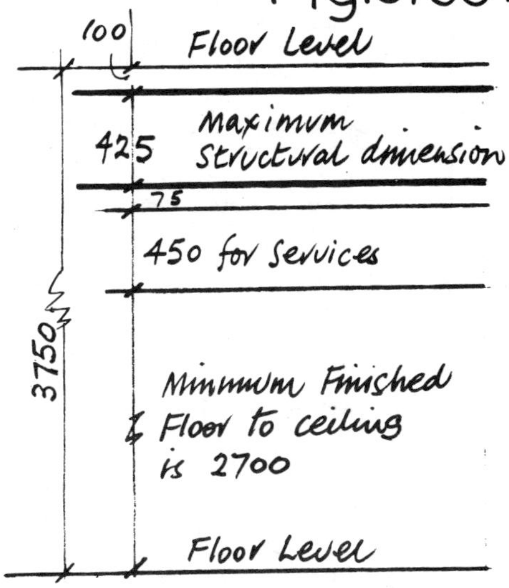

Structural Constraints

Fig.8/002

Ref. (B.S. 8110)

Typical floor calculations: wide beams (1000 wide × 425 deep) & one-way spanning slabs (125 tk.): q_k = 6.0 kN/m²

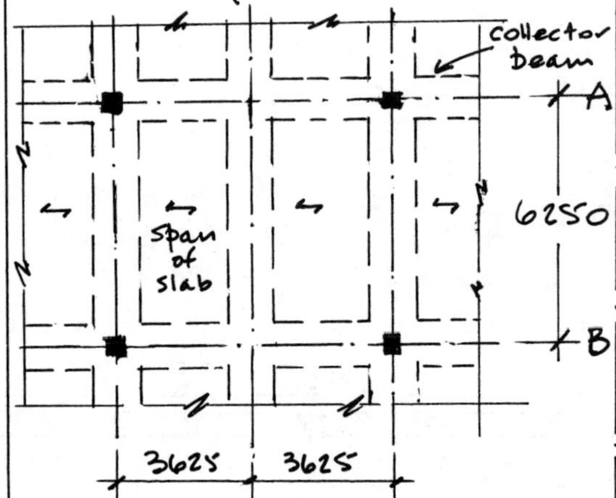

Typical bay

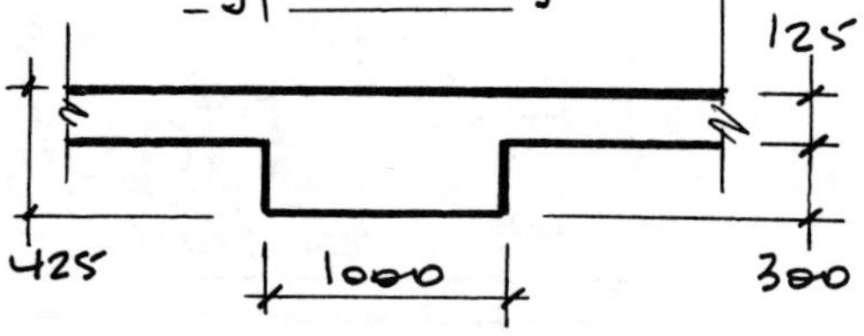

Typical beam section

Fig. 8/003

Average floor dead loading (g_k) :-

Slab (125 mm deep)

$g_k = 0.125 \times 24 = 3$ kN/m²

Volume of 300 deep beam rib in a floor area 6.25 × 3.625 (22.65 m²)

$= 2 \times 3.625 \times 0.3 \times 0.5 + 2 \times 6.25 \times 0.3 \times 0.5 = 3$ m³

∴ Average g_k

$= \dfrac{3 \times 24}{22.65} = 3.2$ kN/m²

Ceiling & floor = 1.0 kN/m²

∴ Total g_k = 7.2 kN/m²

Table 2.1 — ∴ Design loading = 1.6 × 6 + 1.4 × 7.2 = 19.7 kN/m²

Cl. 3.5.2.3 (c) — Take 3 spans of slab + end cantilever; (with relaxing of limit of q_k = 5 kN/m²)

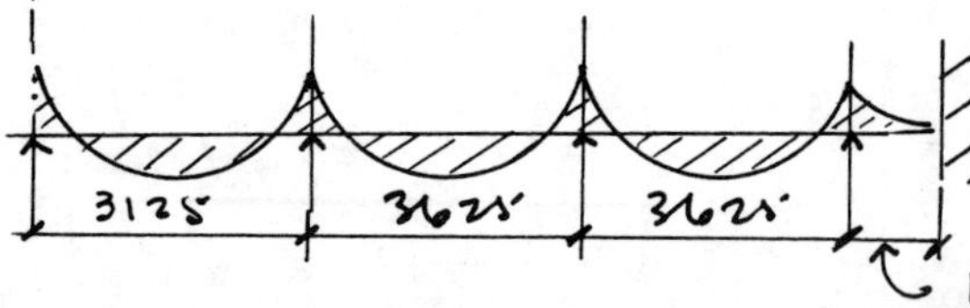

Fig. 8/004

F = 19.7 × 3.625 = 71.4 kN.

Table 3.13 — Hogging Moment = Sagging moment = 0.086 × 71.4 × 3.625 = 22.3 kN.m/m.

Cl. 3.4.4.4, Chart A1

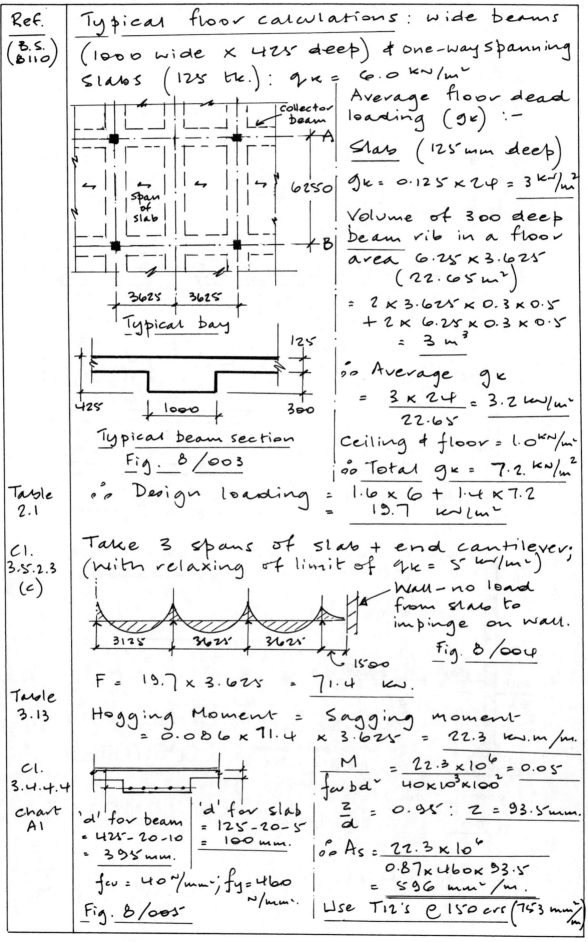

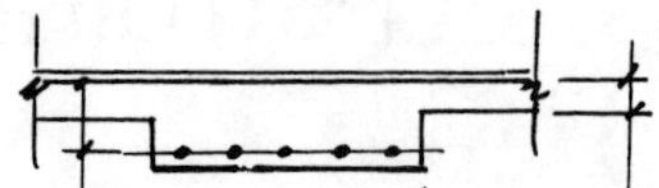

'd' for beam = 425 − 20 − 10 = 395 mm.

'd' for slab = 125 − 20 − 5 = 100 mm.

f_{cu} = 40 N/mm²; f_y = 460 N/mm².

Fig. 8/005

$\dfrac{M}{f_{cu} b d^2} = \dfrac{22.3 \times 10^6}{40 \times 10^3 \times 100^2} = 0.05$

$\dfrac{z}{d} = 0.95$: z = 93.5 mm.

∴ $A_s = \dfrac{22.3 \times 10^6}{0.87 \times 460 \times 93.5} = 596$ mm²/m.

Use T12's @ 150 crs (753 mm²/m)

Ref.	
(B.S. 8110)	**Check minimum steel:**
Table 3.27	Minimum steel = 0.13%
	$= \frac{0.13}{100} \times 1000 \times 125 = 163\ mm^2/m$
	This is < steel provided ($753\ mm^2/m$) ✓ o.k.
	∴ Use T12's @ 150 crs. hogging & sagging
	Check slab deflection:
Cl. 3.5.7 & Tables 3.10 & 3.11	span/d ≯ 26 × (modification factor)
	$span/d = \frac{3625}{100} = 36.25$
	$\frac{M}{bd^2} = \frac{22.2 \times 10^6}{10^3 \times 100^2} = 2.22$
Table 3.11 Note 2	sagging steel provided = $753\ mm^2/m$
	$f_s = \frac{5}{8} \times 460 \times \frac{596}{753} = 228\ N/mm^2$
Table 3.11 Note 1	$\therefore m.f. = 0.55 + \frac{(477 - 228)}{120(0.9 + 2.22)} = 1.22$
	∴ Allowable = 26 × 1.22 = 31.7 < 36.25 Not o.k.
	Increase sagging steel to T16's @ 150 crs.
	(Other alternative is to increase 'd')
	sagging steel provided = $1340\ mm^2/m$
Table 3.11 Note 2	$f_s = \frac{5}{8} \times 460 \times \frac{596}{1340} = 127\ N/mm^2$
	$\therefore m.f. = 0.55 + \frac{(477 - 127)}{120(0.9 + 2.22)} = 1.48$
	∴ Allowable = 26 × 1.48 = 38.6 > 36.25 ✓ o.k.
	Check cantilever

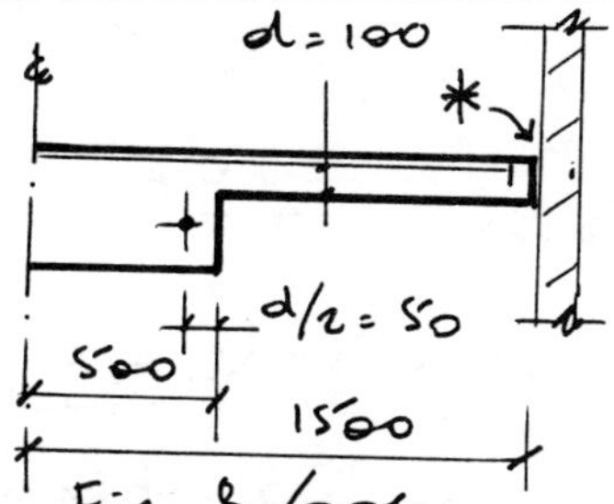

Fig 8/006
Cantilever effective span

Cl. 3.4.1.3

Effective span
= 1500 − 500 + 50 = 1050 mm

Design loading = $19.7\ kN/m^2$
= 19.7 kN/m of slab

$M = \frac{WL^2}{2} = \frac{19.7 \times 1.05^2}{2} = 10.8\ kNm$

(* Horizontal support to wall provided by slab – see Figure 8/020)

Ref.	
(BS 8110) Cl 3.4.4.4 Chart A1	$\frac{M}{f_{cu}bd^2} = \frac{10.8 \times 10^6}{40 \times 10^3 \times 100^2} = 0.03$
	$z/d = 0.95$: $z = 95$ mm.
	$\therefore A_s = \frac{10.8 \times 10^6}{0.87 \times 460 \times 95} = 284$ mm²/m
	(Min. A_s = 162 mm²/m – previous calc.)
	Use T12's @ 150 crs (753 mm²/m.)
Cl. 3.5.7 & Tables 3.10 & 3.11	Check cantilever deflection:
	span/d ≯ 7 : Actual = $\frac{1050}{100} = 10.5$
	$\frac{M}{bd^2} = \frac{10.8 \times 10^6}{10^3 \times 100^2} = 1.08$
Table 3.11 Note 2	$f_s = \frac{5}{8} \times \frac{460 \times 284}{753} = 108$ N/mm²
Table 3.11 Note 1	$m.f = 0.55 + \left(\frac{477 - 108}{120(0.9 + 1.08)}\right) = 2.03$
	But m.f. limited to 2.0
	$\therefore$ Allowable = 2.0 × 7 = 14 > 10.5 ✓ o.k.
	Design of wide floor beams
	A: Beams supporting slabs (6.25m. span) (see Fig. 8/003)
	'F' from slab calculations = 71.4 kN/m.
Table 3.6	$\therefore$ Hogging moment $= 0.11 FL^2$
	$= 0.11 \times 71.4 \times 6.25^2 = 306.8$ kN.m.
	Sagging moment $= 0.07 FL^2$
	$= 0.07 \times 71.4 \times 6.25^2 = 195.2$ kN.m.
	Effective depth of reinf. = 395 mm [Fig. 8/005]
	Hogging steel:
Cl. 3.4.4.4 Chart A1	$\frac{M}{f_{cu}bd^2} = \frac{306.8 \times 10^6}{40 \times 1000 \times 395^2} = 0.05$
	$\frac{z}{d} = 0.95$: $z = 0.95 \times 395 = 375$ mm.
	$\therefore A_s = \frac{306.8 \times 10^6}{0.87 \times 460 \times 375} = 2044$ mm²
	Use : 7 No. T20's (2198 mm²)

Ref. (BS 8110) Cl. 3.4.1.5

Sagging steel: Tee beam – effective breadth $= \frac{0.7 \times 6250}{5} + 1000$

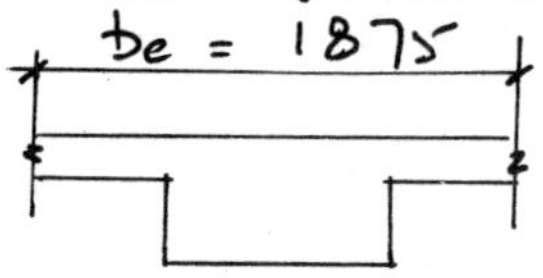

Fig. 8/007

$b_e = 1875$ mm.

Cl. 3.4.4.4

$$\frac{M}{f_{cu} b_e d^2} = \frac{195.2 \times 10^6}{40 \times 1875 \times 395^2} = 0.02$$

Chart A1

$\therefore$ $z/d = 0.95$: $z = 0.95 \times 395 = 375$ mm.

$$\therefore A_s = \frac{195.2 \times 10^6}{0.87 \times 460 \times 375} = 1300 \text{ mm}^2$$

Use : 5 No. T20's (1570 mm^2)

Table 3.27

Check min. steel: Min. steel = 0.13%

$$\therefore \text{Min. steel} = \frac{0.13}{100} \times 425 \times 10^3 = 552 \text{ mm}^2$$

Both hogging & sagging steel > 552 mm² ✓ O.K.

Check shear at supports:

Total 'F' $= 71.4 \times 6.25 = 446$ kN.

Max. shear $= 0.6F = 0.6 \times 446 = 267.8$ kN

$$\therefore v = \frac{267.8 \times 10^3}{1000 \times 395} = 0.68 \text{ N/mm}^2$$

Hogging steel $(A_s) = 2198$ mm²

$$\therefore \frac{100 A_s}{bd} = \frac{100 \times 2198}{1000 \times 395} = 0.56 \; ; \; d = 395 \text{ mm}$$

$\therefore$ $v_c = 0.51$ N/mm²

$$\therefore \underset{(0.25)}{0.5 v_c} < \underset{(0.68)}{v} < \underset{(0.91)}{(v_c + 0.4)}$$

$\therefore$ Use minimum links.

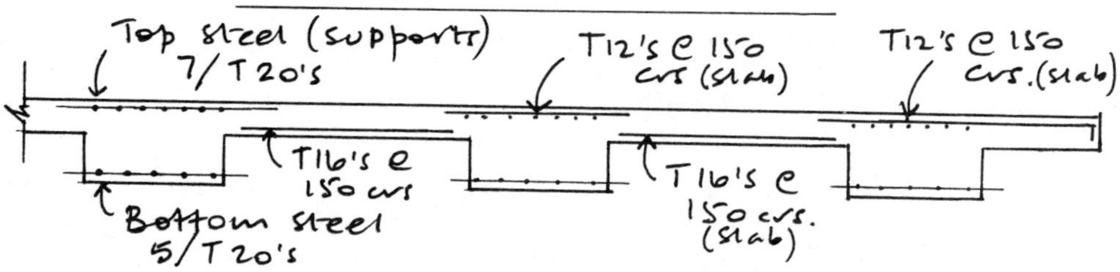

Fig. 8/008: slab & support beam reinforcement.

Ref.	
(BS 8110)	**B: Collector beams – grids A & B (Fig 8/003)**
	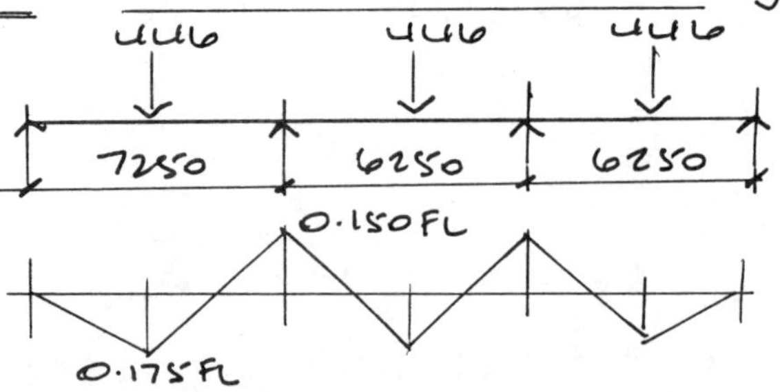
	Fig. 8/009: collector beam loading & moments
	'F' from slab beams = 446 kN (see shear calcs) Take 'L' = maximum span = 7.25 m (conservative)
Table 32 Ref 8/1	From 'Reynolds', Table 32:
	Max. hogging mt = $0.15 \times 446 \times 7.25 = $ **485 kNm**
	Max. sagging mt = $0.175 \times 446 \times 7.25 = $ **566 kNm**
	Hogging steel:
	$\frac{M}{f_{cu}bd^2} = \frac{485 \times 10^6}{40 \times 10^3 \times 395^2} = 0.08$
Cl. 3.4.4.4 Chart A1	$\frac{z}{d} = 0.9$: $z = 0.9 \times 395 = 356$ mm.
	$\therefore A_s = \frac{485 \times 10^6}{0.87 \times 460 \times 355} = 3414\ mm^2$
	Use 7 No. T25's ($3434\ mm^2$)
Cl. 3.4.1.5	**Sagging steel:** $b_e = \left(\frac{0.7 \times 7250}{5}\right) + 1000 = 2015$ mm.
	$\frac{M}{f_{cu}bd^2} = \frac{566 \times 10^6}{40 \times 2015 \times 395^2} = 0.045$
Chart A1	$\frac{z}{d} = 0.94$: $z = 0.94 \times 395 = 371$ mm.
	$\therefore A_s = \frac{566 \times 10^6}{0.87 \times 460 \times 371} = 3812\ mm^2$
	Use 8 No. T25's ($3926\ mm^2$)
Table 3.27	Check min. steel: Min. steel = 0.13%
	$\therefore$ Min. steel $= \frac{0.13}{100} \times 425 \times 10^3 = 552\ mm^2$
	Both hogging & sagging steel > $552\ mm^2$
Table 32 Ref. 8/1	Check shear: $V = 0.6 \times 446 = 268$ kN.
	$v = \frac{268 \times 10^3}{1000 \times 395} = 0.68\ N/mm^2$:
Tables 3.8 & 3.9	$100A_s/bd = (100 \times 3434)/(10^3 \times 395) = 0.87$
	$\therefore v_c = 0.66\ N/mm^2$ $\therefore$ $\underset{(0.33)}{0.5v_c} < \underset{(0.68)}{v} < \underset{(1.04)}{(v_c + 0.4)}$: **Min links**

Ref.	
(B.S. 8110)	**Approximate analysis of column moments and shears in lower columns.**

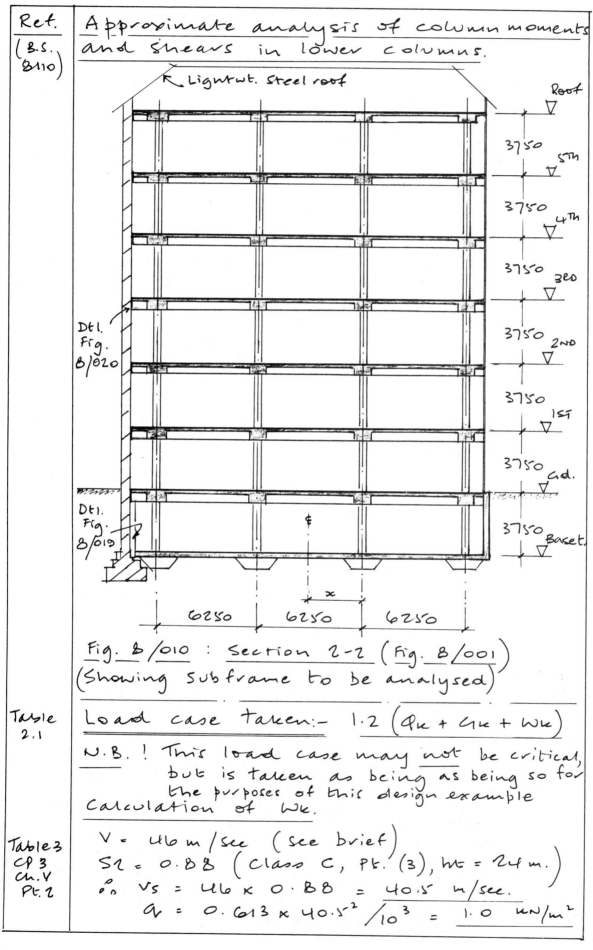

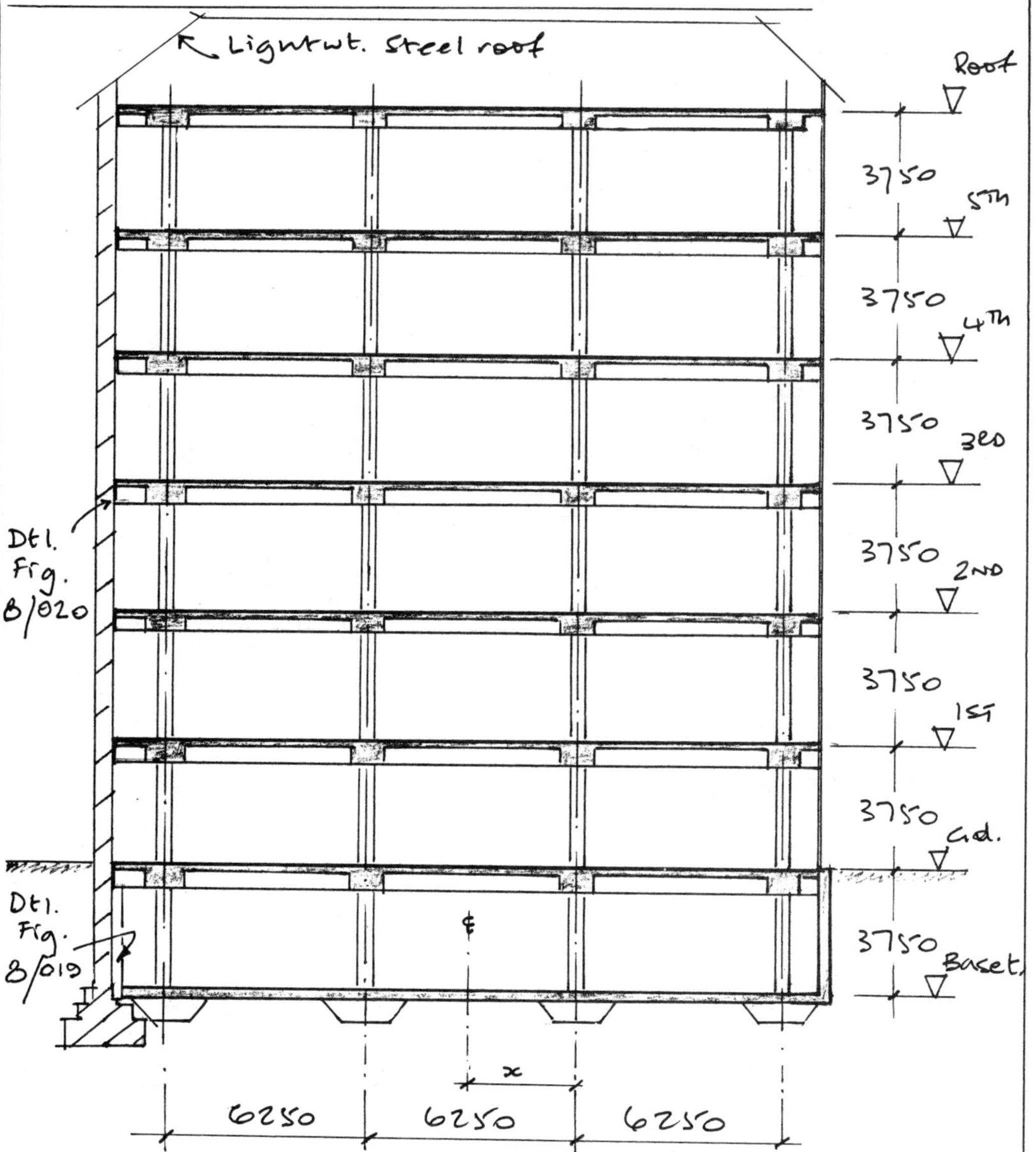

Fig. 8/010 : Section 2-2 (Fig. 8/001)
(Showing subframe to be analysed)

Table 2.1	**Load case taken:- 1.2 ($Q_k + G_k + W_k$)**
	N.B.! This load case may not be critical, but is taken as being as being so for the purposes of this design example
	Calculation of W_k.
Table 3 CP 3 Ch. V Pt. 2	$V = 46$ m/sec (see brief)
	$S_2 = 0.88$ (Class C, Pt. (3), ht = 24 m.)
	$\therefore V_s = 46 \times 0.88 = 40.5$ m/sec.
	$q = 0.613 \times 40.5^2 / 10^3 = 1.0$ kN/m²

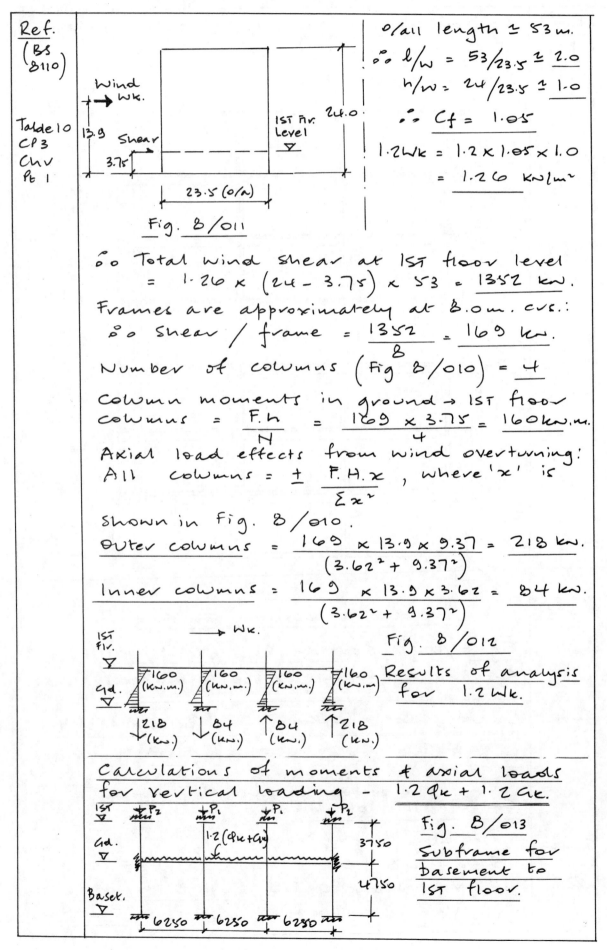

Ref.	
(BS 8110)	
Table 10 CP3 Ch V Pt 1	O/all length ≏ 53 m.
	∴ $l/w = 53/23.5 \doteq \underline{2.0}$
	$h/w = 24/23.5 \doteq \underline{1.0}$
	∴ $\underline{C_f = 1.05}$
	$1.2 W_k = 1.2 \times 1.05 \times 1.0$
	$= \underline{1.26 \text{ kN/m}^2}$

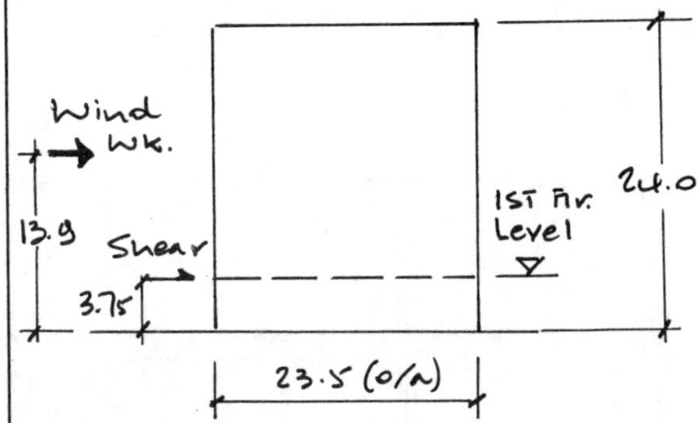

Fig. 8/011

∴ Total wind shear at 1st floor level

$= 1.26 \times (24 - 3.75) \times 53 = \underline{1352 \text{ kN}}.$

Frames are approximately at 8.0 m. crs.:

∴ Shear / frame $= \dfrac{1352}{8} = \underline{169 \text{ kN}}.$

Number of columns (Fig 8/010) $= \underline{4}$

Column moments in ground → 1st floor columns $= \dfrac{F.h}{N} = \dfrac{169 \times 3.75}{4} = \underline{160 \text{ kN.m.}}$

Axial load effects from wind overturning:

All columns $= \pm \dfrac{F.H.x}{\Sigma x^2}$, where '$x$' is shown in Fig. 8/010.

Outer columns $= \dfrac{169 \times 13.9 \times 9.37}{(3.62^2 + 9.37^2)} = \underline{218 \text{ kN}}.$

Inner columns $= \dfrac{169 \times 13.9 \times 3.62}{(3.62^2 + 9.37^2)} = \underline{84 \text{ kN}}.$

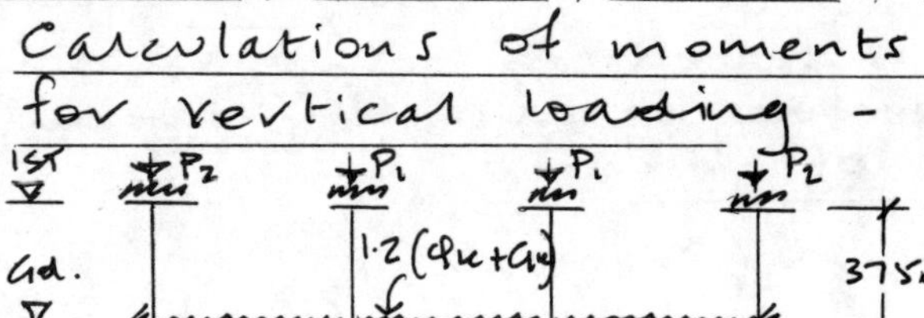

Fig. 8/012 Results of analysis for 1.2 Wk.

Calculations of moments + axial loads for vertical loading – $1.2 Q_k + 1.2 G_k$.

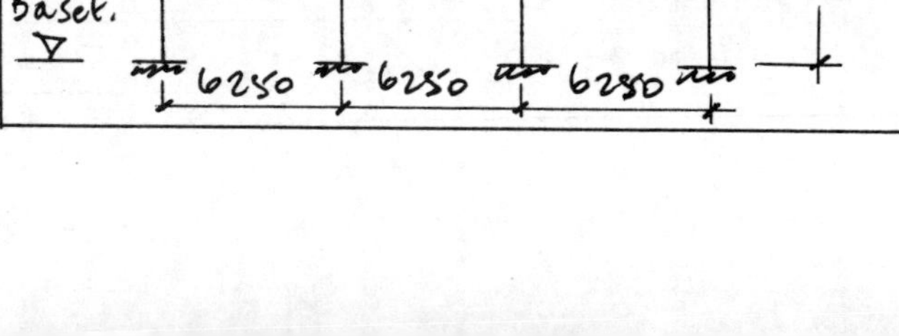

Fig. 8/013 Subframe for basement to 1st floor.

Ref.	
(B.S. 8110)	**Axial loads in columns – loads P_1 & P_2, see Fig. 8/013.**

Loads P_1; 6 storeys of floor loadings + 1 storey of roof loading.

Floors – 2 reactions from collector beams + 2 reactions from slab beams

$= 2 \times 223 + 2 \times 223 = 892$ kN.

So 6 floors $= 6 \times 892 = 5352$ kN

Roof design loading $= 1.6 \times 7.5 + 1.4 \times 6.2$

$= 20.7$ kN/m²

∴ By proportion, roof reactions

$= \frac{20.7}{19.7} \times 892 = 937$ kN.

Total $P_1 = 937 + 5352 = 6289$ kN.

$P_2 \simeq \frac{1}{2} \times P_1 = 3145$ kN.

Axial loads in columns between basement & ground floor:

$P_1 = 5352 + 892 = 6244$ kN.*

$P_2 = 3145 + 446 = 3591$* kN.

* 1.6 Qk + 1.4 Gk

For load case $1.2\,(Q_k + G_k)$, these loads to be proportioned down:–

Floor loading $= 1.2\,(6.0 + 7.2) = 15.8$ kN/m²

Roof loading $= 1.2\,(7.5 + 6.2) = 16.4$ kN/m²

Following same calculations as above:–

P_1 below 1st floor

$= 892 \times 5 \times \frac{15.8}{19.7} + 937 \times \frac{16.4}{20.7} = 4319$ kN.

P_1 below ground floor

$= 4319 + 892 \times \frac{15.8}{19.7} = 5034$ kN.

Fig. 8/014 col. axial loads

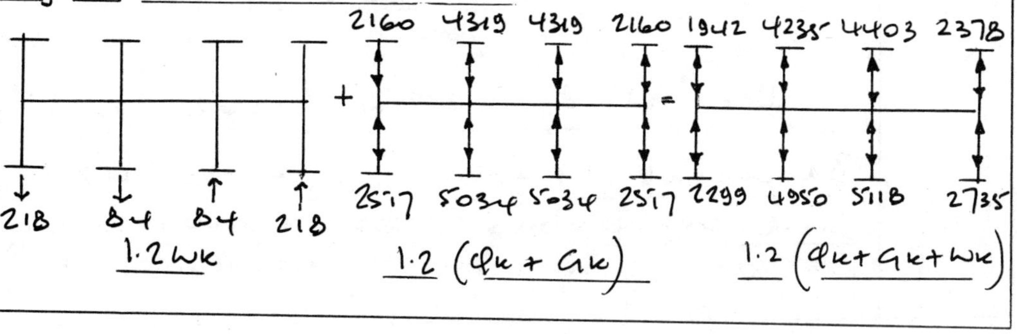

Ref. (BS 8110)

Subframe analysis – due to symmetry of frame & uniformity of loading (all floor beams loaded) – there will be zero bending in the columns.

∴ For a typical column design, taking the right-hand internal column, from Figs. 8/012 & 8/014, we have :–

	M	N
Gd. → 1st Floor	160 kN.m	4403 kN.
Basement → gd.	zero	*5118 kN.

(* also check this lift for 1.6qk + 1.4gk = 6244 kN (previous calcs.))

Gd. → 1st Floor: assume column 500 sq.

Clear height, l_o = 3750 – 425 = 3225 mm.

End conditions: condition 2 each end – beams 425 mm deep, framing on to column 500 mm deep, unbraced frame *(Cl. 3.8.1.6.2)*

∴ l_e = 1.5 × 3225 = 4837 mm

(Table 3.22) l_e/b = 4837/500 = 9.7 < 10

∴ Column is "short"

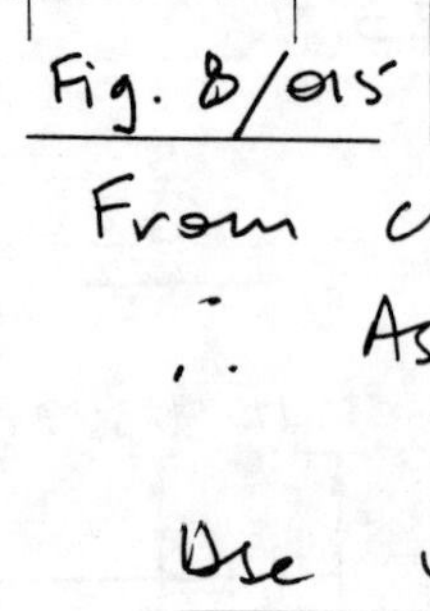

Fig. 8/015

(Pt. 2 Chart 39)

From Fig. 8/015, d/h = $\frac{450}{500}$ = 0.9

∴ Use Chart No. 39

$$\frac{N}{bh} = \frac{4403 \times 10^3}{500^2} = 17.6$$

$$\frac{M}{bh^2} = \frac{160 \times 10^6}{500^3} = 1.28$$

From chart, minimum percentage rules.

$$\therefore\ A_{sc} = \frac{0.4\,bh}{100} = \frac{0.4 \times 500^2}{100} = 1000\ \text{mm}^2$$

Use 4 No. T25's (1962.5 mm²)

Basement to gd. l_o = 3750 – 425 = 3225 mm.

End conditions: top – condition 2, bottom – condition 1 (moment base)

Ref.	
(BS 8110)	∴ $l_e = 1.3 \times 3225 = 4192$ mm $l_e/b = 4192/500 = 8.4 < 10$ ∴ column is 'short'
	Taking more onerous value for 'N' of 6244 kN ($1.6q_k + 1.4G_k$):– $\frac{N}{bh} = \frac{6244 \times 10^3}{500^2} = \underline{25}$
Pt. 2 Chart 39	∴ $\frac{100 A_{sc}}{bh} = 1.0$ ∴ $A_{sc} = \frac{1.0 \times 500^2}{100} = \underline{2500 \text{ mm}^2}$ ∴ Use 6 No. T25's (2945 mm²)
Cl. 3.12.7.1	Links: use 8 mm links (≮ one quarter main bar φ) at a spacing of 300 mm ($\leq 12 \times 25 = \underline{300}$ mm)
	Pilecap design. In the stiff clay, $c = 75$ kN/m². Try a 4-pile cap, diameter (d) = 550 mm., and bored to a depth of 25 m. Formula for a bored pile in clay:– $Q_u = Q_s + Q_b$, where Q_u = ultimate load capacity, Q_s is skin friction & Q_b end bearing. Adopting a safety factor of 2 for skin friction, and 3 for end bearing, the formula becomes:– Allowable $= \frac{Q_s}{2} + \frac{Q_b}{3}$, in which, $Q_s = \pi dh \times 0.45c$ & $Q_b = \frac{\pi d^2}{4} \times 9c$ ∴ For 4 piles, d = 650 & h = 30 m., Capacity $= 4\left[\frac{1}{2}(\pi \times 0.65 \times 30 \times .45 \times 75) + \frac{1}{3}\left(\frac{\pi \times 0.65^2}{4} \times 9 \times 75\right)\right] = \underline{4431 \text{ kN}}$.

Ref. (B.S. 8110)

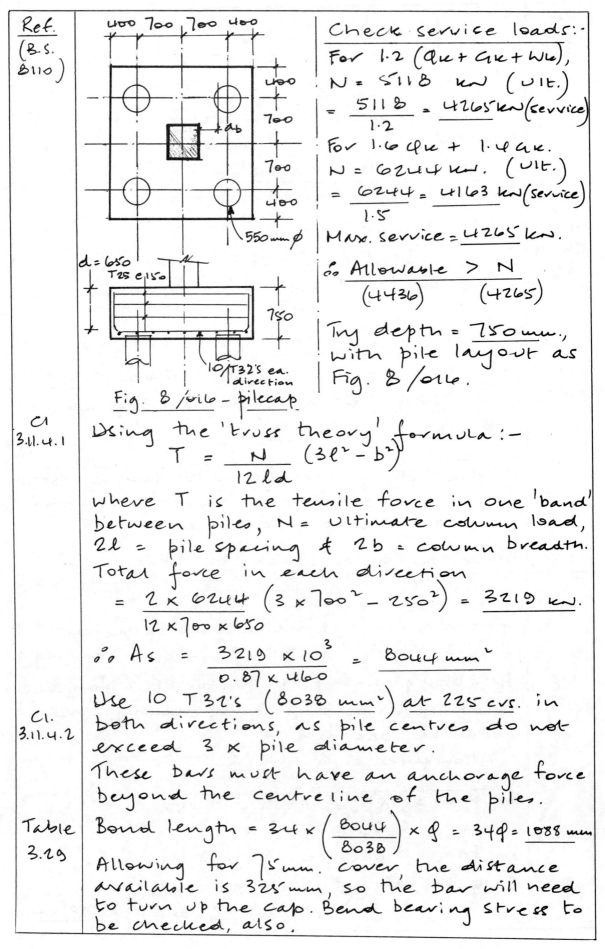

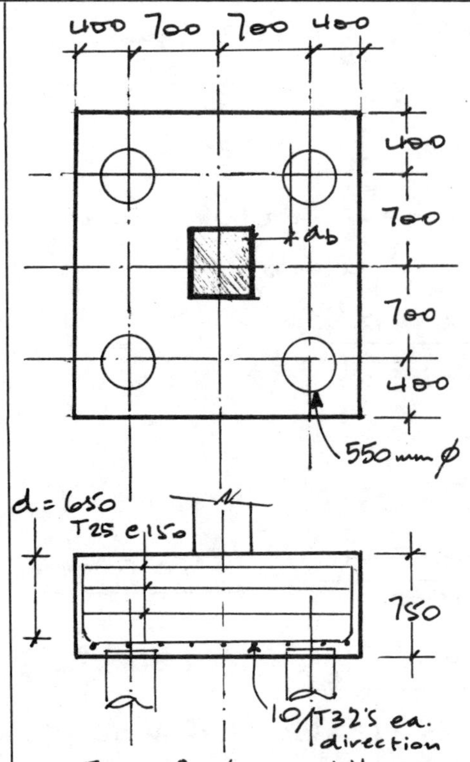

Fig. 8/016 – pilecap

Check service loads:-

For 1.2 ($Q_k + G_k + W_k$),

N = 5118 kN (ult.)

$= \dfrac{5118}{1.2} = 4265$ kN (service)

For 1.6 Q_k + 1.4 G_k.

N = 6244 kN. (ult.)

$= \dfrac{6244}{1.5} = 4163$ kN (service)

Max. service = 4265 kN.

∴ Allowable > N
(4436) (4265)

Try depth = 750 mm., with pile layout as Fig. 8/016.

Cl 3.11.4.1

Using the 'truss theory' formula:-

$$T = \frac{N}{12\,l\,d}\,(3l^2 - b^2)$$

where T is the tensile force in one 'band' between piles, N = ultimate column load, 2l = pile spacing & 2b = column breadth.

Total force in each direction

$$= \frac{2 \times 6244}{12 \times 700 \times 650}\,(3 \times 700^2 - 250^2) = 3219 \text{ kN.}$$

$$\therefore A_s = \frac{3219 \times 10^3}{0.87 \times 460} = 8044 \text{ mm}^2$$

Cl. 3.11.4.2

Use 10 T32's (8038 mm²) at 225 crs. in both directions, as pile centres do not exceed 3 × pile diameter.

These bars must have an anchorage force beyond the centreline of the piles.

Table 3.29

Bond length $= 34 \times \left(\dfrac{8044}{8038}\right) \times \phi = 34\phi = 1088$ mm

Allowing for 75 mm. cover, the distance available is 325 mm, so the bar will need to turn up the cap. Bend bearing stress to be checked, also.

Ref.	
(B.S. 8110)	**Check shear:**
	a) Punching shear around column perimeter
	$= \frac{6244 \times 10^3}{4 \times 500 \times 650} = 4.8\ N/mm^2$
Cl. 3.11.4.5	$< 0.8\sqrt{40}\ (5.0\ N/mm^2)$ ✓ o.k.
	b) Across full width of cap
	$V = \frac{\text{Column load}}{2}$ (2 shear faces)
	$\therefore v = \frac{(6244/2) \times 10^3}{2200 \times 650} = 2.2\ N/mm^2$
Cl. 3.11.4.4 (a)	Shear stress can be enhanced to Clauses 3.5.5 & 3.5.6
	'a_v' is shown in Fig. 8/016 & Fig. 3/23 (code): $a_v = 700 - 250 - 325 + 0.2 \times 650$
	$= 255\ mm.$
	Enhancement factor
	$= \frac{2d}{a_v} = \frac{2 \times 650}{255} = 5.1$
	$\frac{100 A_s}{bd} = \frac{100 \times 8038}{2200 \times 650} = 0.56$
Table 3.9	$\therefore v_c = 0.58\ N/mm^2$
	Allowable shear stress $= 0.58 \times 5.1 = 3\ N/mm^2$
	$\therefore$ Allowable > Actual ✓ o.k.
	(3) (2.2)
	For horizontal binders, allow 25% of main steel.
	Use 3 No. T25's @ 150 crs.
	(6 cross sections $= 2944\ mm^2$)
	25% of main steel $= 2010\ mm^2$ ✓ o.k.
	(see Fig. 8/016 for detail).

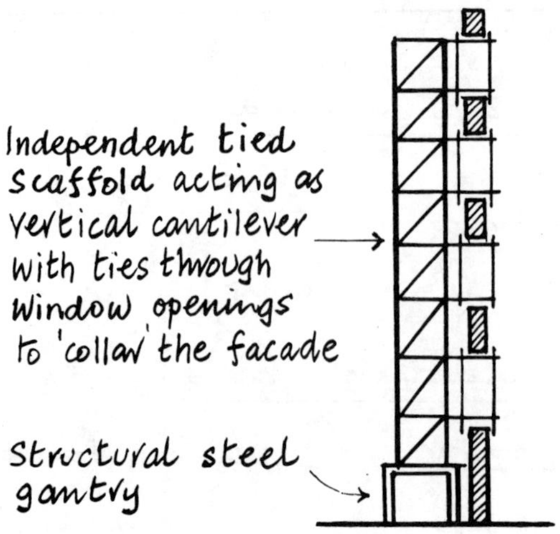

Fig.8/017

Typical example of temporary support to existing Facade

Enlarged Detail

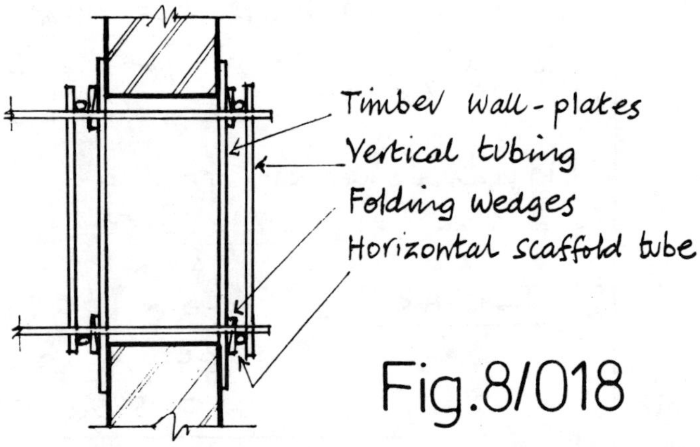

Fig.8/018

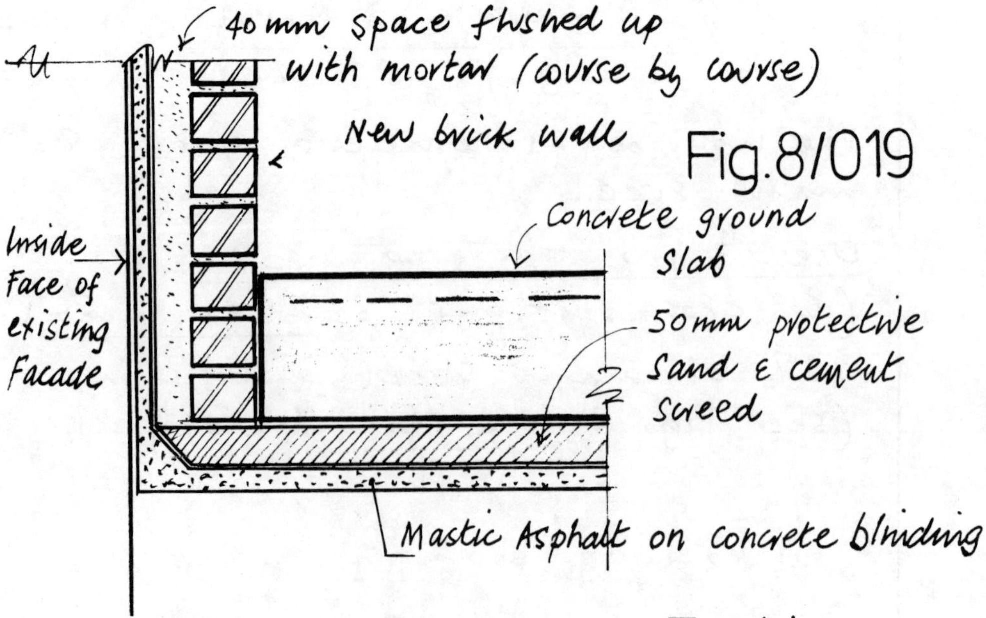

Fig.8/019

Typical Basement Tanking

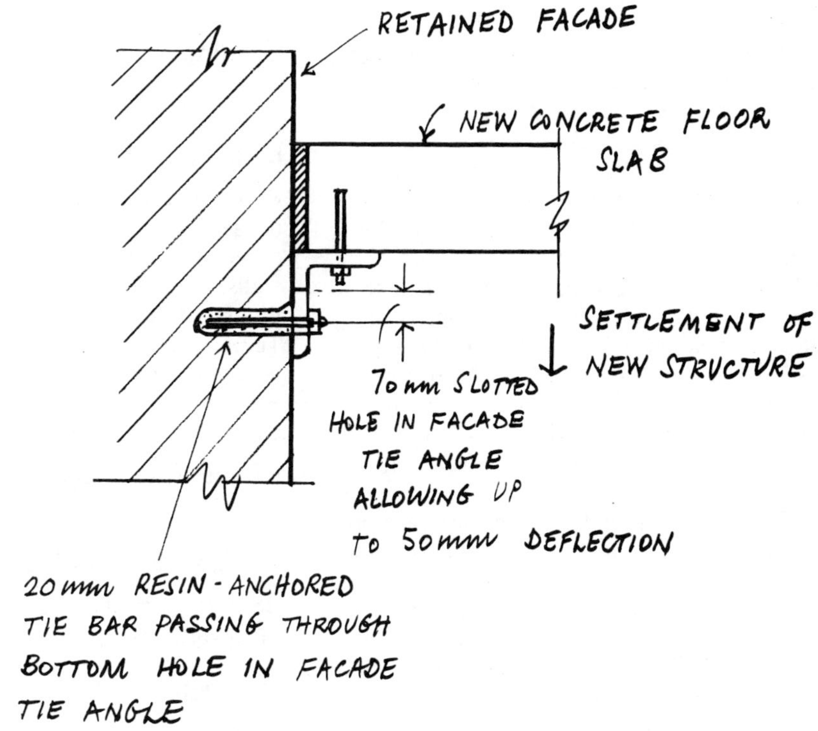

Connection of New Structure to Existing Facade

Fig.8/020

Chapter 9 Part conversion and extension of residential property into a wine bar/restaurant

A client requires the refurbishment, including conversion and extension, of the basement and ground floors of a four-storey end-terrace property. The building is of load-bearing brickwork, with a tile-clad timber roof, and the existing and proposed layouts are shown on the Architect's sketches in Fig. 9/001. The restaurant and wine bar are to be at Ground Floor level and basement level respectively, and are to be free of internal columns. The restaurant is to extend into a purpose-built brickwork extension, with a waterproofed concrete terrace over. The thicknesses of all the existing walls are given on Fig. 9/001, and are considered to be sound in themselves, but stresses from point loads should be restricted to 0.5 N/mm^2. On no account must any lateral loads on the refurbished property impinge on the adjacent properties. The existing ground, first and second floors are timber, and adequate for the intended purpose, with floor joists spanning front-to-rear, whilst the basement floor is of concrete. The proposed floor plans are shown in Figure 9/002, which are based on a two-storey portal frame on grid B, supporting floor and wall loads and part of the 'racking action' load, and simply-supported beams on grid C supporting floor and roof loads only. The remainder of the 'racking action' load is adequately handled by the front elevation of masonry.

Loading – qk

Pitched roof areas (snow)	= 0.75 kN/m^2
First and Second Floors and terrace	= 1.5 kN/m^2
Wine bar and Restaurant Floors	= 4.0 kN/m^2

Loading – 'racking action'

A sideways loading of 0.5% of total loads (Cl. 2.4.2.3 BS 5950)

Loading – wk

Basic wind speed 46 m/sec (outskirts of small town)

Loading – gk

Roof – including tiles, battens, felt and trusses = 1.2 kN/m^2
(Note! survey shows that trusses bear only on outside walls)
Floors – timber – including ceilings = 1.0 kN/m^2
Terrace floor – concrete – including waterproofing = 4.0 kN/m^2

Soil Safe Bearing

Existing strip foundations are on gravel, safe bearing 200 kN/m^2

Design Codes of Practice

BS 5268 – Timber : BS 5628 – Masonry : BS 5950 – Steelwork

Materials

Timber elements – SS grade pitch pine floor joists
Steelwork – EN10025 S275, 1-hour minimum fire resistance
Brickwork – fk = 6.5 N/mm^2: new brickwork – fkx = 1.1 N/mm^2: new blockwork – fx = 3.5 N/mm^2 : fkx = 0.4 N/mm^2, both in mortar (ii). Partial material factor = 3.5

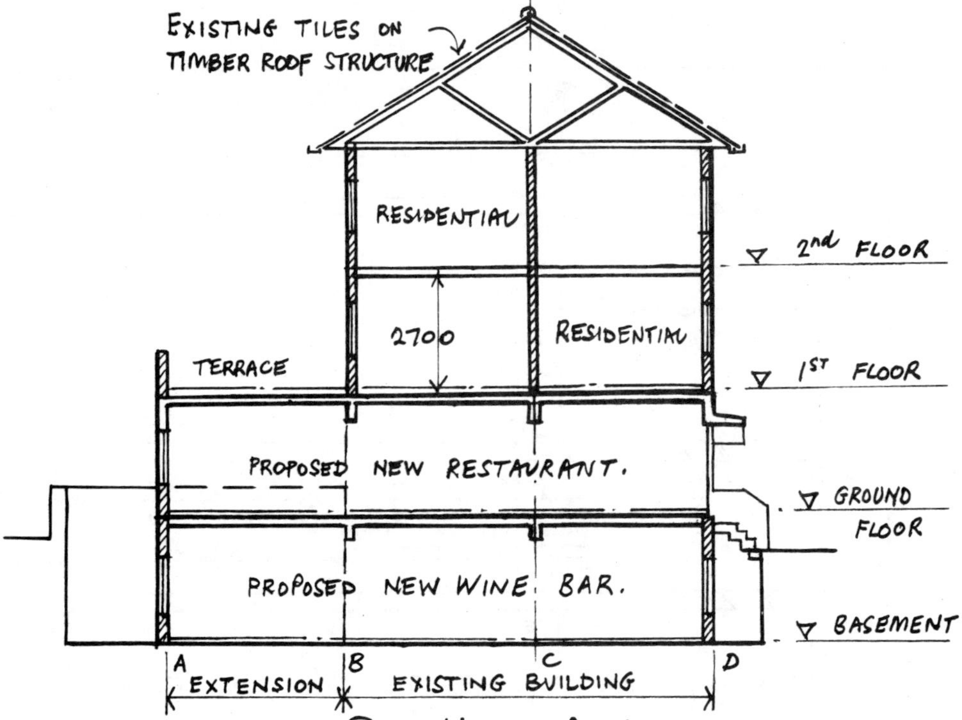

Section A.A.

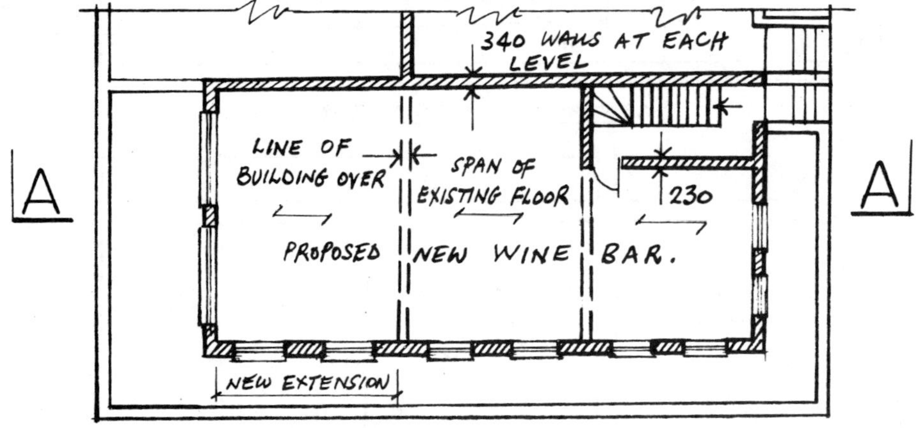

Proposed new Basement Plan

(PROPOSED NEW GROUND FLOOR SIMILAR)

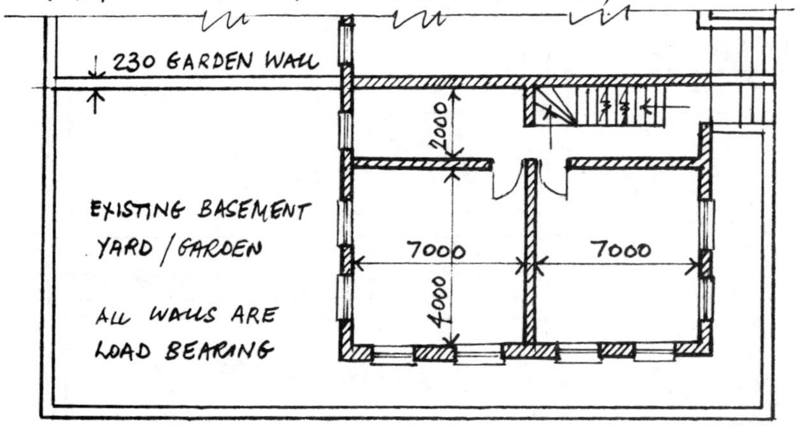

Existing Ground Floor Plan

(BASEMENT AND UPPER FLOOR PLANS SIMILAR)

Fig.9/001

Ref.

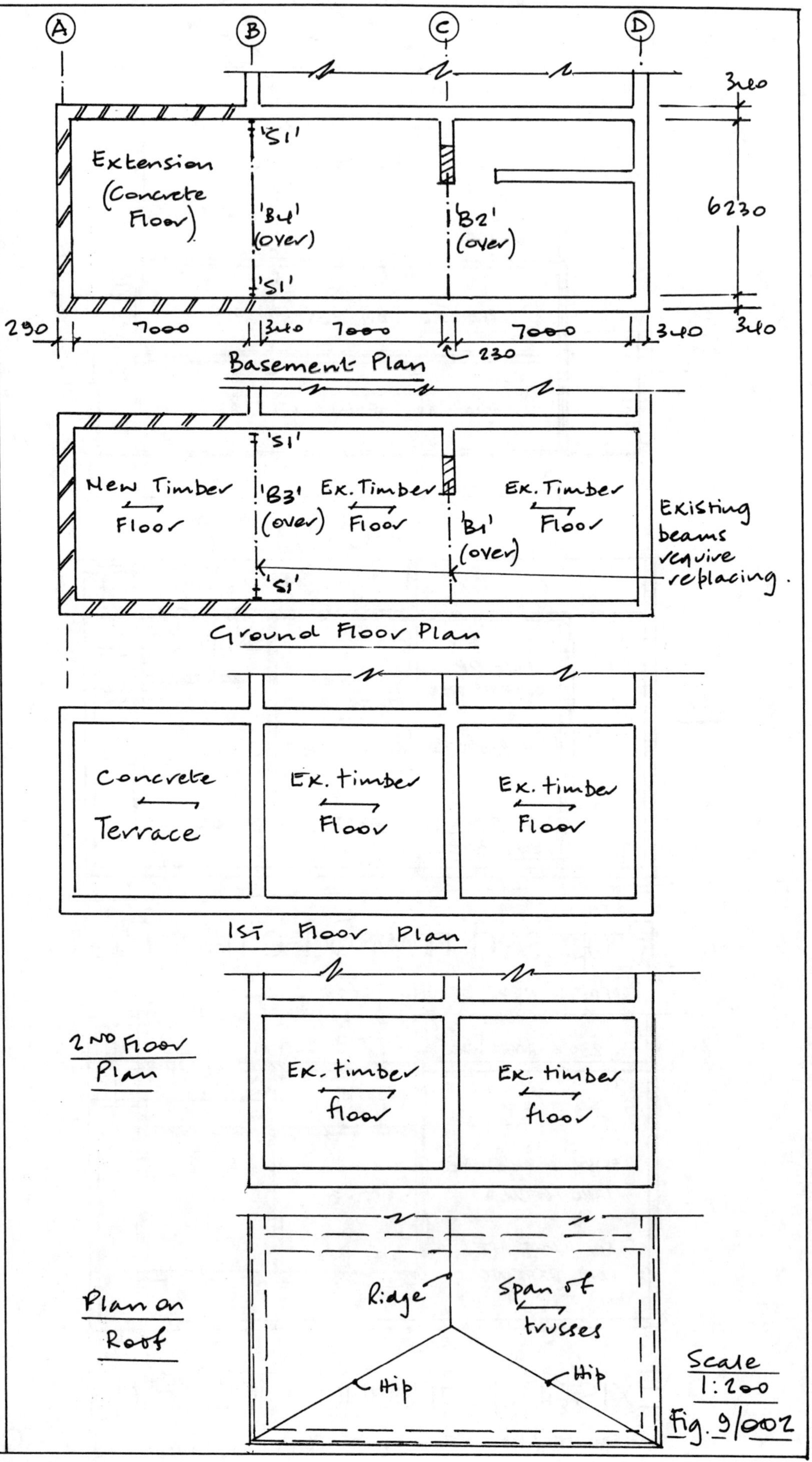
A
B
C
D
340
6230
340
Extension (Concrete Floor)
'S1'
'B4' (over)
'S1'
'B2' (over)
290
7000
340
7000
230
7000
340
Basement Plan
'S1'
New Timber Floor
'B3' (over)
Ex. Timber Floor
'B1' (over)
Ex. Timber Floor
'S1'
Existing beams require replacing.
Ground Floor Plan
Concrete Terrace
Ex. timber Floor
Ex. timber Floor
1st Floor Plan
2ND Floor Plan
Ex. timber floor
Ex. timber floor
Plan on Roof
Ridge
Span of trusses
Hip
Hip

Scale 1:200

Fig. 9/002

Ref.	
(BS 5950)	**Design (factored) loadings:-**
	Roof: Design loading $= 1.6 \times 0.75 + 1.4 \times 1.2$
	$= 2.9\ kN/m^2$
	1ST & 2ND Floors (timber):
Table 2	Design loading $= 1.6 \times 1.5 + 1.4 \times 1.0 = 3.8\ kN/m^2$
	Ground Floor (timber):
	Design loading $= 1.6 \times 4.0 + 1.4 \times 1.0 = 7.8\ kN/m^2$
	Terrace Floor (concrete):
	Design loading $= 1.6 \times 1.5 + 1.4 \times 4.0 = 8.0\ kN/m^2$
	Design of beams "B1" → "B4": Beams "B1" & "B2" are simply supported on padstones: beams "B3" & "B4" are simply-supported on columns "S1".
	A: Beam "B1" - Effective span = 4.3 m.
	U.D.L. (factored):-
	i) **Internal wall**: 230 mm thick; $\gamma = 18\ kN/m^3$; Height = 5.5 m.:
	U.D.L. $= 1.4 \times 0.23 \times 18 \times 5.5 = 31.88\ kN/m.$
	ii) **1ST & 2ND floors**: floor beams span 7.2 m. onto "B1" & onto flank walls.
	U.D.L. $= 2 \times 2 \left[\dfrac{7.2 \times 3.8}{2}\right] = 54.72\ kN/m.$
	iii) Allow 0.5 kN/m for beam self-weight.
	$\therefore$ Total U.D.L. $= 31.88 + 54.72 + (1.4 \times 0.5)$
	$= 87.3\ kN/m.$
	$\therefore M_{MAX} = \dfrac{87.3 \times 4.3^2}{8} = 202\ kN.m.$
	$F_{v\,MAX} = \dfrac{87.3 \times 4.3}{2} = 188\ kN.$
	Try a 356 × 171 × 51 kg/m U.B.
Ref. 9/1	$b/T = 7.46$: $d/t = 42.8$: $D = 355.6\ mm.$:
Table 6	$t = 7.3\ mm.$: $S_{xx} = 895\ cm^3$: $T = 11.5mm < 16$
	$\therefore p_y = 275\ N/mm^2$: $I_{xx} = 14160\ cm^4$

Ref.	
(BS 5950) Table 7	**Classification:** Flange : $b/T = 7.46 < 8.5\varepsilon$ – plastic. Web : $d/t = 42.8 < 79\varepsilon$ – plastic. ∴ Section 'plastic'

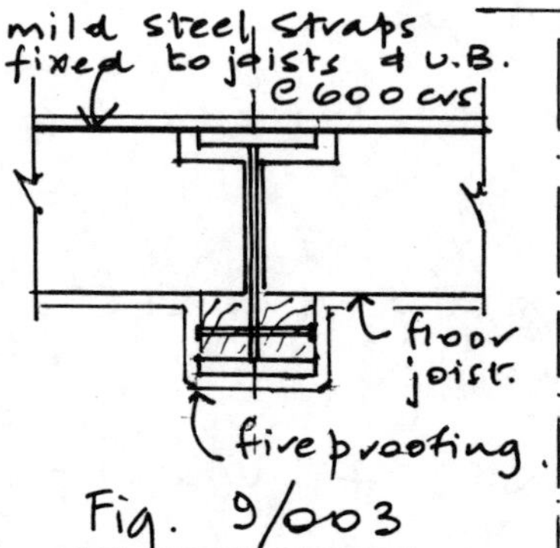

Fig. 9/003
Beam restraint
Detail

Check U.L.S. of flexure:

Consider beam fully restrained (see Fig 9/003)

$$M_{cx} = \frac{275 \times 895}{10^3} = 246 \text{ kN.m.}$$

Cl. 4.2.5

∴ $M_{cx} > M_{MAX}$ ✓ o.k.
(246) (202)

Check shear: $F_v = 188$ kN (previous calcs)

Cl. 4.2.3

$$P_v = \frac{0.6 \times 275 \times 355.6 \times 9.1}{10^3} = 428 \text{ kN.}$$

∴ $P_v > F_v$ ✓ o.k.
(428) (188)

Check Serviceability Limit State – deflection.

$$\text{Unfactored } q_k = 2 \times 2\left[\frac{7.2 \times 1.50}{2}\right] = 21.6 \text{ kN/m.}$$

$$\therefore \Delta = \frac{5 \times 21.6 \times (4300)^4}{384 \times 205 \times 10^3 \times 14160 \times 10^4} = 3.3 \text{mm.}$$

Table 5

Allowable = 4300/360 = 11.9mm ✓ o.k.

'B1' : 356 × 171 × 51 U.B.

B: Beam 'B2' – effective span = 4.3m.

U.D.L. (factored) :–

(i) Ground floor : floor beams span 7.2m. onto "B2" & flank walls.

$$\text{U.D.L.} = 2\left[\frac{7.2 \times 7.8}{2}\right] = 56.2 \text{ kN/m.}$$

(ii) Selfweight = 1.4 × 0.4 = 0.56 kN/m.

Total U.D.L. = 56.76 kN/m

Ref. (BS 5950)

$\therefore M_{MAX} = \dfrac{56.76 \times 4.3^2}{8} = \underline{131.2\ kN.m.}$

$F_v = \dfrac{56.76 \times 4.3}{2} = \underline{122\ kN.}$

Try a 254 x 146 x 37 kg/m U.B.

$b/T = 6.72$: $d/t = 34.2$: $D = 256mm$: $t = 6.4mm$

$S_{xx} = 485\ cm^3$: $I_{xx} = 5560\ cm^4$: $T = 10.9mm < 16$

Table 6 — $\therefore p_y = 275\ N/mm^2$.

Classification:

Table 7 — Flange : $b/T = 6.72 < 8.5\varepsilon$ – plastic

Web : $d/t = 34.2 < 79\varepsilon$ – plastic

$\therefore$ Section 'plastic'

Check U.L.S. of flexure:

Beam fully restrained – see Fig. 9/003

Cl. 4.2.5 — $\therefore M_{cx} = \dfrac{275 \times 485}{10^3} = \underline{133.4\ kN.m.}$

$\therefore \underset{(133.4)}{M_{cx}} > \underset{(131.2)}{M_{MAX}}$ ✓ o.k.

Check U.L.S. of shear : $F_v = 122\ kN.$

Cl. 4.2.3 — $P_v = \dfrac{0.6 \times 275 \times 256 \times 6.4}{10^3} = \underline{270\ kN.}$

$\therefore \underset{(270)}{P_v} > \underset{(122)}{F_v}$ ✓ o.k.

Check Serviceability State of Deflection.

Unfactored $q_k = 2\left[\dfrac{7.2 \times 4.0}{2}\right] = \underline{28.4\ kN/m.}$

$\therefore \Delta = \dfrac{5}{384} \times \dfrac{28.4 \times (4300)^4}{205 \times 10^3 \times 5560 \times 10^4} = \underline{11.1\ mm.}$

Table 5 — Allowable $= 4300/360 = \underline{11.9\ mm}$ ✓ o.k.

'B2' : 254 x 146 x 37 U.B.

C : Beam 'B3' – spans 6.0m, simply-supported, onto portal legs – connections to take moments from 'racking action' only.

U.D.L. (factored)

(i) Roof : trusses span 14.6m. onto flank walls

$U.D.L. = \dfrac{14.6}{2} \times 2.9 = \underline{21.2\ kN/m.}$

Ref.	
(BS 5950)	(ii) External wall – 340 mm thick, 5.5 m high, take $\gamma = 18\ kN/m^3$
	U.D.L. $= 1.4 \times 0.34 \times 18 \times 5.5 = 47.1\ kN/m$
	(iii) 1st & 2nd Floors – span 7.2 m onto internal wall & flank wall.
	U.D.L. $= 2\left[\dfrac{7.2 \times 3.8}{2}\right] = 27.36\ kN/m.$
	(iv) Terrace – spans 7.2 m onto flank wall
	U.D.L $= \dfrac{7.2}{2} \times 8 = 28.8\ kN/m.$
	(v) Selfweight – allow 0.5 kN/m.
	U.D.L. $= 1.4 \times 0.5 = 0.7\ kN/m.$
	Total $= 21.2 + 47.1 + 27.36 + 28.8 + 0.7 = 125.2\ kN/m$
	$\therefore M_{MAX} = \dfrac{125.2 \times 6^2}{8} = 563.2\ kN.m.$
	$F_v = \dfrac{125.2 \times 6}{2} = 375.6\ kN.$
	Try a 457 × 191 × 98 kg/m U.B.
Ref 6/1	$b/T = 4.92$: $d/t = 35.8$: $D = 467.4\ mm$: $t = 11.4\ mm$: $S_{xx} = 2230\ cm^3$: $I_{xx} = 45700\ cm^4$
	$T = 19.6\ mm > 16$: $p_y = 265\ N/mm^2$
Table 6	Classification: $\varepsilon = \sqrt{275/265} = 1.02$
	Flange : $b/T = 4.92 < 8.5\varepsilon$ – plastic.
	Web : $d/t = 35.8 < 79\varepsilon$ – plastic.
	$\therefore$ Section 'plastic'
	Check U.L.S. of flexure :
	Beam fully restrained – see fig 9/003
Cl. 4.2.5	$\therefore M_{cx} = \dfrac{265 \times 2230}{10^3} = 591\ kN.m.$
	$\therefore M_{cx} > M_{MAX}$ ✓ o.k
	(591) (563.2)
	Check U.L.S. of shear : $F_v = 375.6\ kN.$
Cl. 4.2.3	$P_v = \dfrac{0.6 \times 265 \times 467.4 \times 11.4}{10^3} = 847\ kN.$
	$\therefore P_v > F_v$ ✓ o.k.
	(847) (375.6)

Ref.	
(BS 5950)	**Check Serviceability Limit State of Deflection.**
	Unfactored q_k =
	$0.75 \times 14.6/2 + 2\left[\frac{7.2 \times 1.5}{2}\right] + \frac{7.2}{2} \times 1.5 = 21.7$ kN/m
	$\Delta = \frac{5}{384} \times \frac{21.7 \times (6000)^4}{205 \times 10^3 \times 45700 \times 10^4} = 3.91$ mm.
Table 5	Allowable = 6000/360 = 16.67 mm ✓ o.k.
	'B3': 457 × 191 × 98 U.B.
	D: Beam 'B4' - eff. span 6.0 m.
	U.D.L. (factored)
	(i) Restaurant floor – as 'B2' = 56.2 kN/m
	(ii) Selfweight = 1.4 × 0.5 = 0.7 kN/m
	total = 56.9 kN/m.
	∴ $M_{MAX} = \frac{56.9 \times 6^2}{8} = 256$ kN.m.
	$F_v = \frac{56.9 \times 6}{2} = 170.7$ kN.
	Try a 356 × 171 × 57 kg/m U.B.
Ref. 6/1	b/T = 6.62 : d/t = 39 : D = 358.6 mm :
	t = 8 mm : S_{xx} = 1010 cm^3 : I_{xx} = 16100 cm^4 :
Table 6	T = 13 mm < 16 : $p_y = 275$ N/mm^2.
	Classification:
Table 7	Flange: b/T = 6.62 < 8.5ε – plastic
	Web: d/t = 39 < 79ε – plastic.
	∴ Section 'plastic'
	Check U.L.S. of flexure:
	Beam fully restrained – see Fig. 9/003
Cl. 4.2.5	∴ $M_{cx} = \frac{275 \times 1010}{10^3} = 278$ kN.m.
	∴ $M_{cx} > M_{MAX}$ ✓ o.k. (278) (256)
	Check U.L.S. of shear: F_v = 170.7 kN.
Cl. 4.2.3	$P_v = \frac{0.6 \times 275 \times 358.6 \times 8}{10^3} = 473$ kN.
	∴ $P_v > F_v$ ✓ o.k. (473) (170.7)

Ref.	
(BS 5950)	**Serviceability Limit State of Deflection.**
	W_k = 28.4 kN/m (as 'B2')
	∴ $\Delta = \frac{5}{384} \times \frac{28.4 \times (6000)^4}{205 \times 10^3 \times 16100 \times 10^4} = \underline{14.5\text{mm}}$.
Table 5	Allowable = 6000/360 = $\underline{16.67\text{mm}}$ ✓ o.k.
	'B4' : $\underline{356 \times 171 \times 57 \text{ kg/m U.B.}}$

Design of padstones:

Max. factored reaction = $\underline{188 \text{ kN}}$ ('B1')

Width of crosswall under 'B1' = $\underline{230\text{mm}}$.

Make padstones of concrete, 230 wide × 400 long × 150 deep. From brief, take $f_k = 6.5 \text{ N/mm}^2$; $\gamma_m = 3.5$.

BS5628 Cl.34

∴ Capacity = $\frac{1.25 \times 6.5 \times 230 \times 400}{3.5 \times 10^3} = \underline{213.6\text{kN}}$.

$\underline{\text{Capacity}}$ (213.6 kN) > $\underline{\text{Reaction}}$ (188 kN) ✓ o.k.

Design of portal columns on grid 'B'

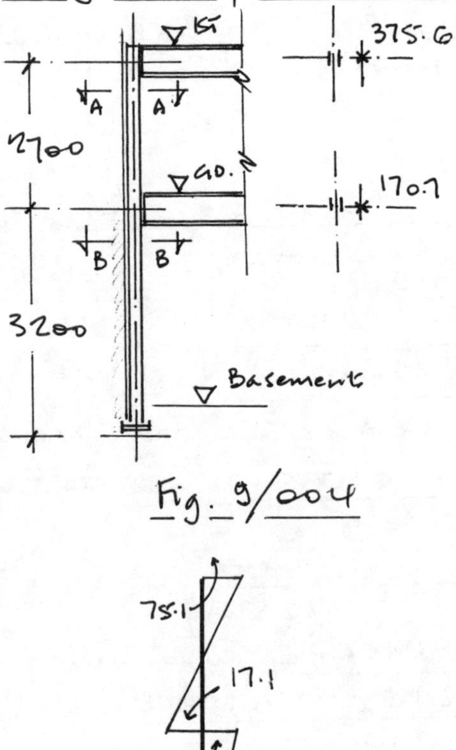

Fig. 9/004

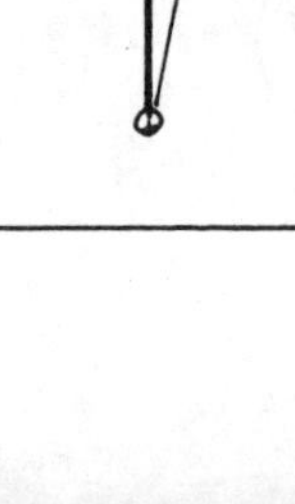

Fig. 9/005 (B.M.D.)

Axial loads (allow 0.5 kN/m for factored self weight).

Top length = 375.6 + 2.7 × 0.5 = $\underline{377}$ kN.

Bottom length = 377 + 170.7 + 3.2 × 0.5 = $\underline{549}$ kN.

Moments:

1st Floor:-

M_{xx} = 375.6 × 0.2 = $\underline{75.1 \text{ kN.m}}$

GD. Floor:-

M_{xx} = 170.7 × 0.2 = $\underline{34.2 \text{ kN.m.}}$

Stiffness ratio = 3.2 : 2.7 = 1.2 : 1 (< 1.5 : 1)

∴ M_{xx} at B–B = $\frac{34.2}{2}$ = $\underline{17.1 \text{ kN.m.}}$

Ref.	
Ref 6/1	Try a 203 × 203 × 46 U.C. ($L_e = 0.85L$, both lengths)
	$r_{yy} = 5.11$ cm. : $T = 11.0$ mm < 16 : $p_y = 275\ \text{N/mm}^2$
Table 6	$b/T = 9.24$: $d/t = 22$: $A = 58.8\ \text{cm}^2$: $S_{xx} = 497\ \text{cm}^3$
Table 7	Classification:
	Flange : $b/T = 9.24 < 15\varepsilon$ – not 'slender'
Cl. 3.5.4	Web – general – see Cl. 3.5.4
	'Squash' load $= \dfrac{275 \times 5880}{10^3} = 1617$ kN.
	For top length: $R = \dfrac{376}{1617} = 0.23$ (positive)
	$\dfrac{120\varepsilon}{1 + 1.5R} = \dfrac{120 \times 1.0}{1 + 1.5 \times 0.23} = 89 > d/t\ (22)$
	$\dfrac{41}{R} - 2 = \dfrac{41}{0.23} - 2 = 176 > d/t\ (22)$
	∴ Not 'slender'
	∴ Section not 'slender'
Cl. 4.7.7	M_b : Top length: $\lambda_{LT} = \dfrac{0.5 \times 2700}{51.1} = 26$
	∴ $p_b = 275\ \text{N/mm}^2$
	$M_b = \dfrac{275 \times 497}{10^3} = 136.7$ kN.m.
	Bottom length: $\lambda_{LT} = \dfrac{0.5 \times 3200}{51.1} = 31.3$
	∴ $p_b = 275\ \text{N/mm}^2$
	$M_b = \dfrac{275 \times 497}{10^3} = 136.7$ kN.m.
	P_c : Top length: $\lambda = \dfrac{0.85 \times 2700}{51.1} = 45$
Table 27(c)	∴ $p_c = 229.5\ \text{N/mm}^2$
	$P_c = \dfrac{229.5 \times 5880}{10^3} = 1349$ kN.
	Bottom length: $\lambda = \dfrac{0.85 \times 3200}{51.1} = 53$
	∴ $p_c = 215\ \text{N/mm}^2$
	$P_c = \dfrac{215 \times 5880}{10^3} = 1264$ kN.
Cl. 4.8.3.3.1	Interaction formulae :-
	Section A–A (Fig. 9/004) : $\dfrac{377}{1349} + \dfrac{75}{136.7} = 0.83 < 1.0$ ✓ o.k
	Section B–B (Fig. 9/004) : $\dfrac{549}{1264} + \dfrac{17.1}{136.7} = 0.56 < 1.0$ ✓ o.k
	'S1' : 203 × 203 × 46 U.C.

Ref.	
	Column base
	'Service Load' ≃ U.L.S /1.5
	∴ 'Service load = $\frac{549}{1.5}$ = 366 kN.
	Try a base 1.6 m. sq. × 0.5 m. deep :
	Selfweight = 0.5 × 1.6^2 × 24 = 30.7 kN.
	Total load = 365 + 30.7 = 400 kN. (say)
	Pressure under base = $\frac{400}{1.6^2}$ = 156.2 kN/m²
	Safe bearing (see brief) = 200 kN/m² ✓ o.k.
	∴ use bases 1.6 m² × 0.5 m. deep.
	(Check later for eccentricity of column load when extent of existing foundations known: base needs checking for shear, etc.)

Check portal for 'racking' action lateral loads & design connections.

Cl. 2.4.2.3

Take a second load case of :-

a) 0.5% of total beam loads as lateral loads – moments taken on joints.

b) shears on joints from $1.3 g_k + 1.4 q_k$.

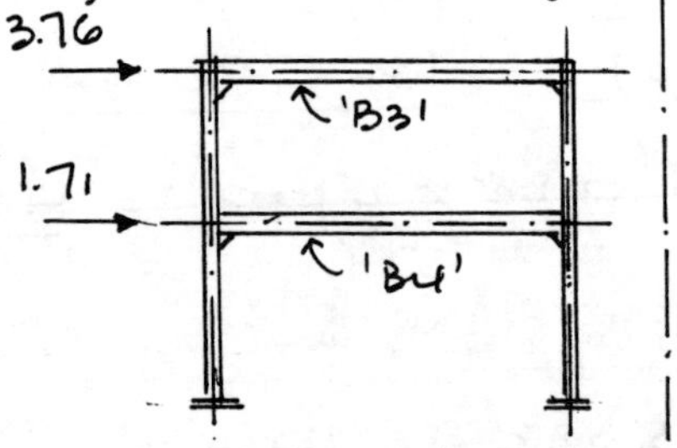

Fig. 9/006 (Loads)

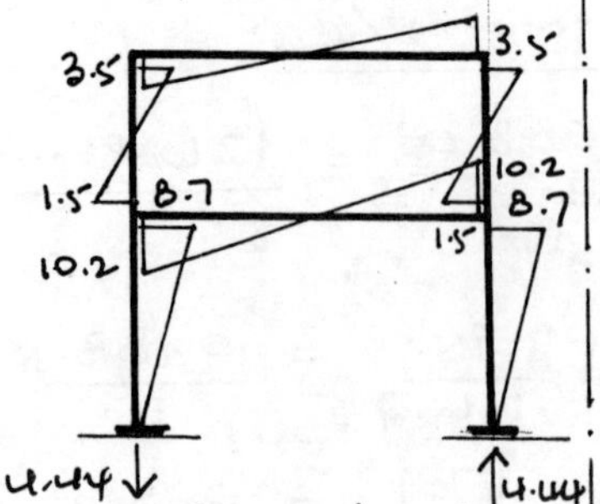

Fig. 9/007 Planeframe Prog. Results. (kN.m.)

Lateral loads :–

L.C. 1 U.D.L. on 1st floor – beam 'B3' = 124.8 kN./m.

∴ Load = $\frac{0.5 \times 125.2 \times 6}{100}$ = 3.76 kN.

L.C. 1 U.D.L. on Gd. floor – beam 'B4' = 56.9 kN/m.

∴ Load = $\frac{0.5 \times 56.9 \times 6}{100}$ = 1.71 kN.

Ref.	
	Vertical loads – 1.3 q_k + 1.4 g_k
	1st Floor – 'B3' – see Fig. 9/006
	Design loading = 1.3 × 1.5 + 1.4 × 1.0 = 3.35 kN/m²
	Design loading from roof = 1.3 × 0.75 + 1.4 × 1.2 = 2.65 kN/m²
	Design loading from Terrace = 1.3 × 1.5 + 1.4 × 4.0 = 7.55 kN/m²
	U.D.L. (see also 'B3' calcs):–
	Roof = 14.6 × 2.65 / 2 = 19.3 kN/m
	External wall (as before) = 47.1 kN/m.
	1st & 2nd floors = $2\left[\frac{7.2 \times 3.35}{2}\right]$ = 24.1 kN/m.
	Terrace = 7.2 × 7.55 / 2 = 27.2 kN/m.
	(+ swt = 0.7 kN/m) : Total = 118.4 kN/m.
	Beam reaction = 118.4 × 6/2 = 355.2 kN.
	Gd. Floor – 'B4' – see Fig. 9/006
	Design loading = 1.3 × 4.0 + 1.4 × 1.0 = 6.6 kN/m.
	U.D.L. (see also 'B4' calcs):–
	Ground floor = $2\left[\frac{7.2 \times 6.6}{2}\right]$ = 47.52 kN/m
	(+ swt. = 0.56 kN/m) : Total = 48.08 kN/m.
	Beam reaction = 48.08 × 6/2 = 144.2 kN.
	Member check:
	Check lower column : M_{xx} = 8.7 kN.m.
	Axial = 355.2 ('B3') + 144.2 ('B4') + 5.9 × 0.5 (swt.) + 4.44 (Notional)
	= 507 kN.
	Compare with design load case 1:–
	M_{xx} = 34.2 kN.m : Axial = 548 kN.
	∴ L.C. 11 not critical for member design.
	Connection design.
	1st Floor: L.C. 1 ↓ 374.4 kN ↻ zero
	L.C. 11 ↓ 355.2 kN ↻ 3.5 kN.m
	Gd. Floor: L.C. 1 ↓ 170.7 kN ↻ zero
	L.C. 11 ↓ 144.2 kN ↻ 10.2 kN.m.

Ref.	Bolt group design (only)
Ref 6/1	**1st Floor Connection – 'B3'**

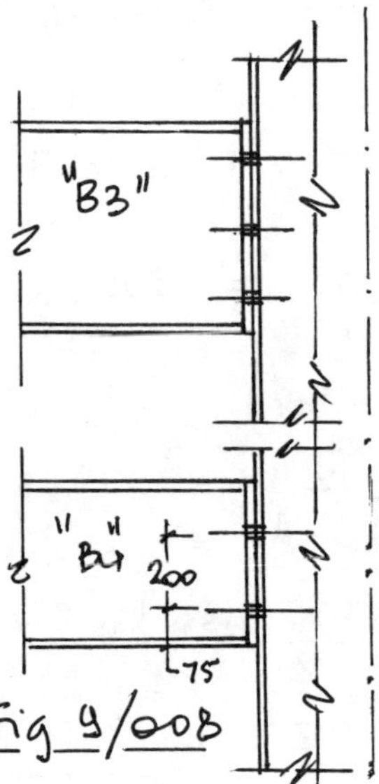

Fig 9/008

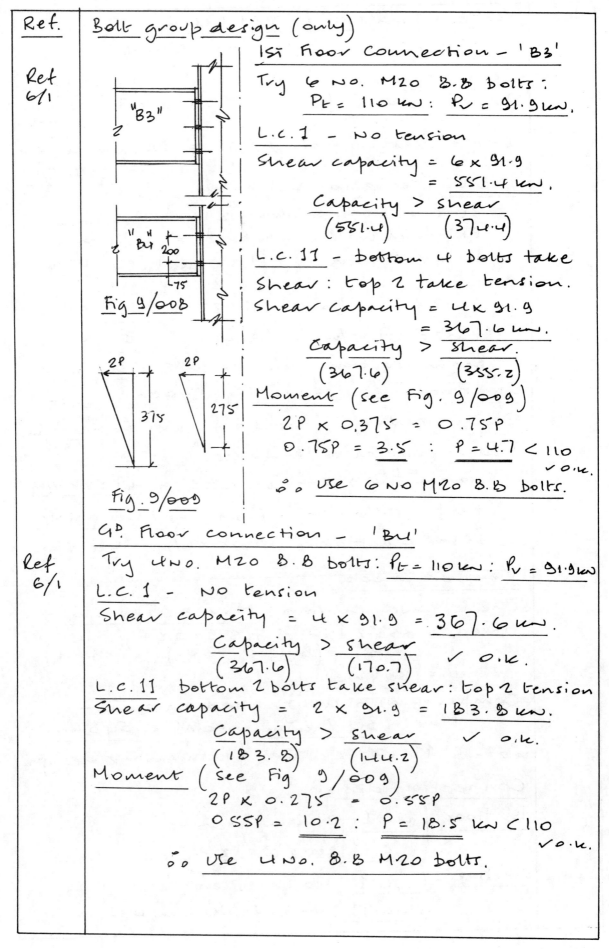

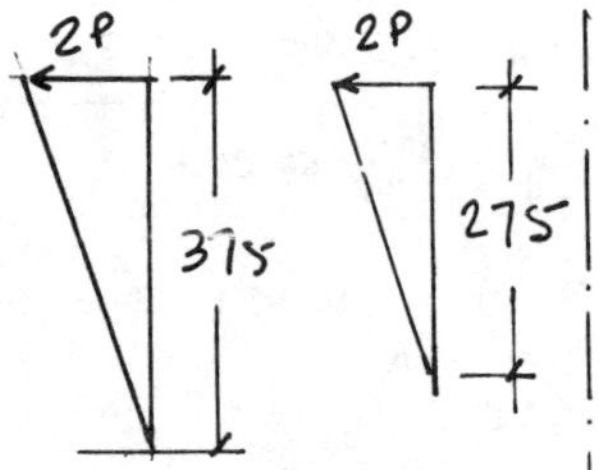

Fig. 9/009

Try 6 No. M20 8.8 bolts:
$P_t = 110$ kN : $P_v = 91.9$ kN.

L.C. I – No tension

Shear capacity = 6 × 91.9
= 551.4 kN.

Capacity > shear
(551.4) (374.4)

L.C. II – bottom 4 bolts take
Shear : top 2 take tension.
Shear capacity = 4 × 91.9
= 367.6 kN.

Capacity > shear.
(367.6) (355.2)

Moment (see Fig. 9/009)
2P × 0.375 = 0.75P
0.75P = 3.5 : P = 4.7 < 110 ✓ o.k.

∴ Use 6 No M20 8.8 bolts.

Gd. Floor connection – 'B4'

Ref 6/1

Try 4 No. M20 8.8 bolts: $P_t = 110$ kN : $P_v = 91.9$ kN

L.C. I – No tension

Shear capacity = 4 × 91.9 = 367.6 kN.

Capacity > shear ✓ o.k.
(367.6) (170.7)

L.C. II bottom 2 bolts take shear: top 2 tension

Shear capacity = 2 × 91.9 = 183.8 kN.

Capacity > shear ✓ o.k.
(183.8) (144.2)

Moment (see Fig 9/009)
2P × 0.275 = 0.55P
0.55P = 10.2 : P = 18.5 kN < 110 ✓ o.k.

∴ Use 4 No. 8.8 M20 bolts.

Ref.	Check on wall panel in new extension for lateral loading

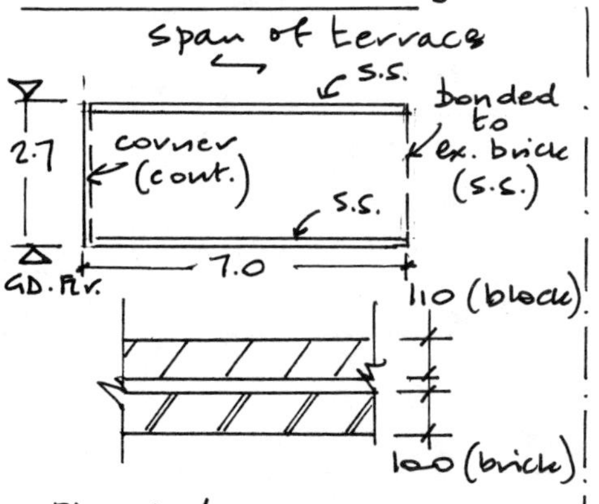

Fig. 9/010 – Wall panel & wall section

Check wall panel dimensions:–

$t_{ef} = \frac{2}{3}(100 + 110)$

$t_{ef} = 140\text{mm}.$

$50 t_{ef} = \frac{50 \times 140}{10^3} = 7.0 \text{ m}.$

$\therefore 50 t_{ef} = \text{max. dim}^n$
(7.0) (7.0) ✓ o.k.

$2025 t_{ef}^2 = \frac{2025 \times 140^2}{10^6} = 39.7\text{m}^2$

$7.0 \times 2.7 = 18.9\text{m}^2$ ✓ o.k.

Check resistance to wind loading:–

CP3 ChV Pt 2 Table 3 and Cl. 6

q: $V = 46 \text{ m/sec}$ (see brief)
$S_2 = 0.55$ (4) – Class B, H = 5

$\therefore V_s = 0.55 \times 46 = 25.3 \text{ m/sec.}$

$q = \frac{0.613 \times 25.3^2}{10^3} = 0.4 \text{ kN/m}^2$

Cf: $h/w = 5.4/9 = 0.6 \; \{ 1/2 < h/w < 3/2$

$l/w = 21/9 = 2.3 \; \{ 3/2 < l/w < 4$

For $\alpha = 0$: $C_{pe} = 0.7$. : for internal suction, total $C_f = 0.7 + 0.3 = 1.0$

Wk: $\therefore W_k = 1.0 \times 0.4 = 0.4 \text{ kN/m}^2$

BS5628 Table 9

Wind moment: section F, Table 9, BS 5628, for support condition given in Fig 9/010:–

$h/L = 2.7/7 = 0.4 : \mu = 0.35 : \alpha = 0.023$

BS5628 Cl. 36.4.1

$M = \alpha W_k \gamma_f L^2 = 0.023 \times 0.4 \times 1.2 \times 7^2$
$= 0.54 \text{ kN.m/m.}$

Calculate Moment of Resistance:

'Z' of block (Fig 9/010) $= \frac{1000 \times 110^2}{6} = 2 \times 10^6 \text{mm}^3$

'Z' of brick (Fig. 9/010) $= \frac{1000 \times 100^2}{6} = 1.67 \times 10^6 \text{mm}^3$

f_{kx} (block) $= 0.45 \text{ N/mm}^2$: f_{kx} (brick) $= 1.1 \text{ N/mm}^2$

Ref.	
BS5628 Cl. 36.4.3	$\therefore M_R = \frac{0.45 \times 2 \times 10^6}{3.5 \times 10^6} + \frac{1.1 \times 1.67 \times 10^6}{3.5 \times 10^6} = 0.78\ \text{kN.m/m}$ $\therefore \underset{(0.78)}{M_R} > \underset{(0.54)}{M}$ ✓ panel o.k.
	Design of wine bar floor extension timber joists. (SS grade p. pine) $q_k = 4.0\ \text{kN/m}^2$, $g_k = 1.0\ \text{kN/m}^2$ Span of joists = 7.2 m. Try 97 × 294 joists @ 300 crs. :- U.D.L. = 0.30 (4.0 + 1.0) = 1.5 kN/m.
BS 5268 Table 9, Cl. 13(a) Cl. 14.6	For S.S. grade pitch pine – $\sigma_{grade} = 10.5\ \text{N/mm}^2$ $\sigma_{par\ adm.} = \sigma_{grade} \times K_3 \times K_8 \times K_7$ $K_7 = \left[\frac{300}{294}\right]^{0.11} = 1.0$ $\therefore \sigma_{par\ adm.} = 10.5 \times 1.25 \times 1.1 \times 1.0 = 14.44\ \text{N/mm}^2$ For a 97 × 294 joist, $Z_{xx} = 1400\ \text{mm}^3 \times 10^3$ $\therefore M_R = \frac{14.44 \times 1400 \times 10^3}{10^6} = 20.16\ \text{kN.m.}$ $M_{MAX} = 1.5 \times 7.2^2/8 = 9.72\ \text{kN.m.}$ $\therefore \underset{(20.16)}{M_R} > \underset{(9.72)}{M_{MAX}}$ ✓ o.k. Deflection: $E_{mean} = 13500\ \text{N/mm}^2$ For a 97 × 294 joist, $I_{xx} = 205 \times 10^6\ \text{mm}^4$
BS 5268 Cl. 14.7	$\Delta = \frac{5}{384} \times \frac{1.5}{13500} \times \frac{(7200)^4}{205 \times 10^6} = 19\ \text{mm}$ Allowable = 0.003 × 7200 = 21.6 mm ✓ ok. $\therefore$ Use 97 × 294 SS grade pitch pine joists at 300 crs.

Chapter 10 Pre-cast concrete rigid framed office building

A Health Service Area Headquarters, comprising of a 3-storey building, approximately 40 m. long x 10 m. wide, is required for a city suburb. The client requires a very short construction period, and pre-cast concrete, in favour of structural steelwork, has been chosen, as shown in Figure 10/001. The client also requires an open plan office and a maximum amount of natural light, and so the Architect has stipulated that no solid shear walls are available. This means that the most common pre-cast joints (pin-joints), shown on Figure 10/001, cannot be used in this instance, as the structure will be a 'sway frame'. In the calculations provided, rigid joints, of a type specified in Ref 10/1 and as shown in Figs. 10/023 & 10/024, are used as a means to rigid frame stability. The client requires as much of the width of the building as possible to be clear of columns. Therefore, in this instance, the Architect has elected to place the columns, which are to be as slender as possible, outside the building, as can be seen in Figure 10/001. In the solution offered, selected elements have been designed. Namely, the rigid frame, including columns and main beams, connections between column and beams, and a pad foundation. Two load cases are offered for the sway frame, and only bending steel has been designed for the footing, as shear checks are not necessary.

Loading – qk

Ground and 1st Floors	– 5.0 kN/m^2
Second Floor	– 3.0 kN/m^2
Roof	– 1.5 kN/m^2

Loading – gk

Hollowcore prestressed pre-cast floor units are to be used. No design for these is provided, and so for the spans given (5.0 m.) and the maximum loading given (5.0 kN/m^2), 150 mm. thick units have been selected from a typical manufacturer's catalogue. These have a self-weight of 2.4 kN/m^2, and together with a 50 mm. structural screed, this makes **gk** = 3.8 kN/m^2 (inc. finishes and service loads).

Loading – wk

Basic wind speed, v = 46 m/sec.

Site Conditions

A thorough site investigation of the site has revealed that there is 0.8 m. of soft clay and topsoil lying over 1.7m. of water-bearing compact sand and gravel, over boulder clay (c = 80 kN/m^2) to a substantial depth. It was decided to use reinforced concrete pad foundations on the compact gravel.

Design Code of Practice

The building is to be designed to BS 8110 – Structural Concrete

Materials

Concrete – fcu = 40 N/mm^2
Reinforcement – fy = 460 N/mm^2

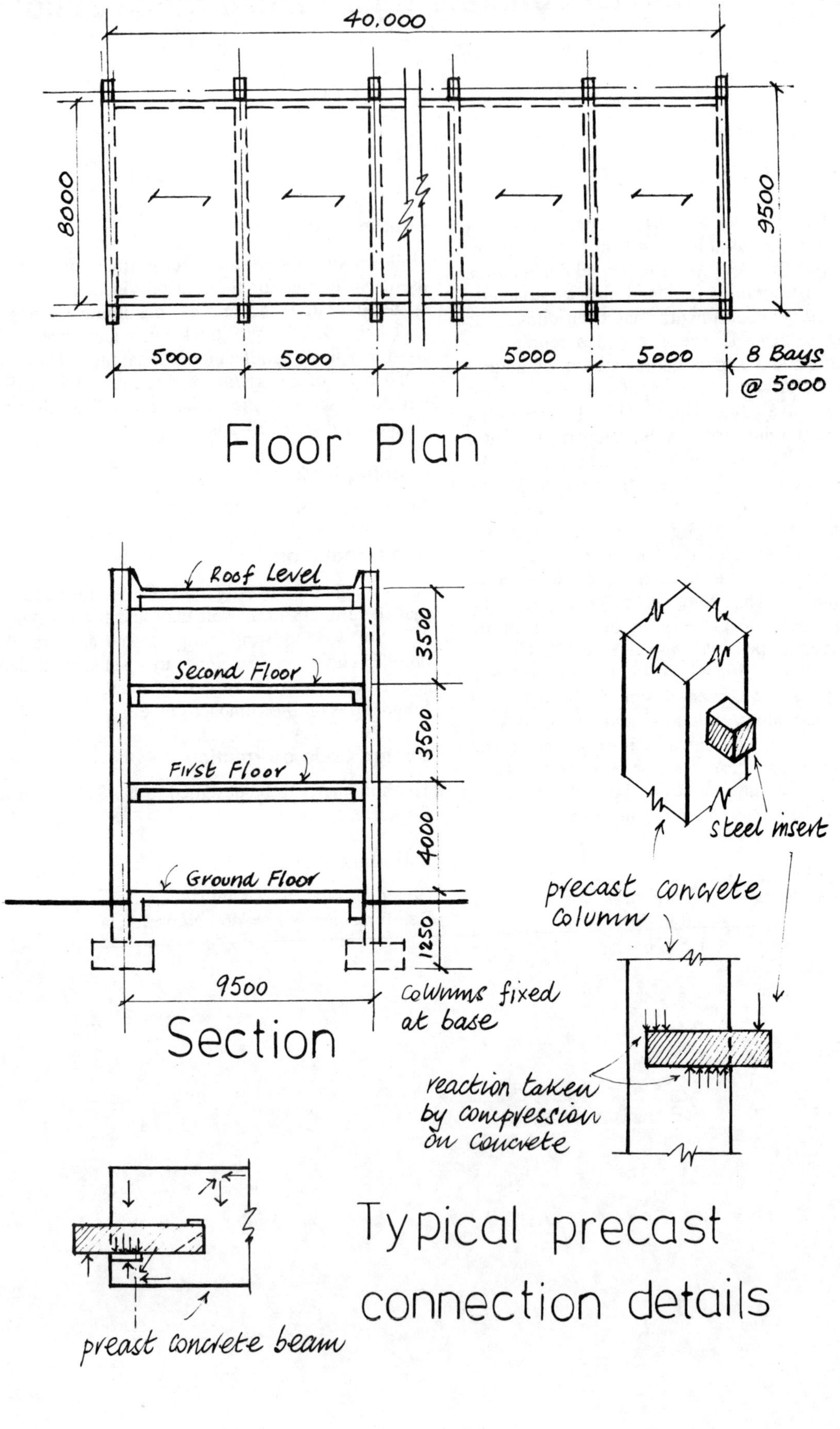

40,000
8000
9500
5000
5000
5000
5000
8 Bays
@ 5000
Floor Plan
Roof Level
Second Floor
First Floor
Ground Floor
3500
3500
4000
1250
9500
columns fixed
at base
Section
steel insert
precast concrete
column
reaction taken
by compression
on concrete
Typical precast
connection details
preast concrete beam

Fig.10/001

Ref.	
(B.S. 8110)	**Structural Design Assumptions :-** 1) Frame will be designed as an unbraced frame, with pre-cast columns with fixity at beam / column connections. 2) Frame to be designed for the 2 load cases :- I : $1.6 Q_k + 1.4 G_k$ II : $1.2 (Q_k + G_k + W_k)$ 3) Try columns 350 sq. & beams (non-composite precast) 600 deep × 400 wide.
Table 2.1	
	Load Case I : Design loadings Slab g_k loading = 3.8 kN/m² (see brief) Roof design loading = 1.6 × 1.5 + 1.4 × 3.8 = 7.72 kN/m² Frames at 5 m. crs. U.D.L. = 5 × 7.72 = 38.6 kN./m. 2ND Floor design load = 1.6 × 3.0 + 1.4 × 3.8 = 10.12 kN/m² ∴ U.D.L. = 5 × 10.12 = 50.6 kN./m. 1ST Floor design load = 1.6 × 5.0 + 1.4 × 3.8 = 13.12 kN/m² ∴ U.D.L. = 5 × 13.12 = 66.6 kN./m.

Fig.10/002

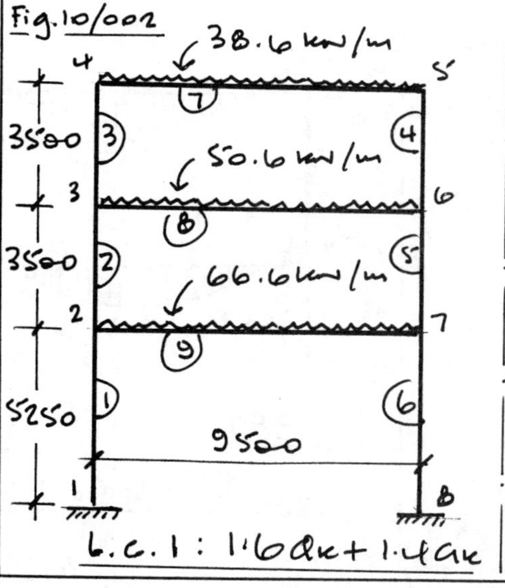

L.C.1 : $1.6 Q_k + 1.4 G_k$

Member properties :

Columns : 350 sq.

$I_{xx} = \frac{350^4}{12} = 125 \times 10^7 \text{ mm}^4$

$A = 350^2 = 122500 \text{ mm}^2$

Beams : 600 deep × 400 wide.

$I_{xx} = \frac{400 \times 600^3}{12} = 720 \times 10^7 \text{ mm}^4$

$A = 400 \times 600 = 240000 \text{ mm}^2$

Ref.		
(B.S 8110)	**Stiffness Factors (I/ℓ)**	**Distribution Factors**

Stiffness Factors (I/ℓ)

Member	I/ℓ ($\times 10^6$)
1	0.24
2	0.36
3	0.36
4	0.36
5	0.36
6	0.24
7	0.76
8	0.76
9	0.76

Distribution Factors

Nodes 4 & 5: 0.32 | 0.68

Nodes 3 & 6: 0.245 | 0.51 | 0.245

Nodes 2 & 7: 0.26 | 0.56 | 0.18

Fixed End Moments ($1.6q_k + 1.4g_k$)

Member 4-5 : $\frac{WL^2}{12} = \frac{38.6 \times 9.5^2}{12} = 290.4$ kN.m

Member 3-6 : $\frac{WL^2}{12} = \frac{50.6 \times 9.5^2}{12} = 380.5$ kN.m

Member 2-7 : $\frac{WL^2}{12} = \frac{66.6 \times 9.5^2}{12} = 500.9$ kN.m

Result of Moment Distribution (next page)

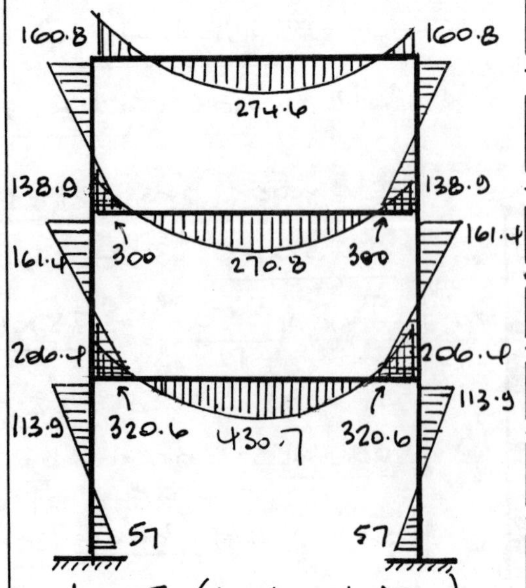

L.C.I ($1.6q_k + 1.4g_k$)
(For sagging mts., $WL^2/8$ hung from hogging mts)

Result of Planeframe Program (good correlation)

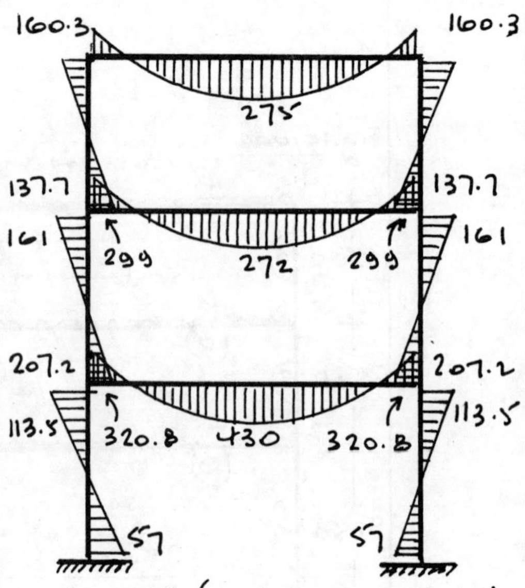

L.C.I ($1.6q_k + 1.4g_k$)

Fig. 10/003

Ref (B.S. 8110)

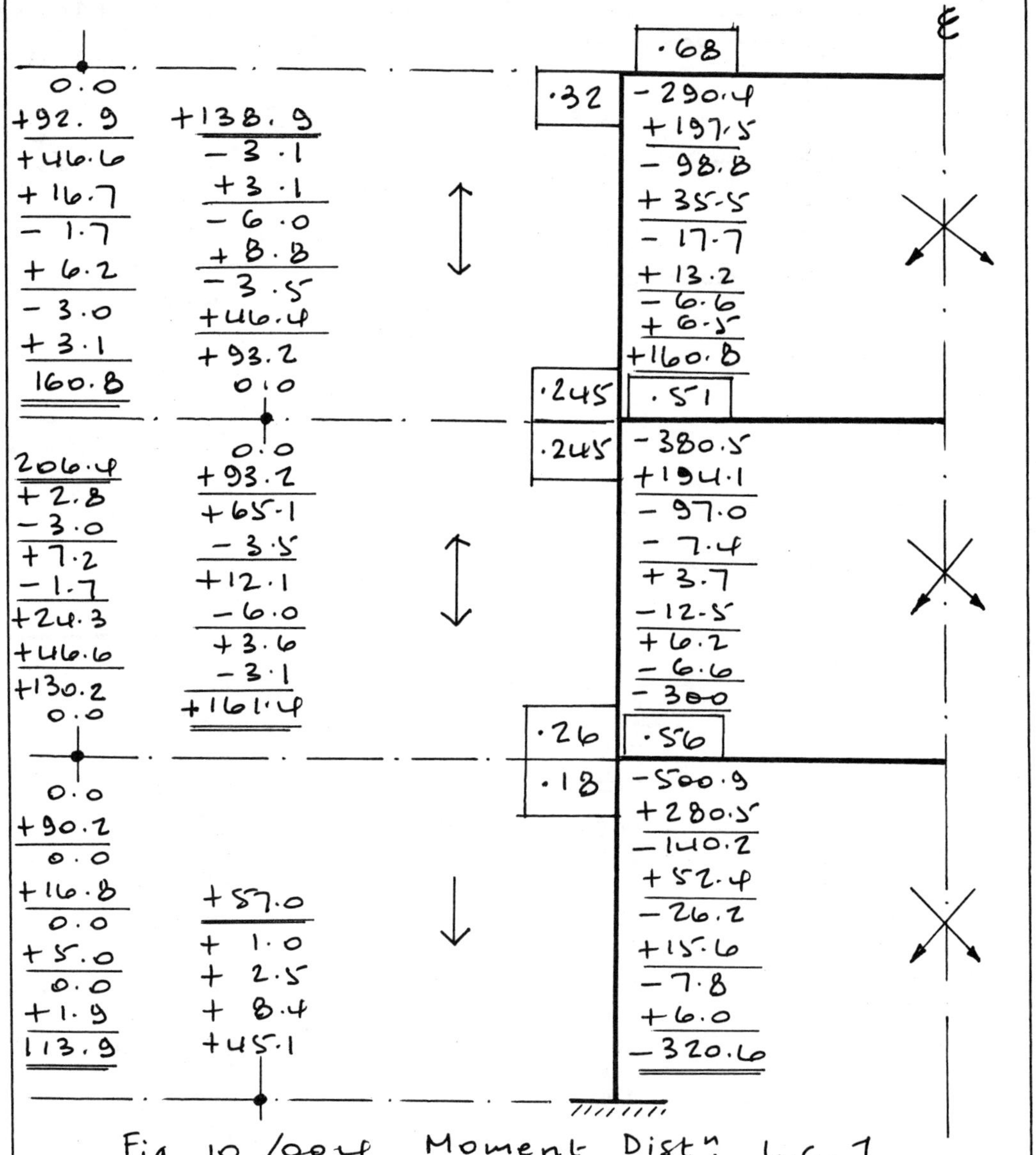

Fig 10/004 Moment Dist^n L.C. 1

Load case 11 : Design loadings.

Roof design loading = 1.2 × 1.5 + 1.2 × 3.8
= 6.36 kN/m²

∴ U.D.L = 5 × 6.36 = 31.8 kN/m.

2ND Floor design loading = 1.2 × 3.0 + 1.2 × 3.8
= 8.16 kN/m²

∴ U.D.L. = 5 × 8.16 = 40.8 kN/m.

1ST Floor design loading = 1.2 × 5.0 + 1.2 × 3.8
= 10.56 kN/m²

∴ U.D.L. = 5 × 10.56 = 52.8 kN/m.

Ref.

(B.S. 8110)

Fig. 10/005

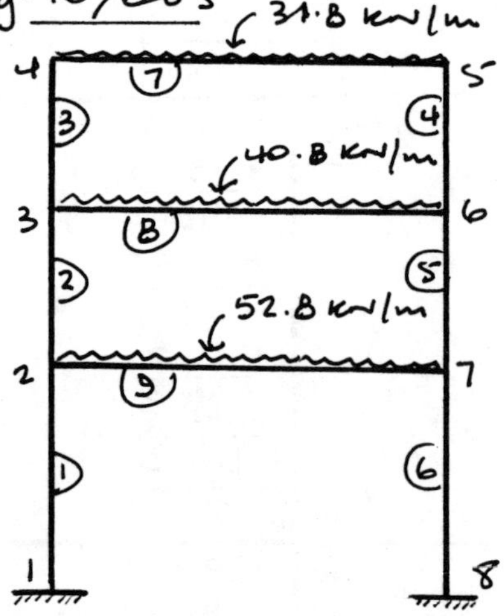

L.C. 11 : 1.2qk + 1.2ak
(Vertical loads)

Fixed End Moments

(1.2 qk + 1.2 ak)

Member 4-5:

$$\frac{WL^2}{12} = \frac{31.8 \times 9.5^2}{12} = 239 \text{ kN.m}$$

Member 3-6:

$$\frac{WL^2}{12} = \frac{40.8 \times 9.5^2}{12} = 307 \text{ kN.m}$$

Member 2-7:

$$\frac{WL^2}{12} = \frac{52.8 \times 9.5^2}{12} = 397 \text{ kN.m}$$

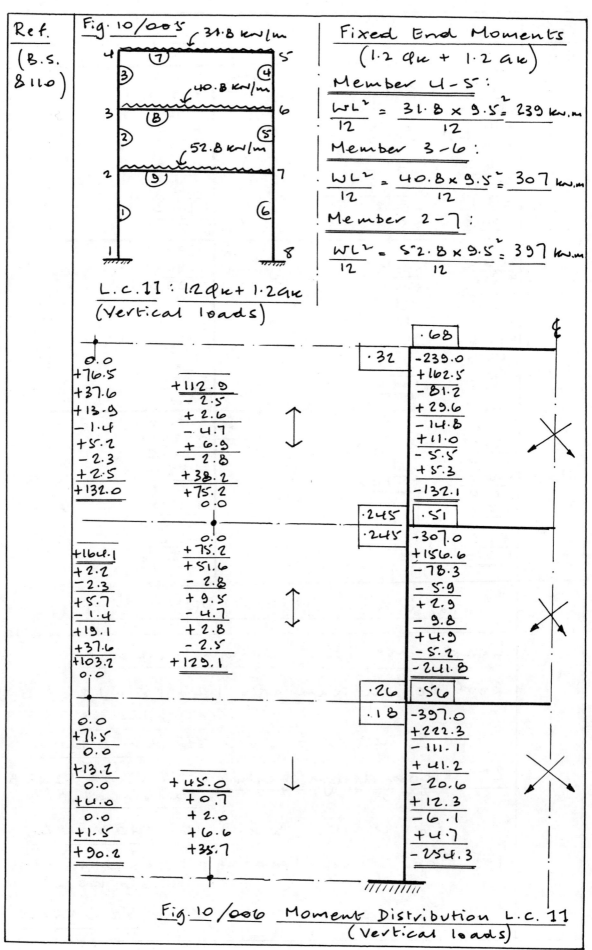

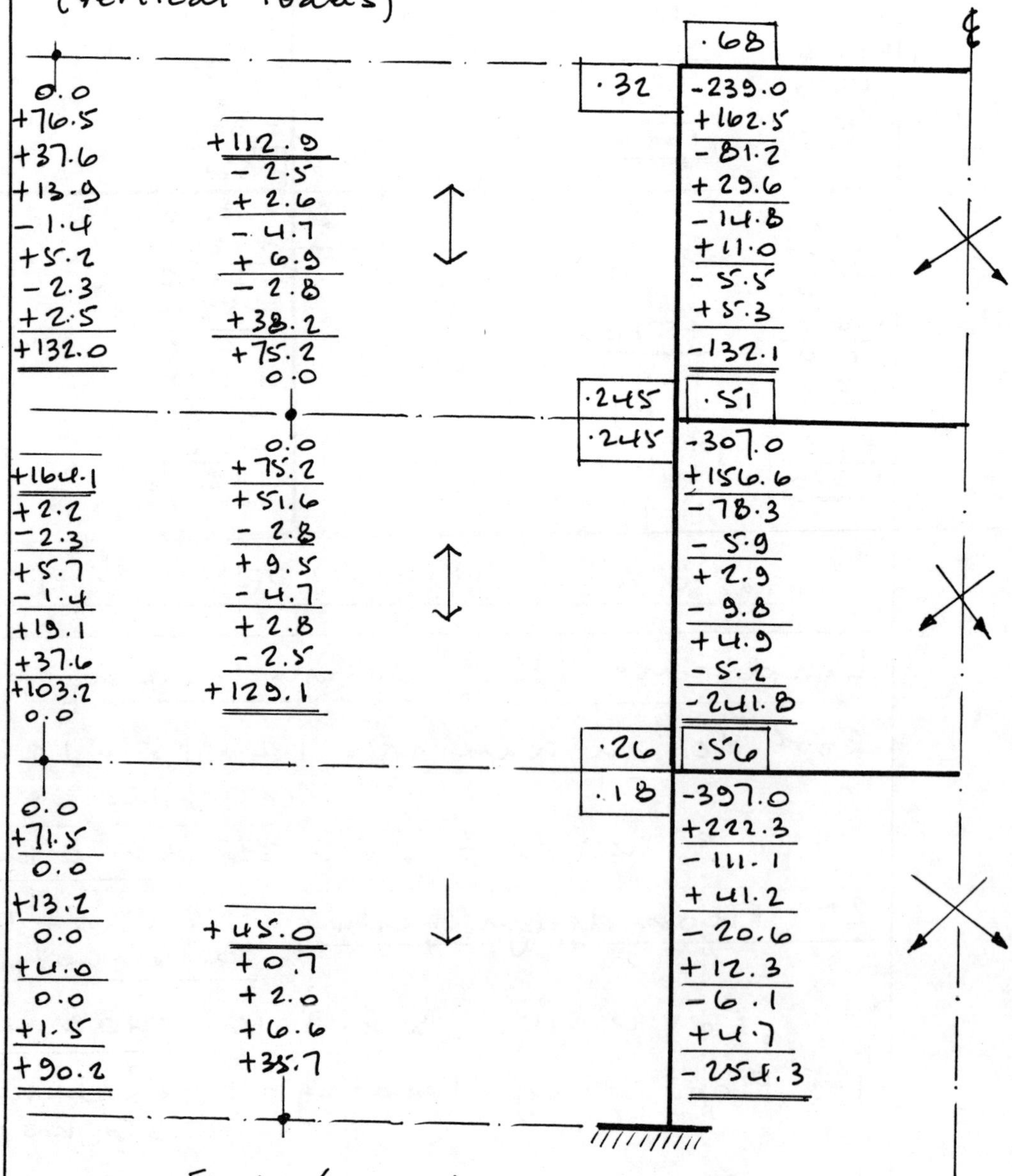

Fig. 10/006 Moment Distribution L.C. 11
(Vertical loads)

Ref. (B.S. 8110) CP3 ChV Pt. 2

Lateral loading – 1.2 Wk.

From brief, V = 46 m/sec (S1 = S3 = 1.0)

S2 value – 'city suburb', Class B.

1st Floor : H ≏ 4.0 : S2 = 0.64 : V_s = 29.4 m/s

2ND Floor : H ≏ 7.5 : S2 = 0.70 : V_s = 32.2 m/s

Roof : H ≏ 11.0 : S2 = 0.76 : V_s = 35.0 m/s

Pressures: $q = 0.613\, V_s^2$

$q\ (\text{1st Floor}) = \dfrac{0.613 \times 29.4^2}{10^3} = 0.53\ \text{kN/m}^2$

$q\ (\text{2nd Floor}) = \dfrac{0.613 \times 32.2^2}{10^3} = 0.64\ \text{kN/m}^2$

$q\ (\text{Roof}) = \dfrac{0.613 \times 35.0^2}{10^3} = 0.75\ \text{kN/m}^2$

Cf values: (w = 10m. (say)) $l/w = 40/10 = 4$:

$h/w = 11/10 = 1.1$: $\tfrac{1}{2} < h/w < 1.5$

Cp

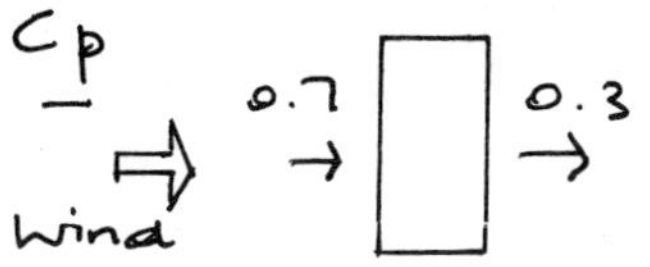

→ Total Cp = 1.0 Fig. 10/007

1.2 Wk node loads at nodes 2, 3 & 4

(Frames at 5.0 m. crs)

Node 2 : 1.2 × 1.0 × [(0.53 × 2) + (0.64 × 1.75)] × 5 = 13.08 kN

Node 3 : 1.2 × 1.0 × [(0.64 × 1.75) + (0.75 × 1.75)] × 5 = 14.58 kN

Node 4 : 1.2 Wk = 1.2 × 1.0 × 0.75 × 5 × 3.5 = 15.75 kN

Column shears = 'storey shear'/2 (see Fig. 10/008)

Column shears, 3 & 4 = $\dfrac{15.75}{2}$ = 7.9 kN.

—"— —"— , 2 & 5 = $\dfrac{(15.75 + 14.58)}{2}$ = 15.2 kN.

—"— —"— , 1 & 6 = $\dfrac{(13.08 + 14.58 + 15.75)}{2}$ = 21.7 kN

Column moments = col. shear × storey ht./2

3 & 4 : Mt. = 7.9 × 3.5/2 = 13.8 kN.m.

2 & 5 : Mt. = 15.2 × 3.5/2 = 26.6 kN.m.

1 & 6 : Mt. = 21.7 × 5.25/2 = 57.0 kN.m.

Ref. (B.S. 8110)

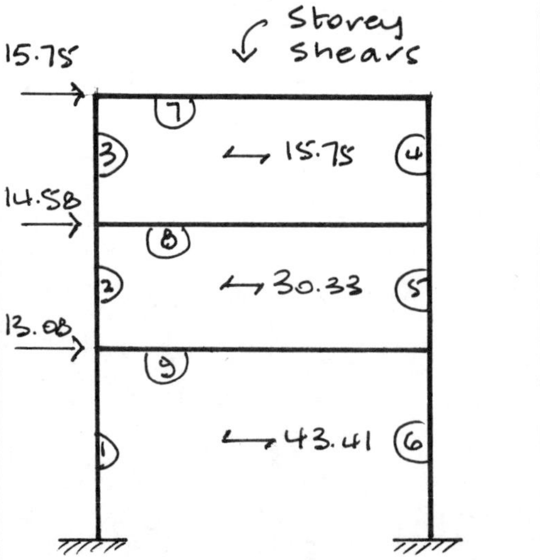

Fig. 10/008 : Lateral Loads

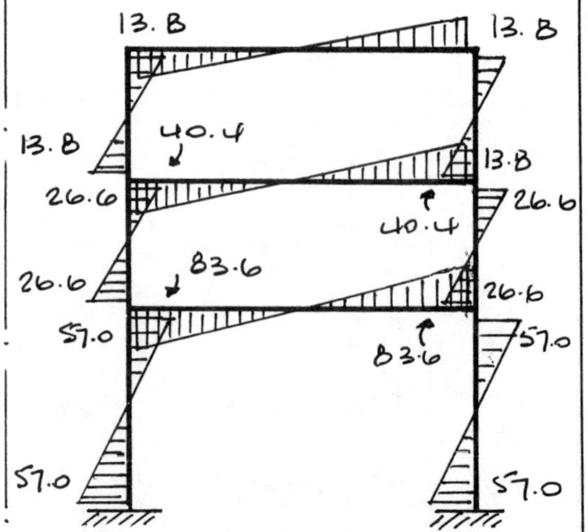

Fig. 10/009 : Moments from 1.2Wk.

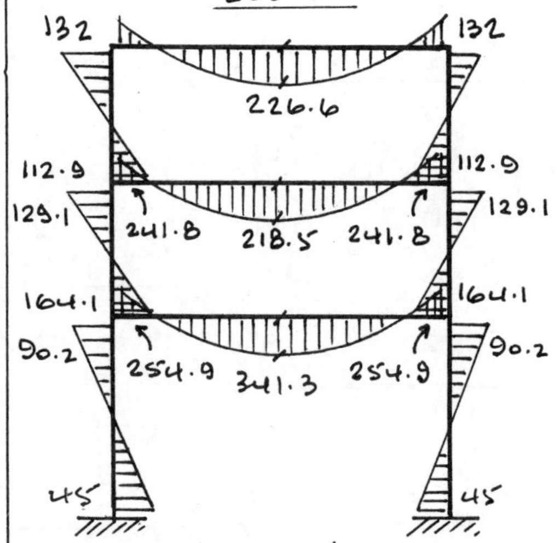

Fig. 10/010: Moments from 1.2 (Qk + Gk)

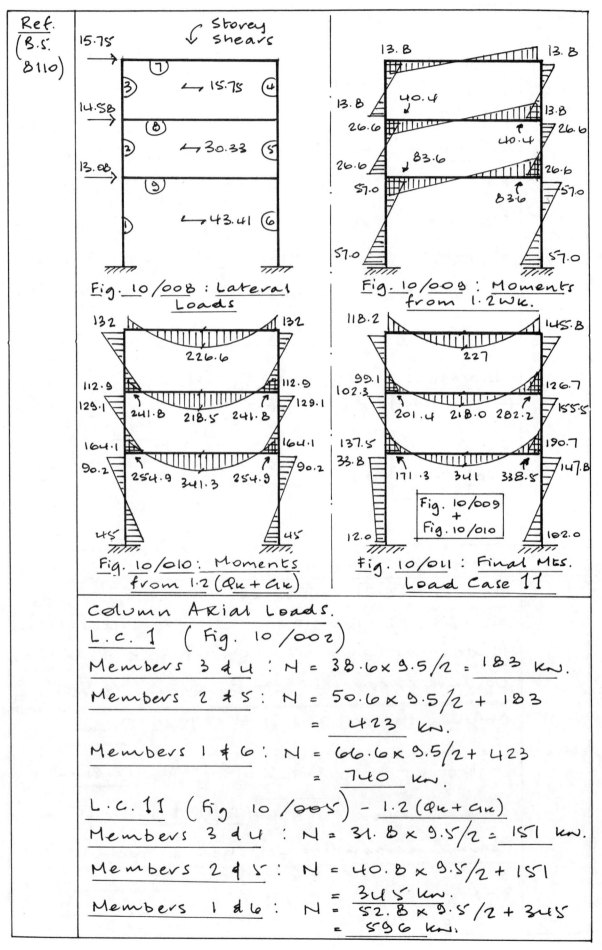

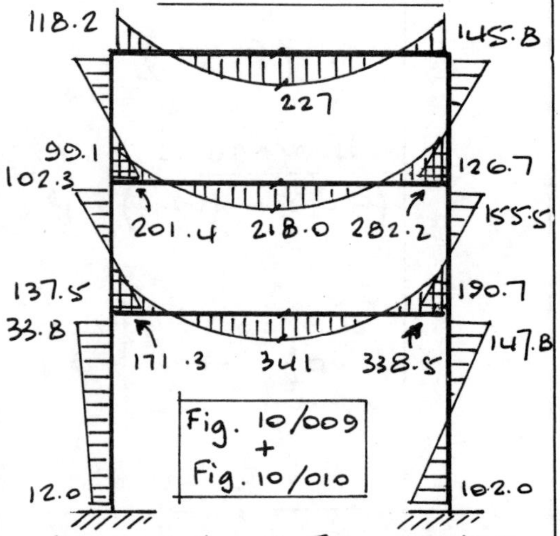

Fig. 10/011 : Final Mts. Load Case 11

Column Axial Loads.

L.C. 1 (Fig. 10/002)

Members 3 & 4 : N = 38.6 × 9.5/2 = 183 kN.

Members 2 & 5 : N = 50.6 × 9.5/2 + 183
= 423 kN.

Members 1 & 6 : N = 66.6 × 9.5/2 + 423
= 740 kN.

L.C. 11 (Fig 10/005) – 1.2 (Qk + Gk)

Members 3 & 4 : N = 31.8 × 9.5/2 = 151 kN.

Members 2 & 5 : N = 40.8 × 9.5/2 + 151
= 345 kN.

Members 1 & 6 : N = 52.8 × 9.5/2 + 345
= 596 kN.

Ref. (B.S. 8110)

L.C. II (Fig. 10/008) – 1.2 Wk

Members 3 & 4

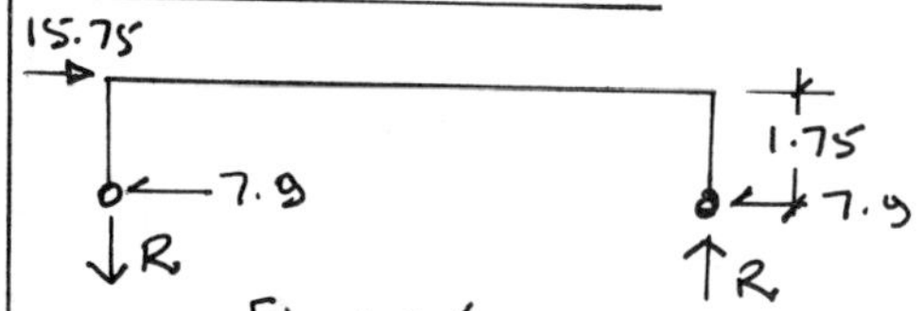

Fig. 10/012

Taking moments:–

$R \times 9.5 = 15.75 \times 1.75$

$\therefore R = 2.9$ kN.

Members 2 & 5

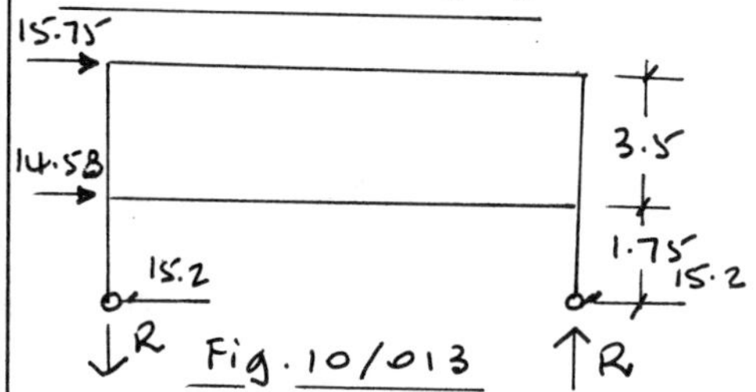

Fig. 10/013

Taking moments:–

$R \times 9.5 = 15.75 \times 5.25 + 14.58 \times 1.75$

$\therefore R = 11.4$ kN

Members 1 & 6

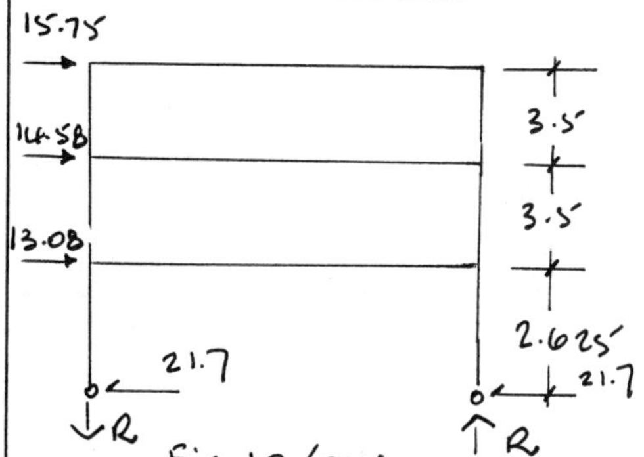

Fig. 10/014

Taking moments:–

$R \times 9.5 = 15.75 \times 9.625 + 14.58 \times 6.125 + 13.08 \times 2.625$

$\therefore R = 29.0$ kN.

Total column loads – L.C. II (reversible)

Member 1	=	596 – 29	=	567 kN.
〃 2	=	345 – 11.4	=	334 kN.
〃 3	=	151 – 2.9	=	148 kN.
〃 4	=	151 + 2.9	=	154 kN.
〃 5	=	345 + 11.4	=	356 kN.
〃 6	=	596 + 29	=	625 kN.

Member Design.

A: Columns – members 1 & 6

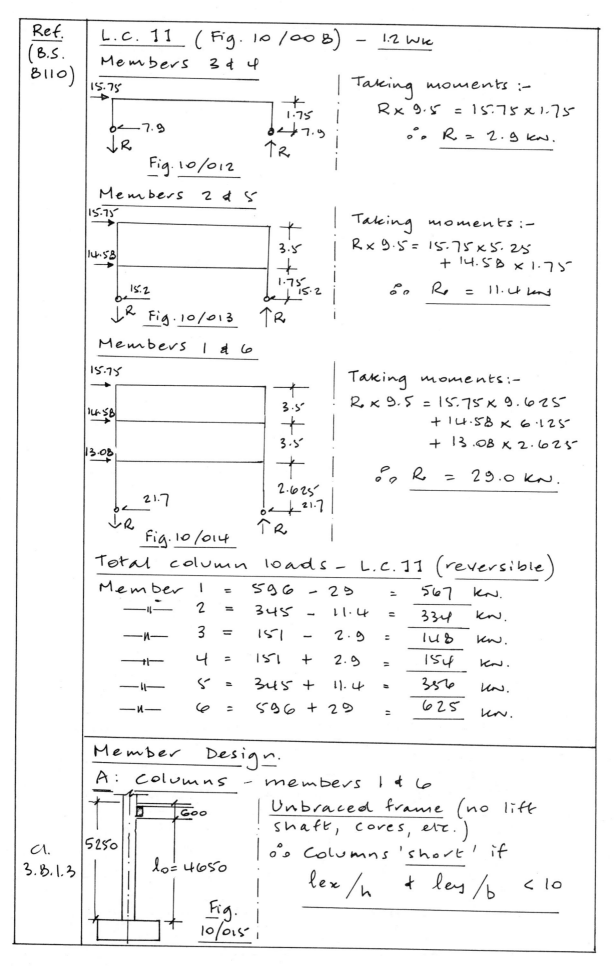

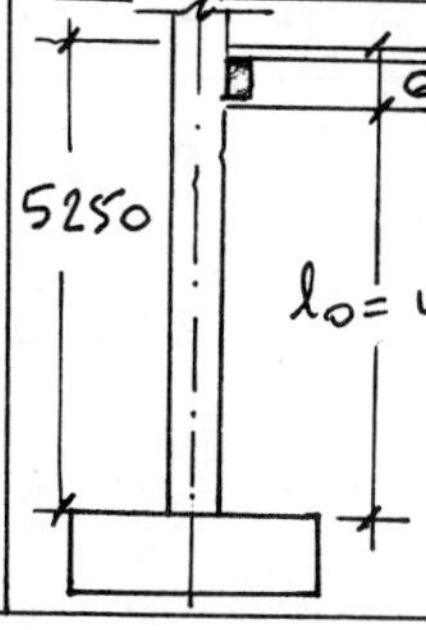

Fig. 10/015

Unbraced frame (no lift shaft, cores, etc.)

Cl. 3.8.1.3 $\therefore$ Columns 'short' if

l_{ex}/h & $l_{ey}/b < 10$

Ref. (B.S. 8110)

Table 3.22

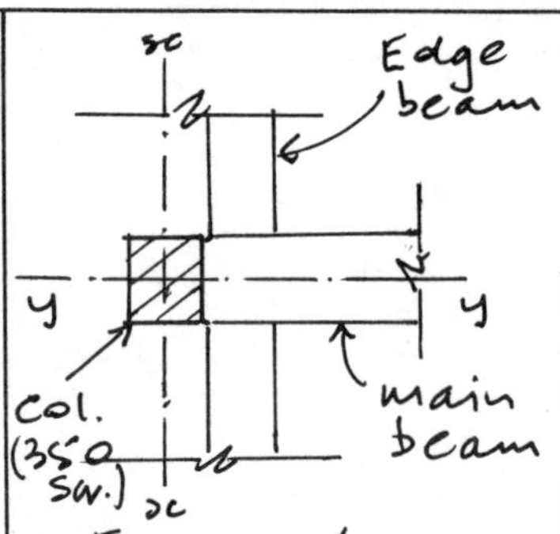

Fig. 10/016 (axes)

Bottom condition ① (Moment base)

Top condition

x–x axis – condition ① (main beam > col. dimⁿ)

y–y axis – condition ① (edge beam moment connected to main beam)

Table 3.22

∴ For both axes, $\beta = 1.2$

∴ $l_{ex} = l_{ey} = 1.2 \times 4650 = 5580\,mm$

∴ $l_{ex}/h = l_{ey}/b = 5580/350 = 16 > 10$

∴ Bottom length 'slender'

Eqⁿ 34

Eq. 34 : $\beta_a = \frac{1}{2000}(16)^2 = 0.128$

Eqⁿ 32

∴ $a_u = 0.128 \times 350 = 45\,mm.$

For member 6 :-

L.C. I	L.C. II

Fig. 10/017

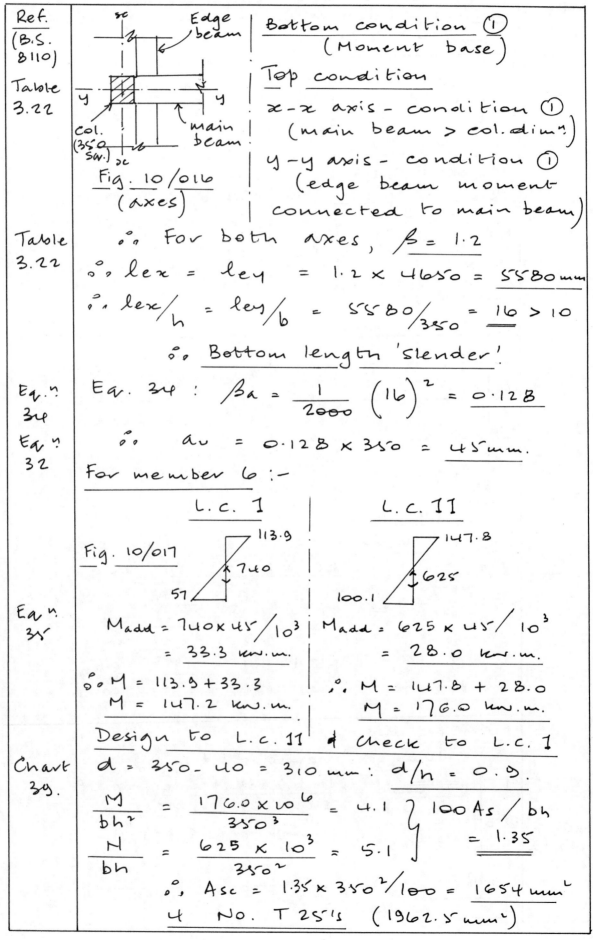

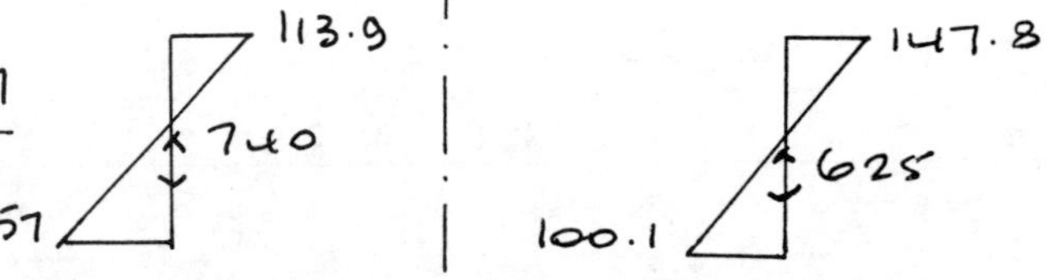

Eqⁿ 35

L.C. I	L.C. II
$M_{add} = 740 \times 45/10^3 = 33.3\,kN.m.$	$M_{add} = 625 \times 45/10^3 = 28.0\,kN.m.$
∴ $M = 113.9 + 33.3$ $M = 147.2\,kN.m.$	∴ $M = 147.8 + 28.0$ $M = 176.0\,kN.m.$

Design to L.C. II & check to L.C. I

Chart 39.

$d = 350 - 40 = 310\,mm$: $d/h = 0.9.$

$\frac{M}{bh^2} = \frac{176.0 \times 10^6}{350^3} = 4.1$

$\frac{N}{bh} = \frac{625 \times 10^3}{350^2} = 5.1$

$\Rightarrow 100A_s/bh = 1.35$

∴ $A_{sc} = 1.35 \times 350^2/100 = 1654\,mm^2$

4 No. T25's $(1962.5\,mm^2)$

Ref. (B.S. 8110)

Check to L.C. I :

$$\frac{M}{bh^2} = \frac{147.2 \times 10^6}{350^3} = \underline{3.43}$$

$$\frac{N}{bh} = \frac{740 \times 10^3}{350^2} = \underline{6.04}$$

$$\frac{100 A_{sc}}{bh} = \underline{\underline{0.8\%}}$$

∴ Not critical.

Members 2 & 5 $l_o = 3500 - 600 = \underline{2900\text{mm}}$

Table 3.22 ∴ $l_{ex} = l_{ey} = 1.2 \times 2900 = \underline{3480 \text{ mm}}$.

$l_{ex}/h = l_{ey}/b = 3480/350 = \underline{\underline{9.94}} < 10$

∴ Column 'short'

For member 5 :-

L.C. I | L.C. II

Fig. 10/018

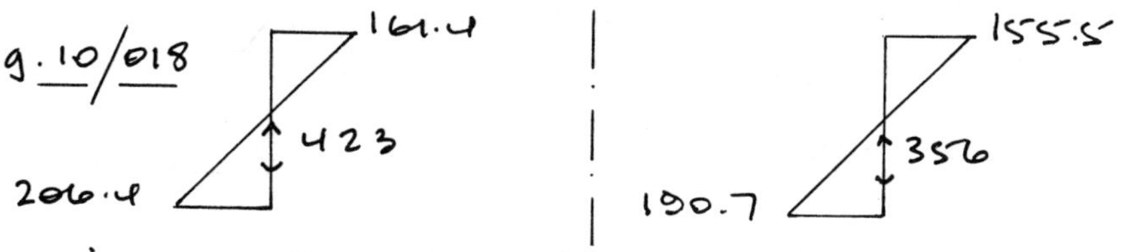

By inspection, L.C. I is critical.

$$\frac{M}{bh^2} = \frac{206.4 \times 10^6}{350^3} = \underline{4.8}$$

$$\frac{N}{bh} = \frac{423 \times 10^3}{350^2} = \underline{3.45}$$

$$\frac{100 A_{sc}}{bh} = \underline{\underline{2.0\%}}$$

$A_s = 2.0 \times 350^2/100 = \underline{2450 \text{ mm}^2}$

Use 4 No. T 32's $= \underline{3215 \text{ mm}^2}$

Checks for members 3 & 4 will show that columns are 'short' & Asc req^d = 2% : Use 4 T 32's full length of column. (max % = 4.0 ✓ o.k., min % = 0.4% ✓ o.k.).

Cl. 3.12.7.1 Links: dia ≮ 32/4 = $\underline{8 \text{ mm}}$.

spacing ≯ 12 × 32 = $\underline{384 \text{ mm}}$.

∴ Use T8 links @ 300 crs.

Ref.
(B.S. 8110)

Pad foundation design.

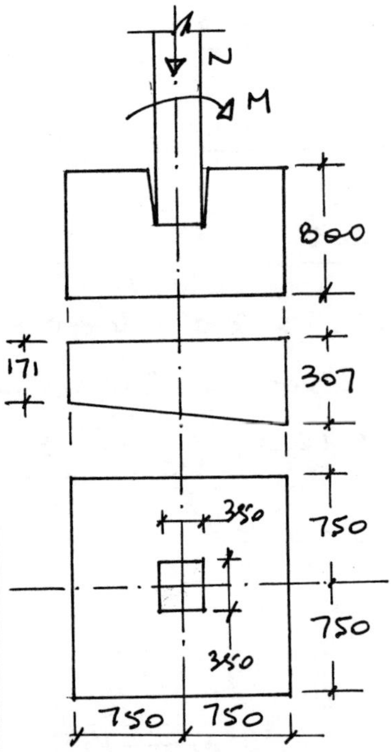

Fig. 10/019
Pad base

Pad bases on 'compact gravel' – safe bearing ≃ 500 kN/m²

Use 'serviceability' loads to determine size ≃ $\frac{\text{U.L.S.}}{1.5}$

For L.C. I :– (see Fig 10/017)

$M = 57/1.5 = \underline{38 \text{ kN/m}^2}$

$N = 740/1.5 = \underline{493 \text{ kN.}}$

Try a base 1.5 m. Sq. x 0.8 m. deep (see Fig 10/019)

Self.weight $= 1.5^2 \times 0.8 \times 25 = \underline{45 \text{ kN.}}$

$\frac{M}{Z} = \pm \frac{38}{1.5^3/6} = \pm 68 \text{ kN/m}^2$

$\frac{P}{A} = + \frac{(493+45)}{1.5^2} = +239 \text{ kN/m}^2$

∴ $\underline{P_{MAX} = 307 \text{ kN/m}^2}$ & $\underline{P_{min} = 171 \text{ kN/m}^2}$ ✓ o.k.

Check with L.C. II :– (see Fig. 10/017)

$M = 102.0/1.5 = \underline{68 \text{ kN/m}^2}$

$N = 625/1.5 = \underline{417 \text{ kN.}}$

$\frac{M}{Z} = \pm \frac{68}{1.5^3/6} = \underline{\pm 121 \text{ kN/m}^2}$

$\frac{P}{A} = + \frac{(417+45)}{1.5^2} = \underline{+205 \text{ kN/m}^2}$

∴ $\underline{P_{MAX} = 326 \text{ kN/m}^2}$ & $\underline{P_{min} = 84 \text{ kN/m}^2}$ ✓ o.k.

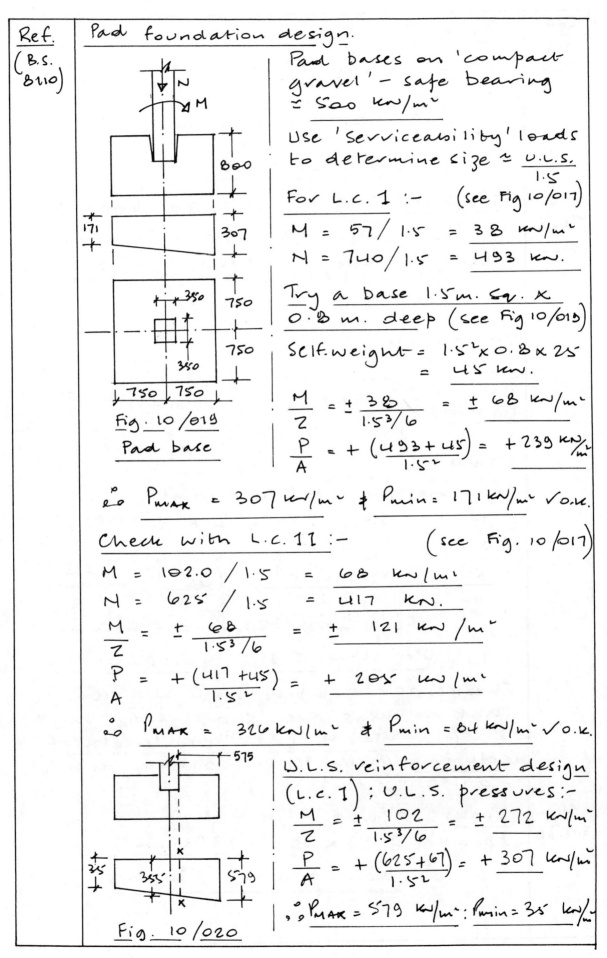

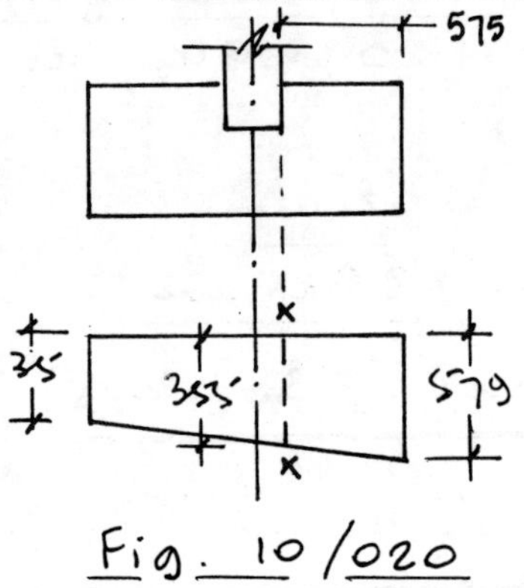

Fig. 10/020

U.L.S. reinforcement design (L.C. I) ; U.L.S. pressures:–

$\frac{M}{Z} = \pm \frac{102}{1.5^3/6} = \underline{\pm 272} \text{ kN/m}^2$

$\frac{P}{A} = + \frac{(625+67)}{1.5^2} = \underline{+307} \text{ kN/m}^2$

∴ $\underline{P_{MAX} = 579 \text{ kN/m}^2}$: $\underline{P_{min} = 35 \text{ kN/m}^2}$

Ref.	
(B.S. 8110)	**Moment per m. of base at X–X (face of column):-**
	$M = 355 \times 0.575^2/2 + 224 \times 0.575^2/3 = \underline{83\ kN.m.}$
	Assuming 16mm bars each way,
	$d = 800 - 50 - 16 = 736\ mm.$
This book App. A1	$\dfrac{M}{f_{cu} b d^2} = \dfrac{83 \times 10^6}{40 \times 10^3 \times 736^2} = \underline{0.004}$
	$\therefore \dfrac{z}{d} = 0.95\ (max.)\ ;\ z = 700\ mm.$
	$\therefore As\ req^d = \dfrac{83 \times 10^6}{0.87 \times 460 \times 700} = \underline{296\ mm^2/m.}$
	$As_{min} = \dfrac{0.13 \times 800 \times 1000}{100} = \underline{1040\ mm^2/m.}$
	$\therefore$ Asmin is critical:-
	Use T16's @ 200crs. ea. way. (1005)
	(Note!! shear checks not applicable).
	C: Main beam design: L.C. 1 critical (see Fig. 10/003 - design member 9 only.

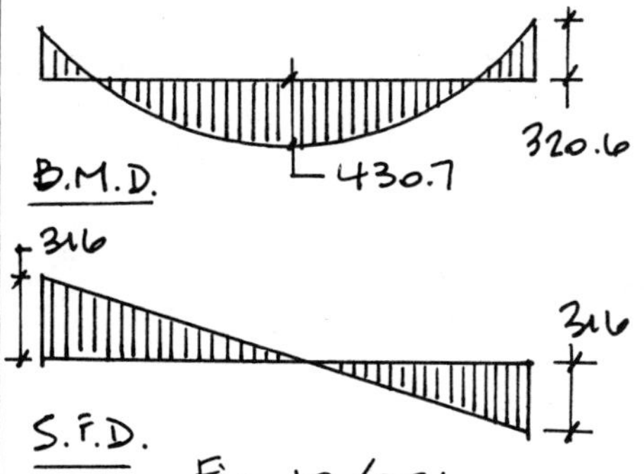

Fig. 10/021

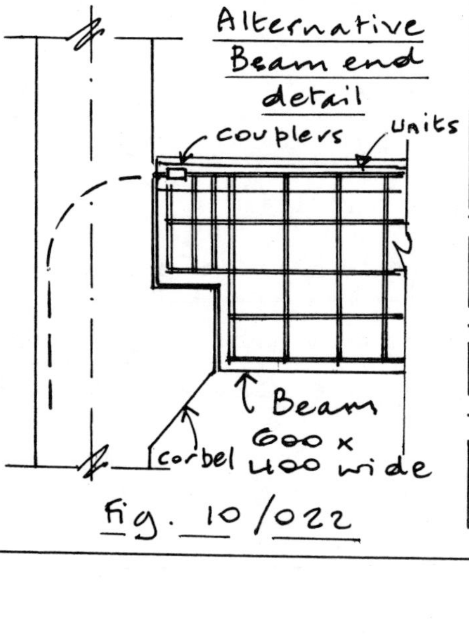

Fig. 10/022

At midspan: $M = 430.7\ kN.m.$

$d = 600 - 50 = 550\ mm.$

1) Check M_u for concrete.

$M_u = 0.156 \times 40 \times 400 \times 550^2$

$= \underline{755\ kN.m.}$

$M_u > M$: no comp^n steel required.

2) Calculate As :

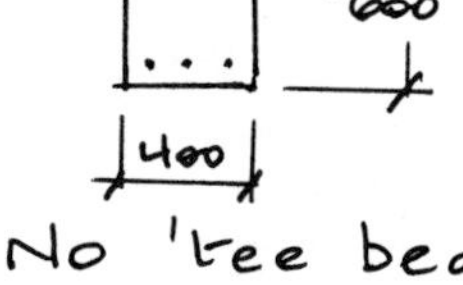

Fig. 10/022a Mid-span Section

No 'tee beam' action - see Fig. 10/022a:-

This book App. A1	
	$\dfrac{M}{f_{cu} b d^2} = \dfrac{430.7 \times 10^6}{40 \times 400 \times 550^2} = 0.09$
	$\dfrac{z}{d} = 0.89\ :\ z = 490\ mm.$

Ref. (B.S. 8110)	
	$\therefore$ As req^d $= \dfrac{430.7 \times 10^6}{0.87 \times 460 \times 490} = 2196\ mm^2$
	Use 3 No. T32's (2411 mm^2)
	Min. steel $= \dfrac{0.13}{100} \times 400 \times 600 = 312\ mm^2$ ✓ o.k.
	At supports: M = 320.6 kN.m.
	1) Check Mu for concrete
	$\underset{(755)}{Mu} > \underset{(320.6)}{M}$: no comp^n steel req^d
	2) Calculate As req^d
	$\dfrac{M}{f_{cu} b d^2} = \dfrac{320.6 \times 10^6}{40 \times 400 \times 550^2} = 0.067$
This book App. A1	$z/d = 0.92$: $z = 506$ mm.
	$\therefore$ As req^d $= \dfrac{320.6 \times 10^6}{0.87 \times 460 \times 506} = 1583\ mm^2$
	Use 2 T32's (1608 mm^2)
	(use couplers to column anchorage bars (see Fig. 10/022))
	Check shear at support: $F_v = 316$ kN.
Cl. 3.4.5.2	$v = \dfrac{316 \times 10^3}{400 \times 550} = 1.44\ N/mm^2$ ($< 5\ N/mm^2$) ✓ o.k.
	$\dfrac{100 As}{bd} = \dfrac{100 \times 1608}{400 \times 550} = 0.73$
Table 3.9	For grade 40 concrete; $v_c = 0.57 \times 1.1 = 0.63\ N/mm^2$
Table 3.8	$\therefore$ $\underset{(1.03)}{v_c + 0.4} < \underset{(1.44)}{v}$: design shear links.
	Try T12 links: $A_{sv} = 2 \times \pi \times 12^2/4 = 226\ mm^2$
	$S_v = \dfrac{226 \times 0.87 \times 460}{400\ (1.44 - 0.63)} = 280$ mm.
	$\therefore$ Use T12 links @ 250 crs.
	Note!! In full design, corbel also would require designing: or use connection shown in Fig. 10/024

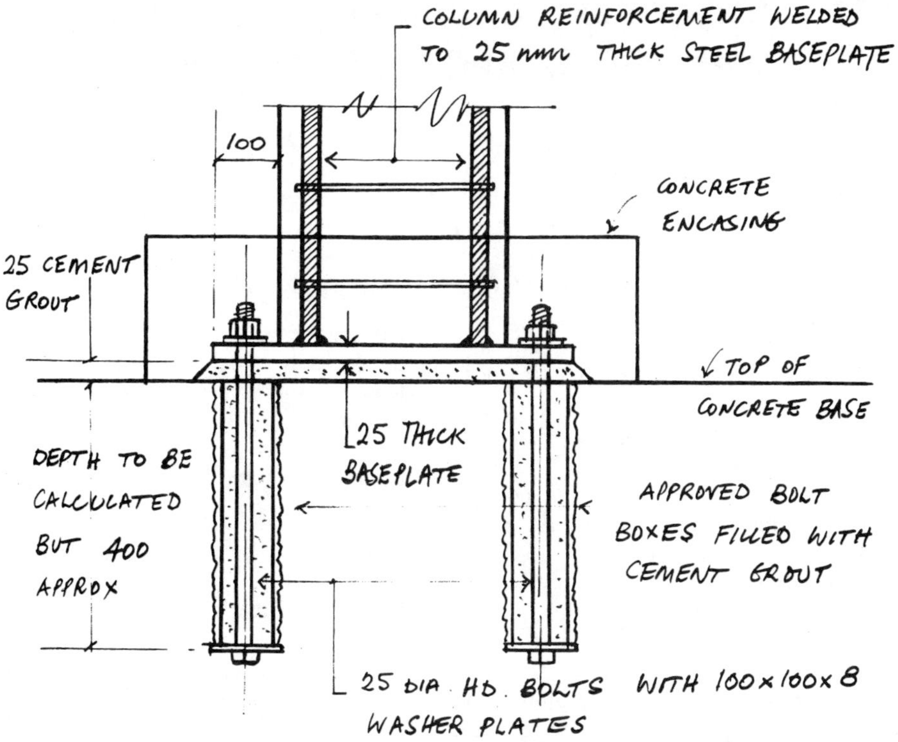

Typical moment connection at base of precast column. Fig.10/023

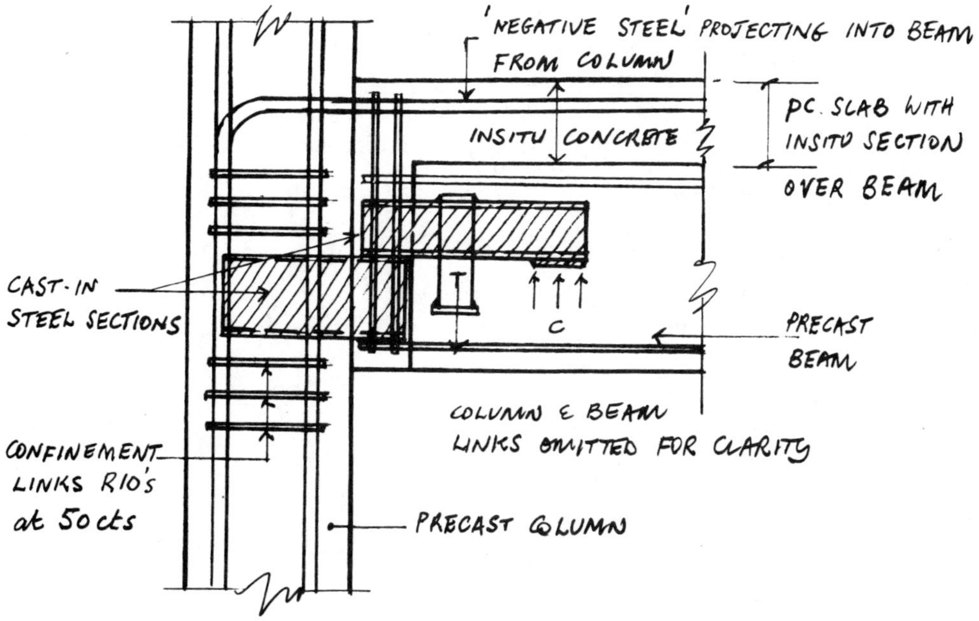

Typical moment connection detail between precast beam and column

Fig.10/024

Chapter 11 Supermarket building in reinforced concrete, with offices and roof car-park

A large supermarket chain require the construction of retail premises on a sloping site, comprising of a ground floor shopping hall, offices above the hall, and a roof car-park. The scheme, shown in Figure 11/001, indicates that the ground floor is partly suspended due to a sloping site. The roof car-park is approached via a ramp, requiring an expansion joint between the ramp span and the building. Thc client requires a column grid spacing of not less than 7.5 m., and a grid of 7.5 m. x 8.0 m. has been chosen (see 'Plan' – Figure 11/001). Only calculations for the main structural elements of the building are provided, which, in this case, comprise of a typical first floor bay of flat slab construction, a typical column and base, a retaining wall bay, a portion of the ramp and support, and various details. With regard to the retaining wall, assumptions and simplifications have been made in order to present a workable design. The panel is monolithic with the frame on all sides, and so a design for a 2-way spanning slab, instead of a traditional cantilever has been used on the assumption that no backfilling will be allowed until the office floor is complete. In addition, the triangular loading has been averaged to an overall UDL, which is an overdesign for much of the panel, but critical for support shear. The justification is that the load factor adopted is conservative. Finally, the ground floor slab has not been designed, and so any uplift on this slab due to water pressure has not been considered.

Loading – qk

Shopping hall and offices	– 10.0 kN/m^2
Car park	– 2.5 kN/m^2
Ramp to car park	– 5.0 kN/m^2

Loading – gk

No requirement for shopping hall ceiling, but for Office ceiling	– 1.0 kN/m^2
Roof (including slab and waterproofing and insulation)	– 8.0 kN/m^2

Loading – wk

Lateral overall stability due to wind loading not considered to be a critical load case.

Site conditions

As is clear from Figure 11/001, the building is to be founded on stiff clay, underlying sandy gravel. The gravel is water-bearing after heavy rainfall, and hence the retaining wall will be designed for resistance to ground water pressure.

Design Code of Practice

The building is to be designed to BS 8110, Structural Concrete.

Materials

Concrete – fcu = 30 N/mm^2 – superstructure: fcu = 40 N/mm^2 – approach ramp
Reinforcement – fy = 460 N/mm^2

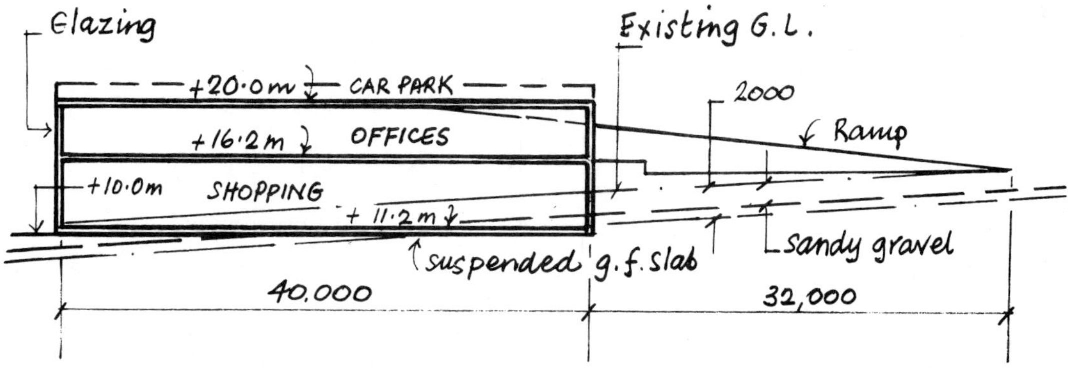

Sectional Elevation

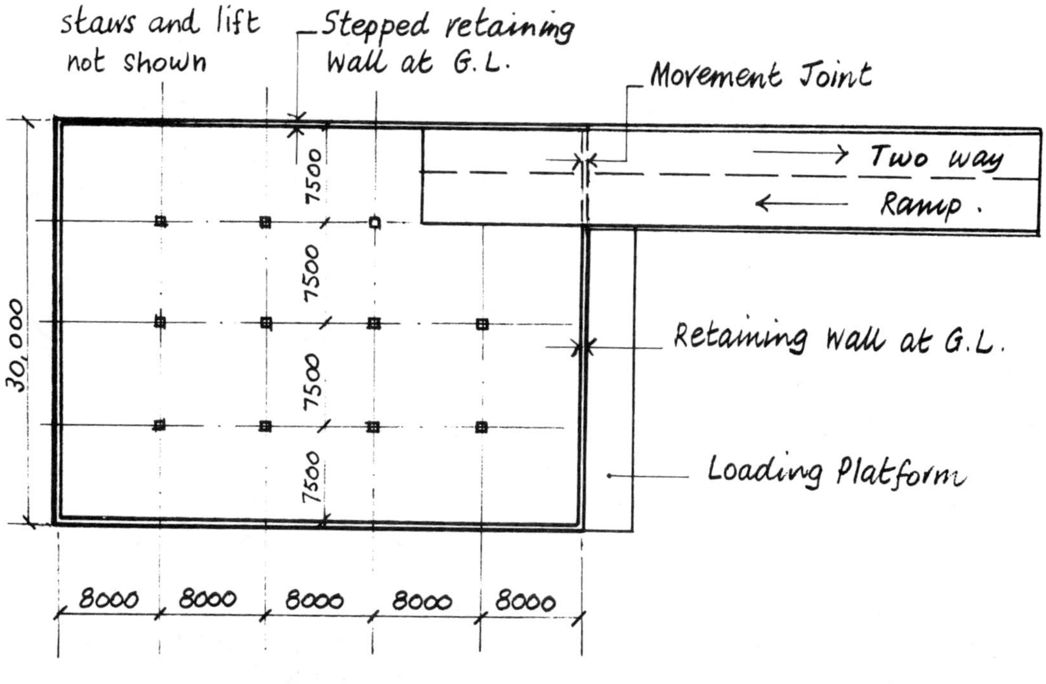

Plan

Fig.11/001

Ref. (BS 8110)

Design of typical first floor slab (under offices) - two-way spanning solid flat slab

Fig. 11/002 Typical floor bay

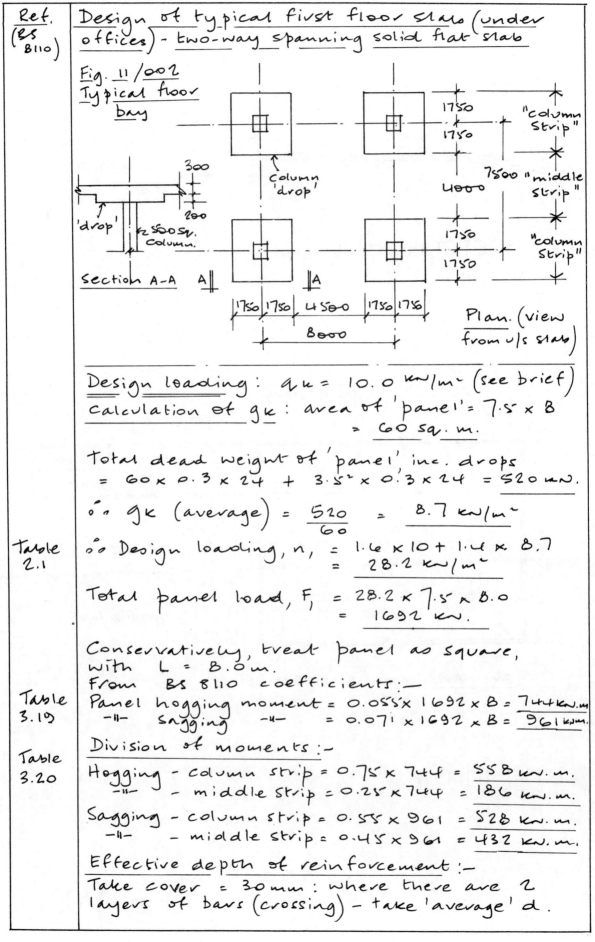

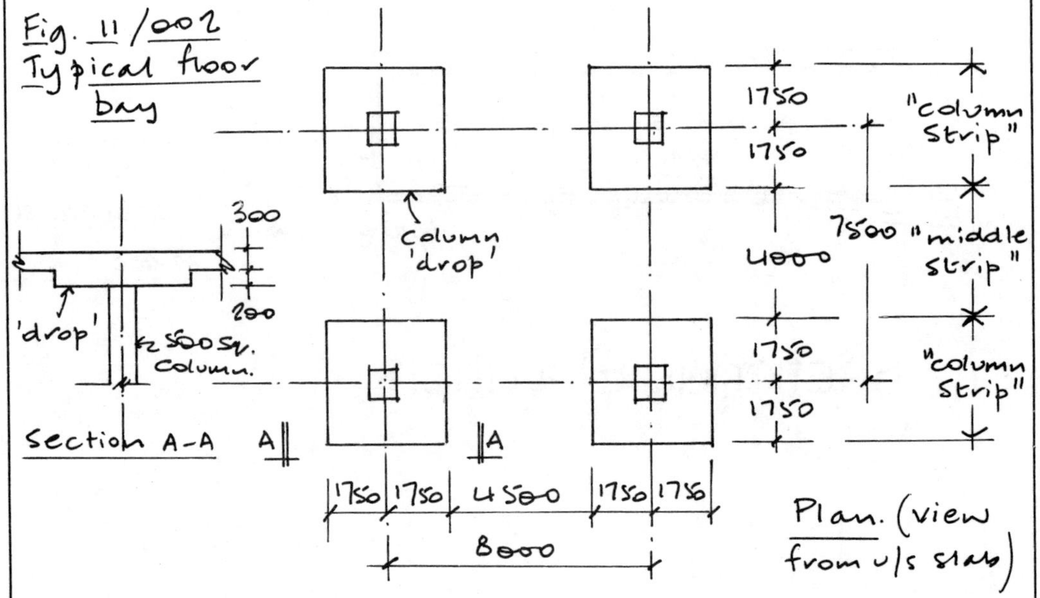

Design loading: $q_k = 10.0\ kN/m^2$ (see brief)

Calculation of g_k: area of 'panel' $= 7.5 \times 8$
$= 60$ sq. m.

Total dead weight of 'panel', inc. drops
$= 60 \times 0.3 \times 24 + 3.5^2 \times 0.3 \times 24 = 520\ kN.$

$\therefore g_k$ (average) $= \frac{520}{60} = 8.7\ kN/m^2$

Table 2.1 $\therefore$ Design loading, n, $= 1.6 \times 10 + 1.4 \times 8.7$
$= 28.2\ kN/m^2$

Total panel load, F, $= 28.2 \times 7.5 \times 8.0$
$= 1692\ kN.$

Conservatively, treat panel as square, with $L = 8.0\ m$.

From BS 8110 coefficients:-

Table 3.19 Panel hogging moment $= 0.055 \times 1692 \times 8 = 744\ kN.m$

— " — sagging — " — $= 0.071 \times 1692 \times 8 = 961\ kN.m.$

Table 3.20 Division of moments:-

Hogging - column strip $= 0.75 \times 744 = 558\ kN.m.$

— " — - middle strip $= 0.25 \times 744 = 186\ kN.m.$

Sagging - column strip $= 0.55 \times 961 = 528\ kN.m.$

— " — - middle strip $= 0.45 \times 961 = 432\ kN.m.$

Effective depth of reinforcement:-

Take cover = 30 mm: where there are 2 layers of bars (crossing) - take 'average' d.

Ref. (BS 8110)	
	Assuming T16 bars:–
	Hogging – column strip: $d = 500 - 30 - 16 = 454$ mm
	Hogging – middle strip: $d = 300 - 30 - 8 = 262$ mm.
	Sagging – column strip: $d = 300 - 30 - 8 = 262$ mm.
	Sagging – middle strip: $d = 300 - 30 - 16 = 254$ mm.
	Reinforcement calculations (Note! area taken is over whole strip width, not per m.)
	Column strip – hogging : $d = 454$ mm.
Cl. 3.4.4.4 This book App. A1	$\frac{M}{f_{cu} b d^2} = \frac{558 \times 10^6}{30 \times 3500 \times 454^2} = 0.04 : \frac{z}{d} = 0.95$
	$\therefore z = 431$ mm.
	A_s req^d. $= \frac{558 \times 10^6}{0.87 \times 460 \times 431} = 3235$. (T16's @ 200 crs) ($3517 \text{ mm}^2$)
	Column strip – sagging : $d = 262$ mm.
Cl. 3.4.4.4 This book App. A1	$\frac{M}{f_{cu} b d^2} = \frac{528 \times 10^6}{30 \times 3500 \times 262^2} = 0.07 : \frac{z}{d} = 0.915$
	$\therefore z = 240$ mm.
	A_s req^d $= \frac{528 \times 10^6}{0.87 \times 460 \times 240} = 5497$ (T16's @ 100 crs) (7035 mm^2)
	Middle strip – hogging : $d = 262$ mm.
Cl. 3.4.4.4 This book App. A1	$\frac{M}{f_{cu} b d^2} = \frac{186 \times 10^6}{30 \times 4000 \times 262^2} = 0.02 : \frac{z}{d} = 0.95$
	$\therefore z = 249$ mm.
	A_s req.^{d} $= \frac{186 \times 10^6}{0.87 \times 460 \times 249} = 1866$ (T16's @ 200 crs) (3517 mm^2)
	Middle strip – sagging : $d = 254$ mm.
Cl. 3.4.4.4 This book – App. A1	$\frac{M}{f_{cu} b d^2} = \frac{432 \times 10^6}{30 \times 4000 \times 254^2} = 0.055 : \frac{z}{d} = 0.935$
	$\therefore z = 237$ mm.
	A_s req^d $= \frac{432 \times 10^6}{0.87 \times 460 \times 237} = 4555$ (T16's @ 150 crs) (5360 mm^2)
Table 3.11	Check slab deflection: $M/bd^2 = (961 \times 10^6)/(8000 \times 254^2) = 1.86 : f_s = \frac{5}{8} \times 460 \times \frac{5026^*}{6198^*} = 233 \text{ N/mm}^2 : \text{m.f.} = 1.29$
	Actual span/d = $8000/254 = 31.5$
Table 3.10	Allowable $= 1.29 \times 26 = 33.5$ ✓ o.k [*av. areas – sagging]
	Note! See Fig 11/005 for rebar layout.

Ref. (BS 8110)

Check column shear:

Cl. 3.7.6.2; Fig. 3.17

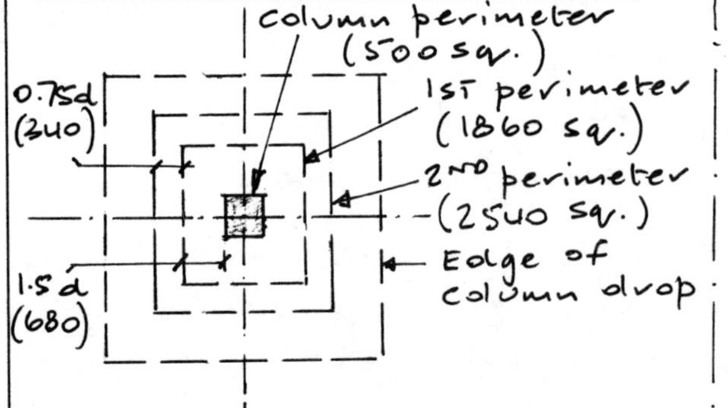

Cl. 3.7.7.3

Fig. 11/003 - column shear

Effective shear around column $= 1.15F$

$\therefore V = 1.15 \times 1692 = 1946$ kN.

At column perimeter :–

$U = 4 \times 500 = 2000$ mm

$v = \dfrac{1946 \times 10^3}{2000 \times 354} = 2.14$ N/mm²

Cl. 3.7.7.2

Value must be < 5 N/mm² ✓ o.k.

or $< 0.8\sqrt{f_{cu}} = 4.4$ N/mm² ✓ o.k.

At 1ST perimeter :–

$U = 4 \times 1860 = 7440$ mm.

$V_{eff} = 1946 - 28.2 \times 1.86^2 = 1848$ kN.

$\therefore v = \dfrac{1848 \times 10^3}{7440 \times 454} = 0.55$ N/mm²

As provided is T16's @ 200 c/s (1005 mm²/m)

Table 3.9

$\therefore \dfrac{100 A_s}{bd} = \dfrac{100 \times 1005}{10^3 \times 454} = 0.22$: $v_c = 0.40$ N/mm²

Table 3.17

$\therefore v > v_c$ (0.55) (0.40) : "shear hoop" reint. required

At 2ND perimeter :–

$U = 4 \times 2540 = 10160$ mm.

$V_{eff} = 1946 - 28.2 \times 2.54^2 = 1764$ kN.

$\therefore v = \dfrac{1764 \times 10^3}{10160 \times 454} = 0.38$ N/mm²

Table 3.17

$\therefore v < v_c$ (0.40) (0.38) : no reint. reqd

$\therefore$ Provide "shearhoops" between column & 2ND perimeter.

Design of column reinforcement using flat slab empirical coefficients.

Table 3.19

Total column moment (biaxial bending)

$= 0.022 \times 1692 \times 8 = 298$ kN.m

Ref. (BS 8110)

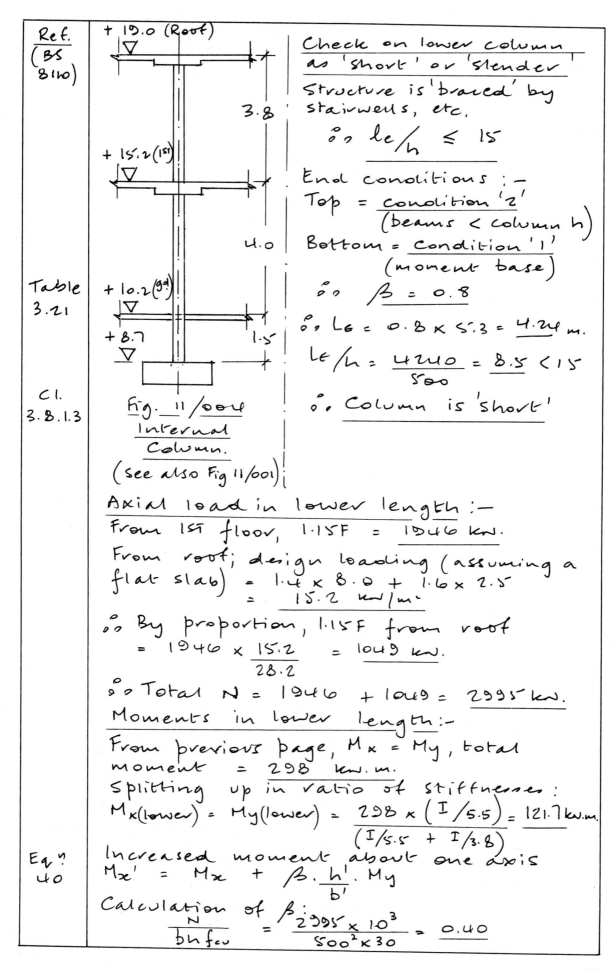

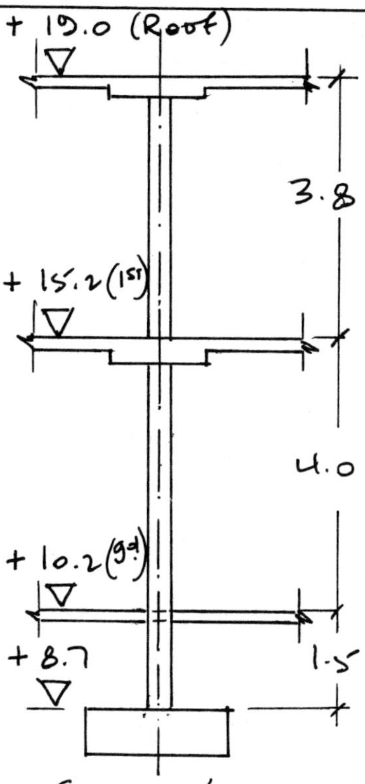

Fig. 11/004 Internal Column.

(see also Fig 11/001)

Check on lower column as 'short' or 'slender'

Structure is 'braced' by stairwells, etc.

$\therefore l_e/h \leq 15$

End conditions :–

Top = condition '2' (beams < column h)

Bottom = condition '1' (moment base)

Table 3.21

$\therefore \beta = 0.8$

$\therefore L_e = 0.8 \times 5.3 = 4.24$ m.

$l_e/h = \dfrac{4240}{500} = 8.5 < 15$

Cl. 3.8.1.3

$\therefore$ Column is 'short'

Axial load in lower length :–

From 1st floor, 1.15F = 1946 kN.

From roof; design loading (assuming a flat slab) $= 1.4 \times 8.0 + 1.6 \times 2.5$

$= 15.2$ kN/m²

$\therefore$ By proportion, 1.15F from roof

$= 1946 \times \dfrac{15.2}{28.2} = 1049$ kN.

$\therefore$ Total N = 1946 + 1049 = 2995 kN.

Moments in lower length :–

From previous page, $M_x = M_y$, total moment = 298 kN.m.

Splitting up in ratio of stiffnesses :

$M_{x(lower)} = M_{y(lower)} = \dfrac{298 \times (I/5.5)}{(I/5.5 + I/3.8)} = 121.7$ kN.m.

Eqn 40

Increased moment about one axis

$M_x' = M_x + \beta . \dfrac{h'}{b'} . M_y$

Calculation of β:

$\dfrac{N}{bhf_{cu}} = \dfrac{2995 \times 10^3}{500^2 \times 30} = 0.40$

Ref.	
(BS 8110)	∴ β = 0.53 (Table 3.24)
	∴ M_x' = 121.7 + 0.53 × 121.7 = 186.2 kN.m.
Table 3.24	h'/h = 450/500 = 0.9 : f_{cu} = 30 N/mm².
	$\frac{M}{bh^2} = \frac{186.2 \times 10^6}{500^3} = 1.5$: $\frac{N}{bh} = \frac{2995 \times 10^3}{500^2} = 12$
Chart 29	From Chart 29 (Pt 2 BS 8110);
	$\frac{100 A_s}{bh} = 0.4$ (min.) : $A_s = \frac{0.4 \times 500^2}{100} = 1000$ mm²
	∴ Use 4 T20's (1256 mm²)
	Base size: founding on stiff clay, c = 125 kN/m² gives a safe bearing ≏ 3c = 375 kN/m²
	Try a base 2.5m. sq. × 0.6m. deep.
	Selfweight = 2.5² × 0.6 × 24 = 90 kN
	column service load = $\frac{2995}{1.5}$ = 1997 kN.
	Total = 2087 kN.
	From a moment distribution analysis, take 50% of moment to base.
	∴ $M_x' = \frac{186.2}{2}$ = 93.1 kN.m. (one axis)
	Stresses under base :-
	$\frac{P}{A} \pm \frac{M}{Z} = \frac{2087}{2.5^2} \pm \frac{93.1}{2.5^3/6}$ = 319.5 ± 35.75
	∴ Max. stress = 355 kN/m² (< 375 ✓ o.k.)
	Min. stress = 284 kN/m² (> 0 ✓ o.k.)
	∴ Base size o.k.

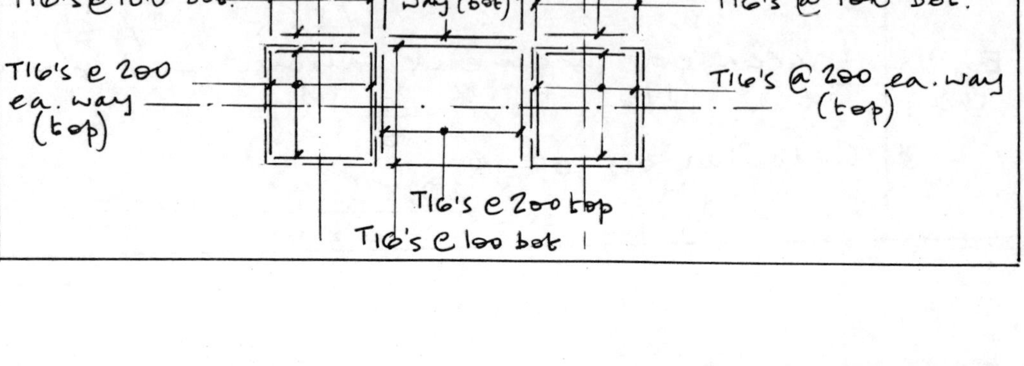

Fig. 11/005 Slab reinf.

Ref. (BS 8110)	Vehicle Ramp Design.
	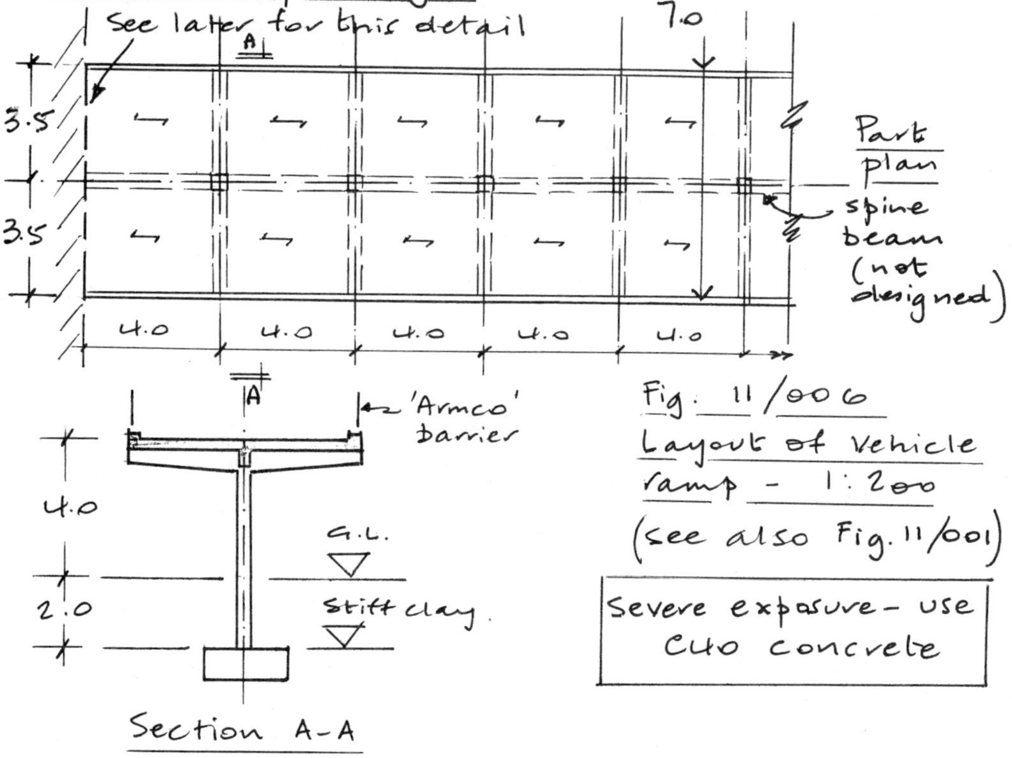
	Fig. 11/006 Layout of vehicle ramp – 1:200 (see also Fig. 11/001)
	Severe exposure – use C40 concrete
	One-way spanning slab:
	Try a 150 mm. thick slab ($d = 105$ mm.)
	$\therefore g_k = 0.15 \times 24 = 3.6$ kN/m²
	Design loading $= 1.6 \times 5.0 + 1.4 \times 3.6 = 13.0$ kN/m²
Table 3.13	Hogging & sagging moment $= 0.063FL$
	$F = 13.0 \times 4 = 52$ kN.
	$\therefore M = 0.063 \times 52 \times 4 = 13.1$ kN.m.
Cl. 3.4.4.4 This book App. A1	$\dfrac{M}{f_{cu} b d^2} = \dfrac{13.1 \times 10^6}{40 \times 10^3 \times 105^2} = 0.03$: $\dfrac{z}{d} = 0.95$
	$\therefore z = 100$ mm
	A_s req^d $= \dfrac{13.1 \times 10^6}{0.87 \times 460 \times 100} = 327$ mm²/m
Table 3.27	Use A 352 mesh (H.Y.S.) top & bottom
	Check min. steel: $A_{s\,min} = \dfrac{0.13 \times 10^3 \times 150}{100} = 195$ mm²/m < 352 ✓ o.k.
	Check deflection: $\dfrac{\text{span}}{d} = \dfrac{4000}{105} = 38$
Table 3.10	Allowable $= 26 \times$ m.f.

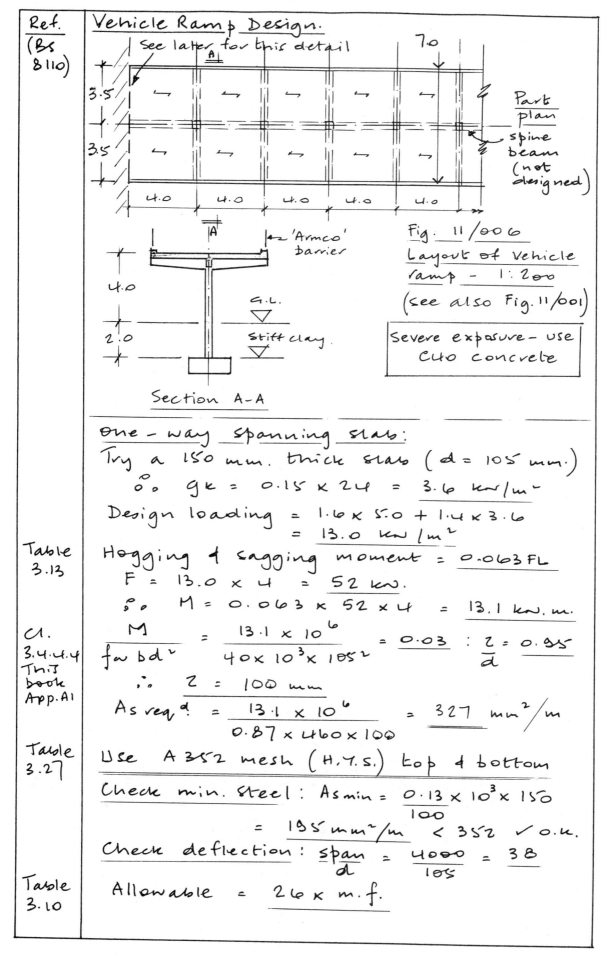

Ref.	Calculations
(BS 8110) Table 3.11 Notes 1 & 2	$f_s = \frac{5}{8} \times 460 \times \frac{327}{352} = \underline{267\ N/mm^2}$
	$\frac{M}{bd^2} = \frac{13.1 \times 10^6}{10^3 \times 105^2} = \underline{1.19}$
	∴ Mod. factor $= 0.55 + \frac{(477-234)}{120\,(0.9 + 1.19)}$
	$= \underline{1.39}\ (< 2.0)$
	∴ Allowable $= 26 \times 1.39 = \underline{36.1}$
	$\frac{\text{Allowable}}{(36.1)} \ngtr \frac{\text{Actual}}{(38)}$: Not o.k. – more steel required
	Design of cantilever beam.
	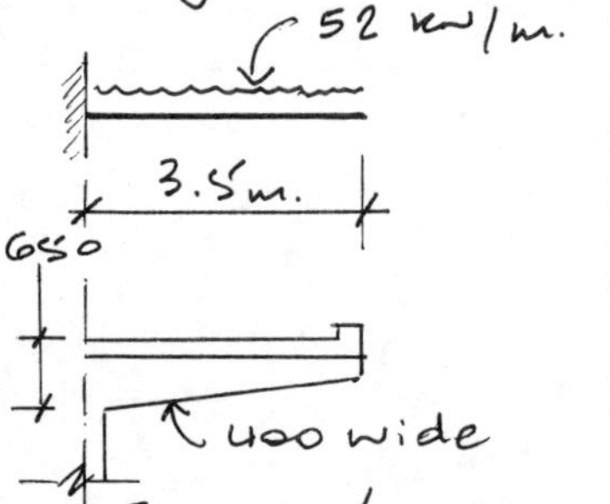 Fig. 11/007 Cantilever beam
	Slab reaction $= \frac{52}{2} \times 2 = 52\ kN/m$
	$M = \frac{52 \times 3.5^2}{2} = 318.5\ kN.m.$
Cl. 3.4.4.4 This book App. A1	$\frac{M}{f_{cu} b d^2} = \frac{318.5 \times 10^6}{40 \times 400 \times 600^2} = \underline{0.055}$
	∴ $\frac{z}{d} = \underline{0.93}$: $\underline{z = 558\ mm.}$
	∴ A_s req^d $= \frac{318.5 \times 10^6}{0.87 \times 460 \times 558}$
	∴ A_s req^d $= \underline{\underline{1426\ mm^2}}$ $(4\ T25\text{'s} = 1962\ mm^2)$
Cl. 3.4.5.2	Check shear : $V = 52 \times 3.5 = 182\ kN.$
	$v = \frac{182 \times 10^3}{400 \times 600} = \underline{0.76\ N/mm^2}$
Table 3.9	$\frac{100 A_s}{bd} = \frac{100 \times 1962}{400 \times 600} = \underline{0.82}$: $\underline{v_c = 0.58\ N/mm^2}$
Table 3.8	∴ $\frac{0.5 v_c}{(0.29)} < \frac{v}{(0.76)} < \frac{(v_c + 0.4)}{(0.98)}$
	Minimum links req^d
	Check Deflection : Actual $\frac{span}{d} = \frac{3500}{600} = \underline{5.83}$
Table 3.10	Allowable $= \underline{7.0}$
	∴ Allowable > Actual ✓ ok.

Ref (BS 8110)

Ramp support column & base.

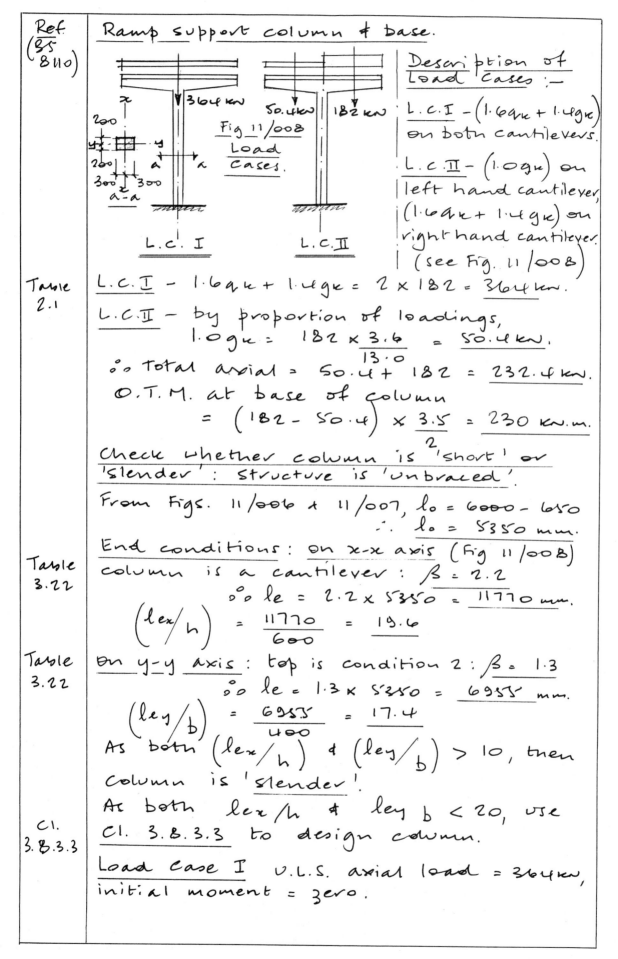

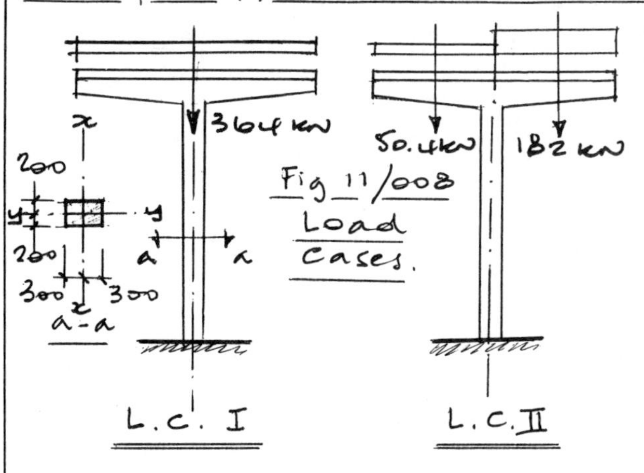

Fig 11/008 Load cases.

Description of Load Cases :-

L.C.I – $(1.6q_k + 1.4g_k)$ on both cantilevers.

L.C.II – $(1.0g_k)$ on left hand cantilever, $(1.6q_k + 1.4g_k)$ on right hand cantilever. (see Fig. 11/008)

Table 2.1 — L.C.I – $1.6q_k + 1.4g_k = 2 \times 182 = 364$ kN.

L.C.II – by proportion of loadings,

$1.0g_k = 182 \times \frac{3.6}{13.0} = 50.4$ kN.

∴ Total axial $= 50.4 + 182 = 232.4$ kN.

O.T.M. at base of column

$= (182 - 50.4) \times \frac{3.5}{2} = 230$ kN.m.

Check whether column is 'short' or 'slender': structure is 'unbraced'.

From Figs. 11/006 & 11/007, $l_o = 6000 - 650$

∴ $l_o = 5350$ mm.

Table 3.22 — End conditions: on x-x axis (Fig 11/008) column is a cantilever : $\beta = 2.2$

∴ $l_e = 2.2 \times 5350 = 11770$ mm.

$\left(l_{ex}/h\right) = \frac{11770}{600} = 19.6$

Table 3.22 — On y-y axis : top is condition 2 : $\beta = 1.3$

∴ $l_e = 1.3 \times 5350 = 6955$ mm.

$\left(l_{ey}/b\right) = \frac{6955}{400} = 17.4$

As both $\left(l_{ex}/h\right)$ & $\left(l_{ey}/b\right) > 10$, then column is 'slender'.

Cl. 3.8.3.3 — As both l_{ex}/h & $l_{ey}/b < 20$, use Cl. 3.8.3.3 to design column.

Load Case I U.L.S. axial load = 364 kN, initial moment = zero.

Ref. (BS 8110) Eq. 32

Because column is 'slender' there is a moment due to the 'PΔ' effect, Madd.

$$a_u = \beta a . h$$

For $le/h = 19.6$, $\beta a = \frac{1}{2000} \times 19.6^2 = 0.192$

$\therefore a_u = 0.192 \times 600 = 115$ mm.

$\therefore$ Madd $= 364 \times \frac{115}{10^3} = 42$ kN.m.

$\therefore$ Design as for a 'short' column, with N = 364 kN., M = 42 kN.m.

Chart 36

$$\frac{N}{bh} = \frac{364 \times 10^3}{400 \times 600} = 1.52$$

$$\frac{M}{bh^2} = \frac{42 \times 10^6}{400 \times 600^2} = 0.30$$

Chart 36: $\frac{100 A_s}{bh} = 0.4$ (min. steel)

Load case II N = 232.4 kN.
M = 230 + 42 = 272 kN.m.

$$\frac{N}{bh} = \frac{232.4 \times 10^3}{400 \times 600} = 0.97$$

$$\frac{M}{bh^2} = \frac{272 \times 10^6}{400 \times 600^2} = 1.89$$

Chart 36: $\frac{100 A_s}{bh} = 1.1$

$\therefore \frac{100 A_{sc}}{bh} = 1.1$: $A_{sc} = \frac{1.1 \times 400 \times 600}{100} = 2640 \text{mm}^2$

Use 6 No. T 25's (2945 mm²)

Base size: founded on stiff clay: c = 125 kN/m²
Safe bearing ≃ 3c = 375 kN/m²

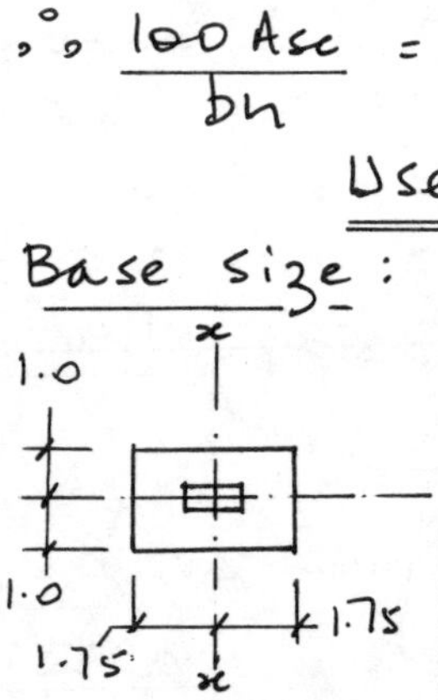

Fig. 11/009 Base.

Try a base 3.5 m × 2.0 m × 1.0 m tk.

Self-weight = 3.5 × 2.0 × 1.0 × 24 = 168 kN.

$$'Z' = \frac{bd^2}{6} = \frac{2 \times 3.5^2}{6} = 4.1 \text{ m}^3$$

$$'A' = b.d = 2 \times 3.5 = 7.0 \text{ m}^2$$

L.C. I service loads:

$$N = \frac{364}{1.5} + 168 = 411 \text{ kN.} : M = \frac{42}{1.5} = 28 \text{ kN.m}$$

$$\therefore \frac{P}{A} \pm \frac{M}{Z} = \frac{411}{7} \pm \frac{28}{4.1} = +65.5 \text{ kN/m}^2 \ \& \ +51.9 \text{ kN/m}^2$$

✓ o.k.

L.C. II service loads: $N = \frac{232.4}{1.5} + 168 = 323$ kN.

$$M = 272/1.5 = 182 \text{ kN.m}$$

$$\therefore \frac{P}{A} \pm \frac{M}{Z} = \frac{323}{7} \pm \frac{182}{4.1} = +90.5 \text{ kN/m}^2 \ \& \ +1.7 \text{ kN/m}^2$$

✓ o.k.

$\therefore$ Base size o.k.

Ref.	
(BS 8110)	Retaining wall design (see Figs 11/001 & 11/010) plus design outline under 'site conditions'

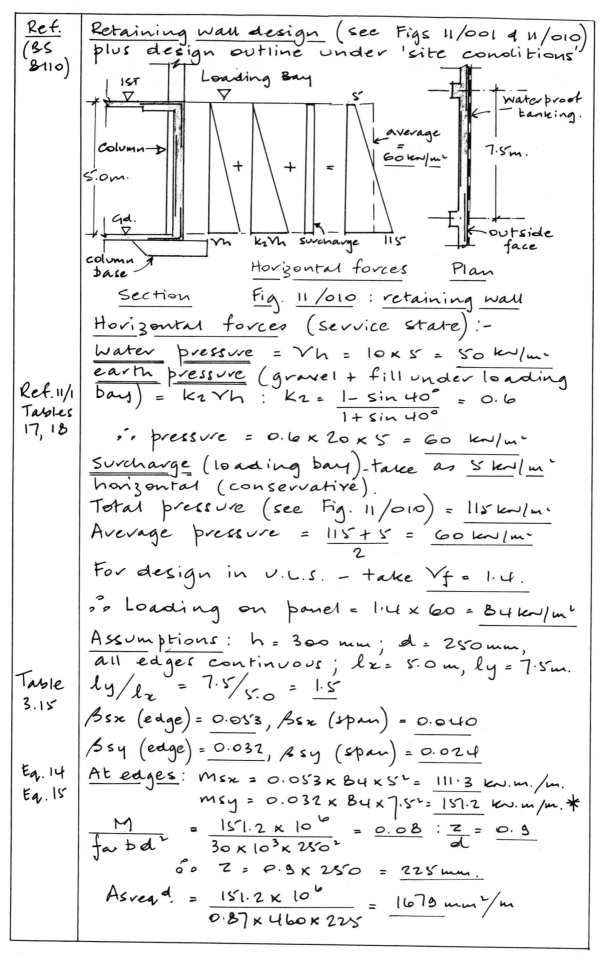

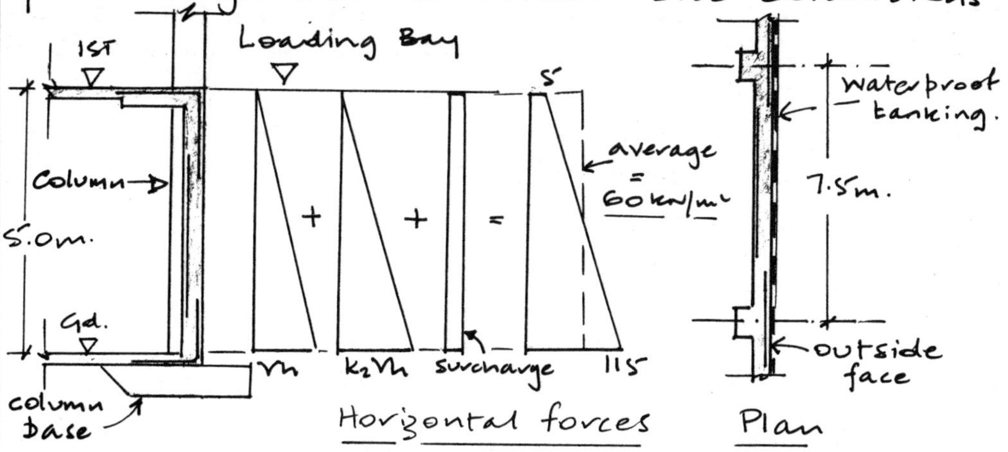

Section

Fig. 11/010 : retaining wall

Horizontal forces (service state):-

Water pressure $= \gamma h = 10 \times 5 = 50\ kN/m^2$

Ref. 11/1 Tables 17, 18 — earth pressure (gravel + fill under loading bay) $= k_2 \gamma h$: $k_2 = \dfrac{1 - \sin 40^\circ}{1 + \sin 40^\circ} = 0.6$

$\therefore$ pressure $= 0.6 \times 20 \times 5 = 60\ kN/m^2$

Surcharge (loading bay) - take as $5\ kN/m^2$ horizontal (conservative).

Total pressure (see Fig. 11/010) $= 115\ kN/m^2$

Average pressure $= \dfrac{115 + 5}{2} = 60\ kN/m^2$

For design in U.L.S. - take $\gamma_f = 1.4$.

$\therefore$ Loading on panel $= 1.4 \times 60 = 84\ kN/m^2$

Assumptions: $h = 300\ mm$; $d = 250\ mm$, all edges continuous; $l_x = 5.0\ m$, $l_y = 7.5\ m$.

Table 3.15 — $l_y / l_x = 7.5/5.0 = 1.5$

β_{sx} (edge) $= 0.053$, β_{sx} (span) $= 0.040$

β_{sy} (edge) $= 0.032$, β_{sy} (span) $= 0.024$

Eq. 14, Eq. 15 — At edges: $m_{sx} = 0.053 \times 84 \times 5^2 = 111.3\ kN.m/m$.

$m_{sy} = 0.032 \times 84 \times 7.5^2 = 151.2\ kN.m/m.$ *

$\dfrac{M}{f_{cu} b d^2} = \dfrac{151.2 \times 10^6}{30 \times 10^3 \times 250^2} = 0.08$: $\dfrac{z}{d} = 0.9$

$\therefore z = 0.9 \times 250 = 225\ mm.$

$As_{req^d} = \dfrac{151.2 \times 10^6}{0.87 \times 460 \times 225} = 1679\ mm^2/m$

Ref. (BS 8110)	
	Use T20's @ 175 crs. (1794 mm²/m)
	Span moments: $m_{sx} = 0.040 \times 84 \times 5^2 = 84$ kN.m/m
	$m_{sy} = 0.024 \times 84 \times 7.5^2 = 113.4$ kN.m/m
	Critical mt. = 113.4 kN.m/m.
Cl. 3.4.4.4 This book - App. A1	$\frac{M}{f_{cu} b d^2} = \frac{113.4 \times 10^6}{30 \times 10^3 \times 250^2} = 0.06$: $\frac{z}{d} = 0.93$
	∴ $z = 0.93 \times 250 = 232$ mm.
	$A_{s\,req^d} = \frac{113.4 \times 10^6}{0.87 \times 460 \times 232} = 1224$ mm²/m
	Use T20's @ 200 crs (1570 mm²/m)
Table 3.16	Check shear at edges:
	$\beta_{vx} = 0.45$; $\beta_{vy} = 0.33$
	∴ V_{sx} (kN/m) = $0.45 \times 84 \times 5 = 189$ kN/m
	V_{sy} (kN/m) = $0.33 \times 84 \times 7.5 = 208$ kN/m*
	∴ $v = \frac{208 \times 10^3}{10^3 \times 250} = 0.83$ N/mm²
Table 3.9	$\frac{100 A_s}{bd} = \frac{100 \times 1794}{10^3 \times 250} = 0.72$: $v_c = 0.64$ N/mm²
Fig. 3.10.	∴ $v_c < v < (v_c + 0.4)$
	(0.64) (0.83) (1.04)
	∴ Minimum single leg links req^d around edges.

Chapter 12 Rooftop swimming pool on an existing multi-storey hotel

A major hotel chain wish to upgrade one of their properties, by incorporating a swimming pool between the 10th floor and the roof, as shown in Figure 12/001. In addition, a canopy in structural steelwork and masonry is required over the pool, and between two existing lift and stairs cores. The swimming pool is to have a depth of 0.9 m. at the shallow end, 1.8 m. at the deep end, and is to have a width of 10.0 m. The scheme, which is a combination of pre-cast concrete, in-situ concrete and structural steelwork, is shown in section in Figure 12/002, together with details in Figures 12/003 and 12/004. The pool and new structure sits on an existing reinforced concrete frame, with main frames at 5.0 m. centres. The structure below the 10th Floor will have to be strengthened for the additional loads. However, this does not form part of this exercise, which gives the design of the swimming pool slab, walls and support beams, together with the pool canopy. With regard to the swimming pool slab, an alternative design, not taken here, would be of taking the pre-cast planks and in-situ concrete as acting compositely. Design then would be as the bridge deck in Chapter 2.

Loading – qk

Pool surround – 3.0 kN/m^2
Pool canopy – 0.60 kN/m^2 (snow)

Loading – gk

Take unit weight of water as 10.0 kN/m^3

Loading – wk

Level site on the coast – basic wind speed = 50 m/sec.

Design Codes of Practice

Concrete elements to be designed to BS 8110 (Structural Concrete)
Water-retaining concrete elements to be designed to BS 8007 (Water-retaining concrete structures)
Structural steelwork elements to be designed to BS 5950 (Structural steelwork)

Materials

Concrete – fcu = 40 N/mm^2
Reinforcement – fy = 460 N/mm^2
Prestressed concrete planks – fci = 40 N/mm^2:
fcu = 50 N/mm^2
Steelwork – EN10025 S275

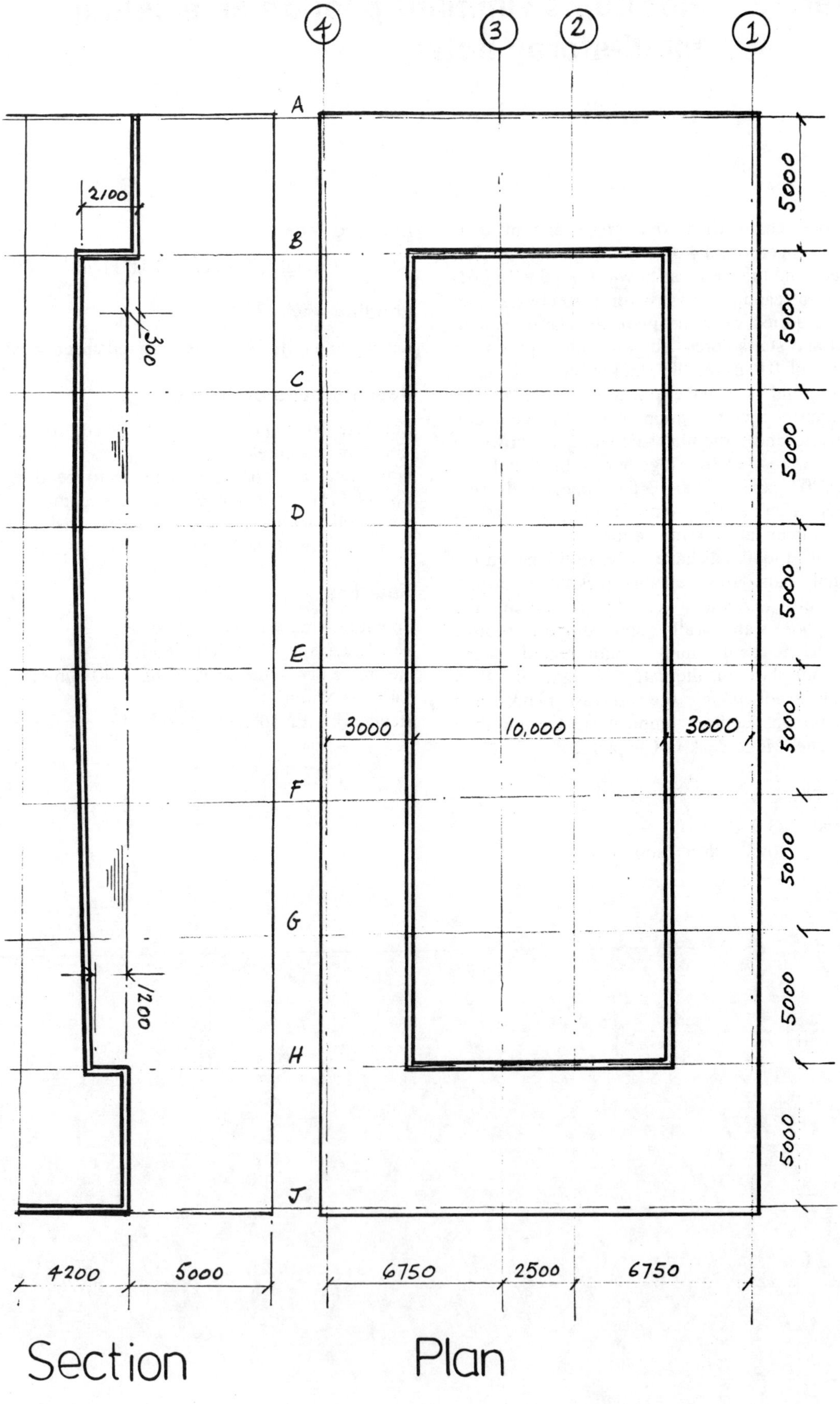

4
3
2
1
A
B
C
D
E
F
G
H
J
2100
300
1200
4200
5000
3000
10,000
3000
5000
5000
5000
5000
5000
5000
5000
5000
6750
2500
6750
Section
Plan

Fig.12/001

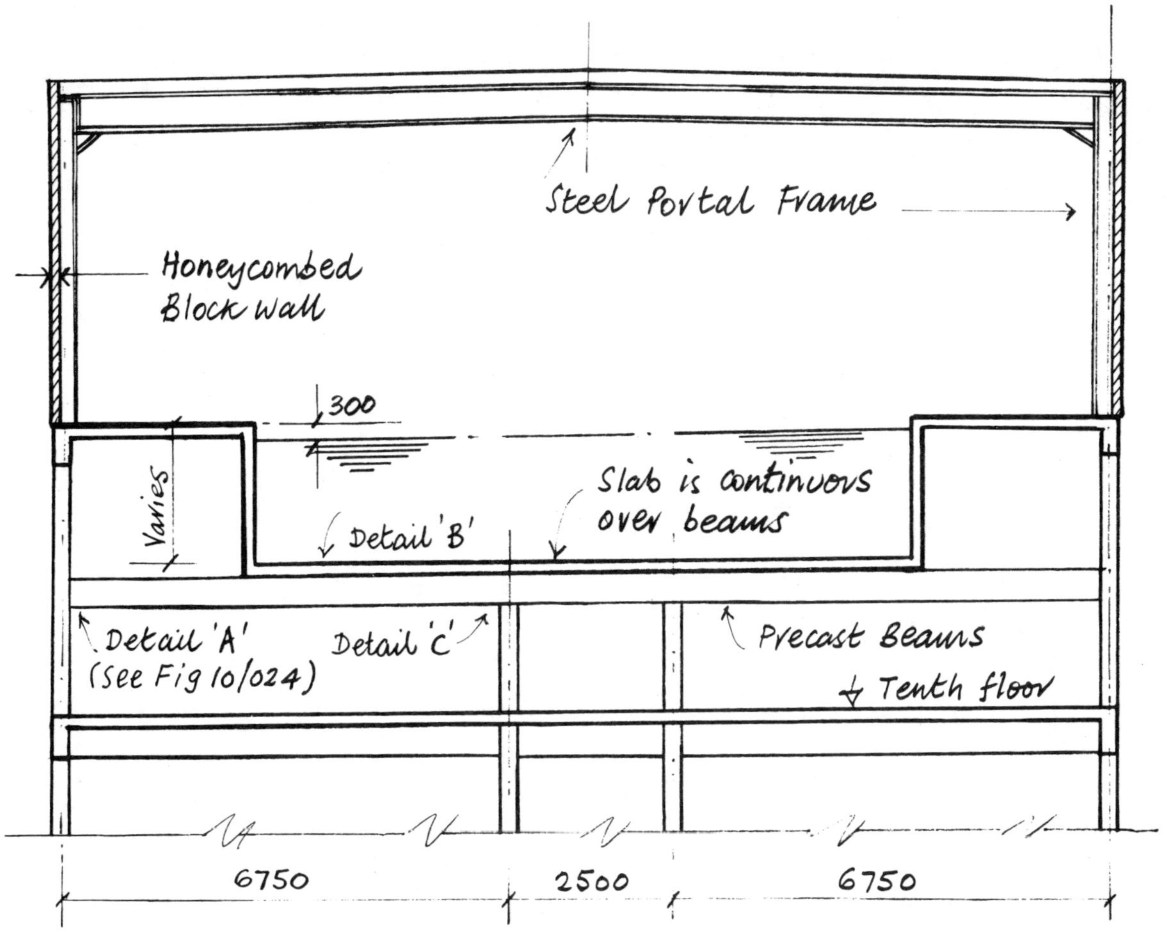

Typical Cross Section Fig.12/002

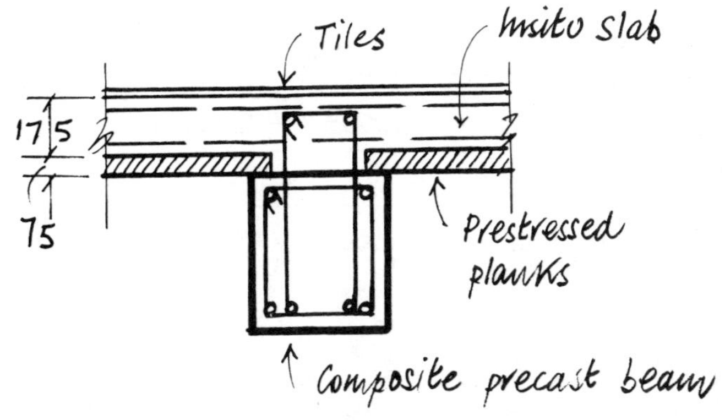

Detail 'B'

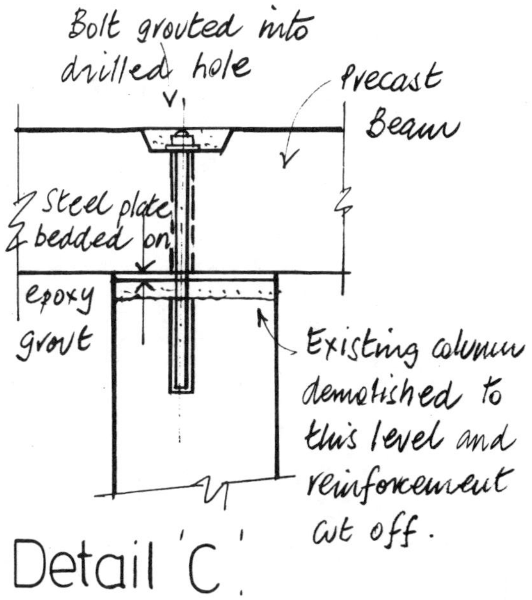

Detail 'C'

Fig.12/003

Fig.12/004

Ref.	A Design of pool floor slab – continuous over 5 m. centres. (Figs. 12/001, 002 & 003)
(BS8110 Pt 1)	1) Design of prestressed, pre-cast concrete planks. – take pool slab dead wt.

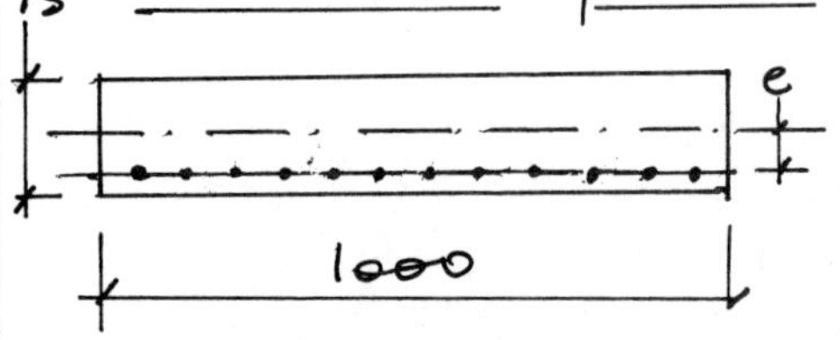

Fig. 12/005
Plank dimensions

Try 75 mm thick × 1000 wide planks with 15 mm cover to tendons (7 mm. ø) (See Fig. 12/005)

∴ Tendon eccentricity
$e = 75/2 - 15 - 3.5 = \underline{19}$ mm.

Tendon force, P_i (losses – take $\beta = 0.8$)

For 7 mm ø wires, $f_{pi} = 1440$ N/mm²

P_i/tendon $= \frac{\pi \times 7^2}{4} \times \frac{1440}{10^3} = \underline{55.4 \text{ kN}}$.

Use 12 tendons / plank :–

$P_i = 12 \times 55.4 = \underline{\underline{665 \text{ kN}}}$.

Allowable stresses (take planks as Class 2)

Cl. 4.3.5 — $f_{ci} = 40$ N/mm² – transfer

∴ $f'_{min} = 0.45\sqrt{40} = \underline{-2.85}$ N/mm² (tens.)

$f'_{max} = 0.50 \times 40 = \underline{+20}$ N/mm² (comp.)

Cl. 4.3.4 — $f_{cu} = 50$ N/mm² – serviceability.

∴ $f_{min} = 0.45\sqrt{50} = \underline{-3.20}$ N/mm²

$f_{max} = 0.33 \times 50 = \underline{+16.5}$ N/mm²

Bending Moments (planks simply supported)

Self weight U.D.L. $= 0.075 \times 24 \times 1.0 = \underline{1.8}$ kN/m

$M_{SWT} = \frac{1.8 \times 5^2}{8} = \underline{5.625 \text{ kN.m.}}$

dead load (service) – pool in-situ slab = 175 mm thick.

U.D.L. $= 0.175 \times 24 \times 1.0 = 4.2$ kN/m.

$M_D = \frac{4.2 \times 5^2}{8} = \underline{13.12 \text{ kN.m.}}$

Section properties :–

$Z_t = Z_b = 1000 \times 75^2/6 = \underline{937500 \text{ mm}^3}$

Area, $A = 75 \times 1000 = \underline{75000 \text{ mm}^2}$

Ref	
(BS8110 Pt1)	

Check stresses at transfer ($\alpha = 0$)

Top level: $P_i/A - P_i.e/Z_t + M_{swt}/Z_t \nless -2.85$

$$\therefore \frac{665\times10^3}{75000} - \frac{665\times10^3\times19}{937500} + \frac{5.625\times10^6}{937500}$$

$$= 8.87 - 13.48 + 6.00 = +1.39$$

Bottom level: $P_i/A + P_i.e/Z_b - M_{swt}/Z_b < 20$

$$\therefore \frac{665\times10^3}{75000} + \frac{665\times10^3\times19}{937500} - \frac{5.625\times10^6}{937500}$$

$$= 8.87 + 13.48 - 6.00 = +16.35$$

Fig. 12/006 transfer stresses

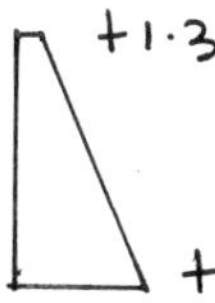

+1.39 ($\nless -2.85$ ✓ o.k.)

+16.35 (< 20 ✓ o.k.)

Check stresses at service ($\beta = 0.8$)

Top level: $\beta P_i/A - \beta P_i.e/Z_t + (M_{swt}+M_D)/Z_t < 16.5$

$$\therefore \frac{0.8\times665\times10^3}{75000} - \frac{0.8\times665\times10^3\times19}{937500} + \frac{(5.625+13.12)\times10^6}{937500}$$

$$= 7.09 - 10.78 + 19.99 = +16.3$$

Bottom level: $\beta P_i/A + \beta P_i.e/Z_b - (M_{swt}+M_D)/Z_b \nless -3.2$

$$\therefore \frac{0.8\times665\times10^3}{75000} + \frac{0.8\times665\times10^3\times19}{937500} - \frac{(5.625+13.12)\times10^6}{937500}$$

$$= 7.09 + 10.78 - 19.99 = -2.12$$

Fig. 12/007 service stresses

+16.30 (< 16.5 ✓ o.k.)

−2.12 ($\nless -3.2$ ✓ o.k.)

$\therefore$ Planks satisfactory – 12 No. 7mm ϕ tendons for pool slab dead wt.

2) **Pool bottom slab** (r.c. 175mm. thick)

Loading (g_k)

Slab – load taken by planks.

tiles (on bed) $= 0.050\times20 = 1.0$ kN/m²

Ref.	
(BS8110 Pt.1)	**Loading (q_k)**
	Water (1.8m. depth) = 1.8 × 10 = 18 kN/m²
	Note! from BS 8007, γ_f = 1.4.
	∴ Factored loading = 1.4 × 18 + 1.4 × 1.0 = 26.6 kN/m²
	Slab is continuous over beams at 5m. crs. & is non-composite with planks. Coefficients from BS 8110 Table 3.13 cannot be used, as $q_k : g_k$ = 18 : 1.0 > 1.25 : 1
Ref 12/1	From 12/1,
	$M_{sagging}$ = 0.046 × 26.6 × 5² = 30.6 kN.m.
	$M_{hogging}$ = 0.080 × 26.6 × 5² = 53.2 kN.m.
	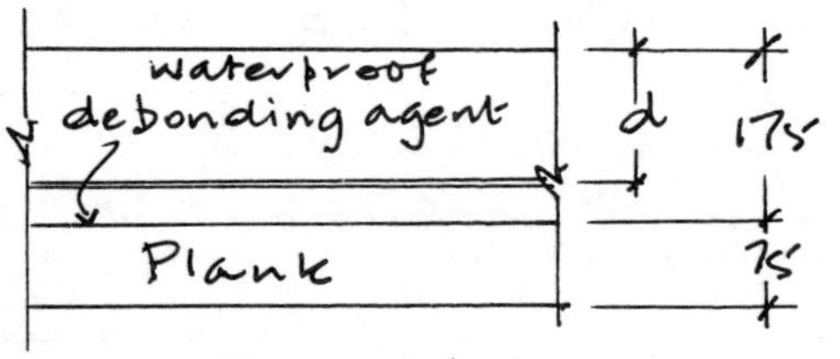 Fig. 12/008
	For sagging steel :- d = 175 − 20 − 8 = 147mm
	$\frac{M}{f_{cu}bd^2} = \frac{30.6 \times 10^6}{40 \times 10^3 \times 147^2}$ = 0.035
This book App. A1	$\frac{z}{d}$ = 0.95 : z = 140 mm.
	∴ $A_{s\,req'd} = \frac{30.6 \times 10^6}{0.87 \times 460 \times 140}$ = 546 mm²/m.
	Use T16's @ 250 crs (804 mm²/m)
	For hogging steel : d = 175 − 40 − 8 = 127mm. (min. cover = 40mm in BS 8007 for surfaces in contact with water)
	$\frac{M}{f_{cu}bd^2} = \frac{53.2 \times 10^6}{40 \times 10^3 \times 127^2}$ = 0.082
This book App. A1	z/d = 0.88 : z = 112 mm.
	∴ $A_{s\,req'd} = \frac{53.2 \times 10^6}{0.87 \times 460 \times 112}$ = 1187 mm²/m.
	Use T16's @ 150 crs (1340 mm²/m)
	Check cracking in top of slab:
BS8110/ BS8007	From BS 8110, w_{cr} ≯ 0.2mm.
	M_{ULT} = 53.2 kN.m. : ∴ $M_{service} = \frac{53.2}{1.4}$ = 38 kN.m

Ref.	
(BS8110 Pt 1)	

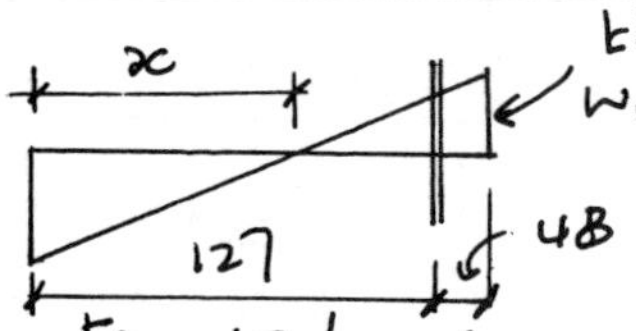

Fig. 12/009
Elastic stresses across slab

Steel ratio, $\rho = \frac{A_s}{bd} = \frac{1340}{10^3 \times 127} = 0.010$

Take modular ratio, $\alpha_e = 15.4$ & $E_s = 200 \times 10^3\ N/mm^2$

$\therefore \rho . \alpha_e = 15.4 \times 0.010 = 0.154$

$$\frac{x}{d} = -0.154 + \sqrt{0.154(2 + 0.154)} = 0.422$$

$\therefore x = 0.422 \times 127 = 53.6\ mm.$

$$Z = \left(1 - \frac{1}{3} \times 0.422\right) \times 127 = 109\ mm$$

Steel service stress, $f_s = \frac{M}{A_s . Z}$

$$f_s = \frac{38 \times 10^6}{1340 \times 109} = 260.2\ N/mm^2$$

(Limit $= 0.8 f_y = 0.8 \times 460 = 368\ N/mm^2$ o.k.)

$$\therefore \varepsilon_s = \frac{260.2}{200 \times 10^3} = 0.0013$$

BS8007 App. B4.

$$\varepsilon_1 = \frac{(175 - 53.6)}{(127 - 53.6)} \times 0.0013 = 0.00215$$

$$\therefore \varepsilon_m = 0.00215 - \frac{10^3 \times (175 - 53.6)^2}{3 \times 200 \times 10^3 (127 - 53.6) \times 1340}$$

$\varepsilon_m = 0.0019$

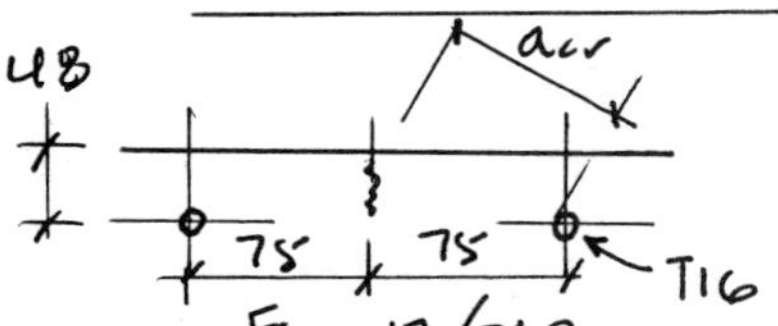

Fig. 12/010
Calculation of acr

From Fig. 12/010 :-

$a_{cr} = \sqrt{48^2 + 75^2} - 8$

$a_{cr} = 81\ mm.$

$C_{min} = 40\ mm.$

BS8007 App. B2

$$\therefore W_{cr} = \frac{3 \times 81 \times 0.0019}{1 + 2\left(\frac{81 - 40}{175 - 53.6}\right)} = 0.276\ mm$$

Wcr slightly greater than 0.20 mm.; but in view of tiled surface & some added support from planks - allow

$\therefore$ T16's @ 150 mm. crs. O.K.

Ref. (BS 8110 Pt. 1)

Swimming pool sides – take as 'free' cantilever

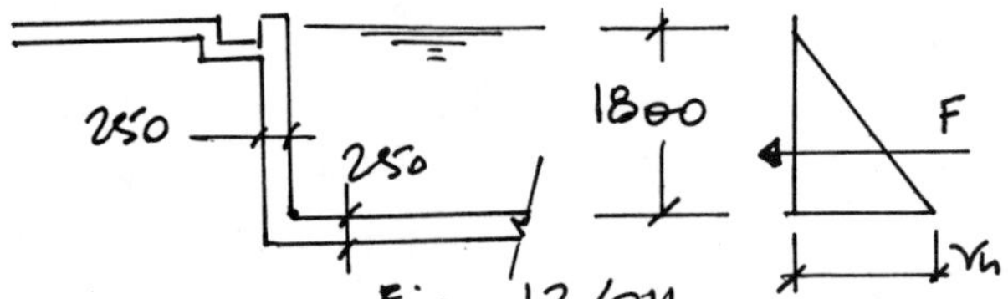

Fig. 12/011

Water pressure, Vh (see Fig 12/011)
$= 10 \times 1.8 = 18\ kN/m^2$

$\therefore F = \frac{18}{2} \times 1.8 = 16.2\ kN/m$

Moment, M, about root of cantilever
$= 16.2 \times 1.8/3 = 9.72\ kN.m/m.$

In the U.L.S., $M_u = 1.4 \times 9.72 = 13.61\ kN.m/m.$

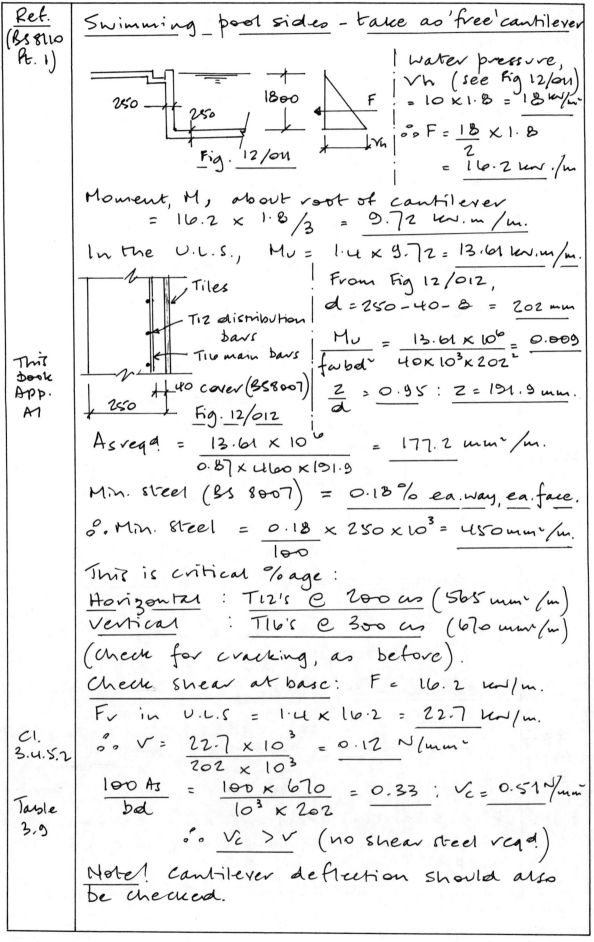

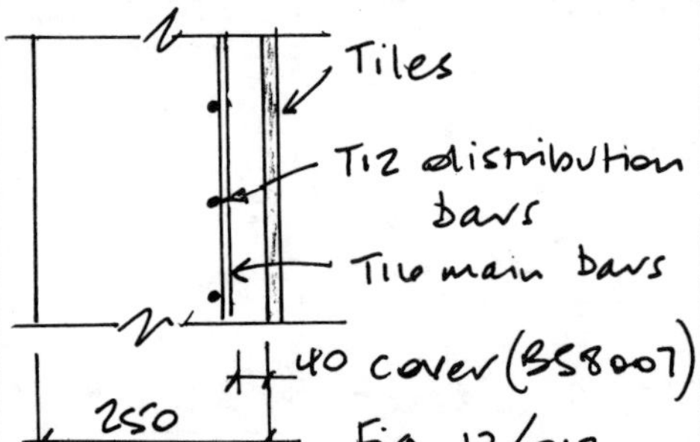

Fig. 12/012

From Fig 12/012,
$d = 250 - 40 - 8 = 202\ mm$

This book App. A1: $\frac{M_u}{f_{cu} b d^2} = \frac{13.61 \times 10^6}{40 \times 10^3 \times 202^2} = 0.009$

$\frac{z}{d} = 0.95 : z = 191.9\ mm.$

$As_{reqd} = \frac{13.61 \times 10^6}{0.87 \times 460 \times 191.9} = 177.2\ mm^2/m.$

Min. steel (BS 8007) = 0.18% ea. way, ea. face.

$\therefore$ Min. steel $= \frac{0.18}{100} \times 250 \times 10^3 = 450\ mm^2/m.$

This is critical %age:

Horizontal : T12's @ 200 c/s (565 mm^2/m)

Vertical : T16's @ 300 c/s (670 mm^2/m)

(Check for cracking, as before).

Check shear at base: $F = 16.2\ kN/m.$

F_v in U.L.S $= 1.4 \times 16.2 = 22.7\ kN/m.$

Cl. 3.4.5.2: $\therefore v = \frac{22.7 \times 10^3}{202 \times 10^3} = 0.12\ N/mm^2$

Table 3.9: $\frac{100 A_s}{bd} = \frac{100 \times 670}{10^3 \times 202} = 0.33 : v_c = 0.51\ N/mm^2$

$\therefore v_c > v$ (no shear steel reqd)

Note! Cantilever deflection should also be checked.

Ref.	
(BS8110 Pt.1)	**Design of pre-cast, composite, propped T beams** (see figs 12/001 → 12/006).

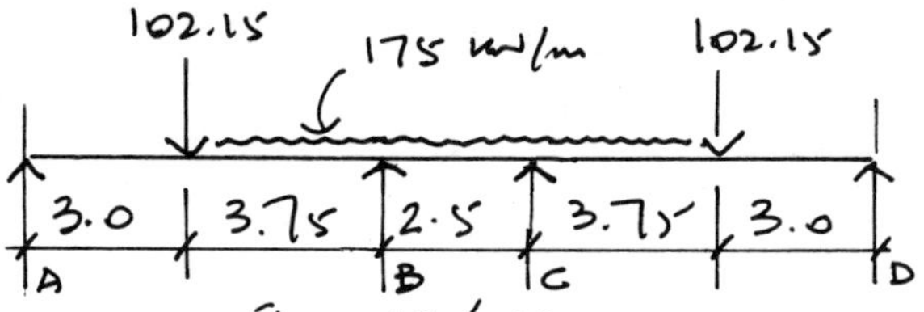

Fig. 12/013

Factored loading of pool slab :-

Planks + slab
$= 1.4(1.8 + 4.2)$
$= 8.4$ kN/m²

Tiles + water
$= 26.6$ kN/m²
(see previous calcs.)

∴ U.D.L. (beams at 5m. c/s.)
$= 5(8.4 + 26.6) = 175$ kN/m.

Point loads from pool wall = wall self weight + reaction from pool sides spanning from wall to side of building.

i) Pool wall $= 1.4 \times 0.25 \times 24 \times 2.1 = 17.64$ kN/m

ii) Allow 200 mm slab for pool surround
$g_k = 0.2 \times 24 = 4.8$ kN/m²
Allow $g_k = 1.5$ kN/m² for tiles + screed
$q_k = 3.0$ kN/m² (see brief)

∴ Factored loading $= 1.6 \times 3.0 + 1.4(4.8 + 1.5)$
$= 13.62$ kN/m²

Reaction $= \frac{3}{2} \times 13.62 = 20.43$ kN/m.

So Point load $= 5 \times 20.43 = 102.15$ kN.
(Loads shown in Fig. 12/013)

Continuous beam properties :-
take as rectangular beam for stiffness factors - 750 deep × 400 wide.

$I_{xx} = 40 \times 75^3/12 = 1406250$ cm⁴

$A = 40 \times 75 = 3000$ cm²

Results of plane frame program :-

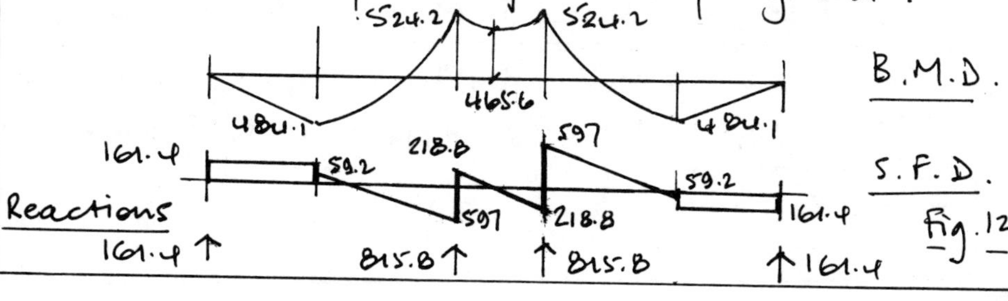

Fig. 12/014

Ref. (BS8110 Pt-1)

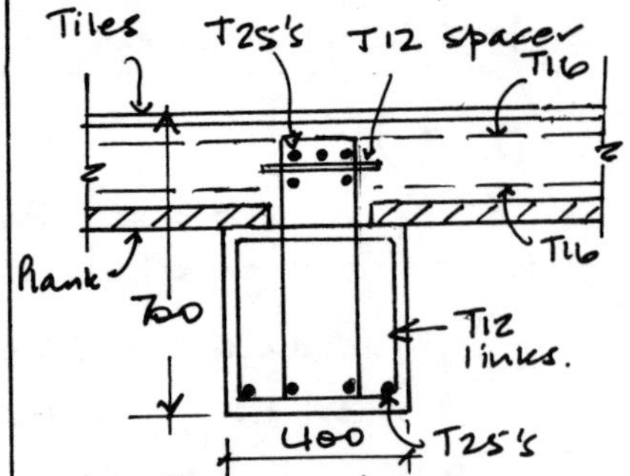

Fig. 12/015
Composite main beam

Effective depth (hogging steel) - see Fig. 12/015 :-

$d = 700 - 40 - 16 - 25 - 6$

$d = 613$ mm

Effective depth (sagging steel) - see Fig. 12/015 :-

$d = 700 - 20 - 12 - 12.5$

$d = 655.5$ mm.

Hogging moment :- $M_u = 524.2$ kN.m

Check concrete compression:

Cl. 3.4.4.4

$M = 0.156 \times 40 \times 400 \times 613^2 / 10^6$

$= 938$ kN.m.

$\therefore M_{ec} > M$ - no compression steel required.

$$\frac{M}{f_{cu}bd^2} = \frac{524.2 \times 10^6}{40 \times 400 \times 613^2} = 0.09$$

This book - App. A1

$z/d = 0.89$: $z = 546$ mm.

$$\therefore A_s \text{ req}^d = \frac{524.2 \times 10^6}{0.87 \times 460 \times 546} = 2399 \text{ mm}^2$$

Use 5 No. T25's (2 layers) (2453 mm²)

Sagging moment :- $M_u = 484.1$ kN.m.

Effective breadth = $\frac{\text{span}}{5}$ + rib

$$b_e = \frac{6750}{5} + 400 = 1750 \text{ mm.}$$

$$\frac{M}{f_{cu}bd^2} = \frac{484.1 \times 10^6}{40 \times 1750 \times 655.5^2} = 0.02$$

This book App. A1

$z/d = 0.95$: $z = 623$ mm.

$$\therefore A_s \text{ req}^d = \frac{484.1 \times 10^6}{0.87 \times 460 \times 623} = 1942 \text{ mm}^2$$

Use 4 No. T25's (1963 mm²)

Table 3.27

Minimum steel (bottom steel critical)

$$= \frac{0.13}{100} \times \left[(1750 \times 250) + (450 \times 400)\right] = 803 \text{ mm}^2 < 1963 \text{ mm}^2$$

✓ o.k.

Ref.	
(BS 8110 Pt 1)	**Check shear at supports B & C. $F_v = 597$ kN.**
	$v = \dfrac{597 \times 10^3}{400 \times 613} = 2.43$ N/mm²
Cl. 3.4.5.2 Table 3.9	$\dfrac{100 A_s}{bd} = \dfrac{100 \times 2453}{400 \times 613} = 1.0$: $v_c = 0.74$ N/mm²
Table 3.8	∴ use designed links as,
	$(v_c + 0.4) < v < 0.8\sqrt{f_{cu}}$
	(1.14) (2.43) (5.0)
	Using 4 legs of T12 links (see Fig. 12/015)
	$A_{sv} = 4 \times \dfrac{\pi \times 12^2}{4} = 452$ mm²
	∴ $S_v = \dfrac{452 \times 0.87 \times 460}{400\,(2.43 - 1.14)} = 350$ mm.
	∴ Use prs. of T12 links @ 300 mm crs.
	Check deflection, midway between A & B.
	span/d ≯ 26 × (mod. factor)
	Actual span/d = 6750/655.5 = 10.3 < 26
	✓ O.K.
	Design of roof canopy over pool – steel portals @ 5 m. crs.
BS 5950 Table 2	Load cases: I): $1.6 q_k + 1.4 g_k$.
	Loading : $q_k = 0.6$ kN/m² (snow)
	$g_k = 0.4$ kN/m² (cladding)
	∴ Factored loading = 1.6 × 0.6 + 1.4 × 0.4
	= 1.52 kN/m²
	U.D.L. on one portal = 5 × 1.52 = 7.6 kN/m
BS 5950 Table 2	II): $1.4 W_k + 1.0 g_k$.
	$V = 50$ m/sec (see brief).
CP3 Ch V Pt 2	S_2 value: h = 11 × 4.2 + 5 = 52 m.
	class C, group (3): $S_2 = 1.00$
	∴ $V_s = V = 50$ m/sec.
	$q = \dfrac{0.613 \times V_s^2}{10^3} = 1.53$ kN/m²

Ref.	
(BS 5950)	Cpe coefficients (wind on side)

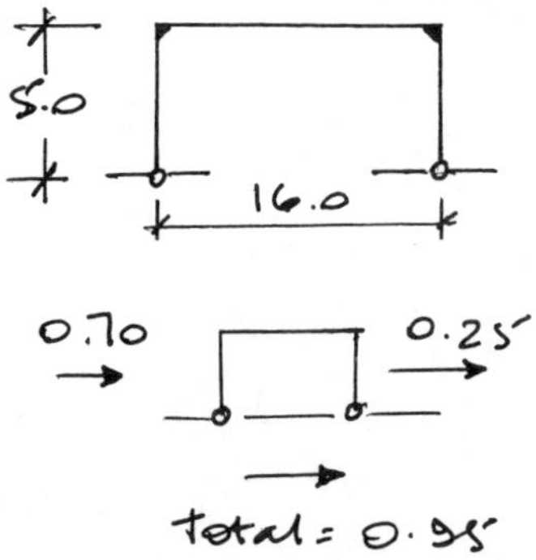

Fig. 12/016

$l/w = 40/16 = 2.5$

$\therefore 3/2 < l/w < 4$

$h/w = 5/16 = 0.31$

$\therefore h/w < 0.5$

$\therefore$ coefficients as shown in Fig. 12/016

CP3 ChV Pt 2 Table 7

$\therefore W_k = 0.95 \times 1.53 = 1.45 \; kN/m^2$

$1.4W_k$ (U.D.L.)

$= 1.4 \times 1.45 \times 5 = 10.2 \; kN/m$

Ref. 12/4

Using 'Kleinlogel Charts' for square portal with dimensions as shown in Fig. 12/016 and a constant inertia :-

$K = h/L = 5/16 = 0.31 \quad : N = 2k + 3 = 3.62$

Load Case 1 $(1.6 q_k + 1.4 g_k)$

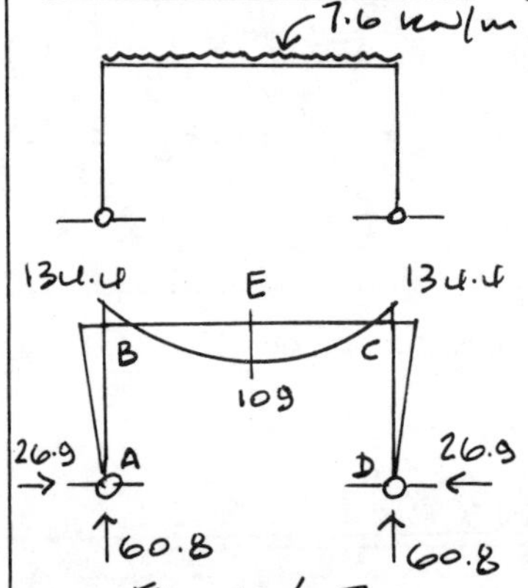

Fig. 12/017

$M_B = M_C = \dfrac{-wL^2}{4N} = \dfrac{7.6 \times 16^2}{4 \times 3.62}$

$= -134.4 \; kN.m$

$M_E = \dfrac{7.6 \times 16^2}{8} - 134.4 = 109 \; kN.m.$

$H_A = -H_B = \dfrac{+134.4}{5} = 26.9 \; kN.$

Values shown in Fig. 12/017

Load Case 11 $(1.4 w_k + 1.0 g_k)$

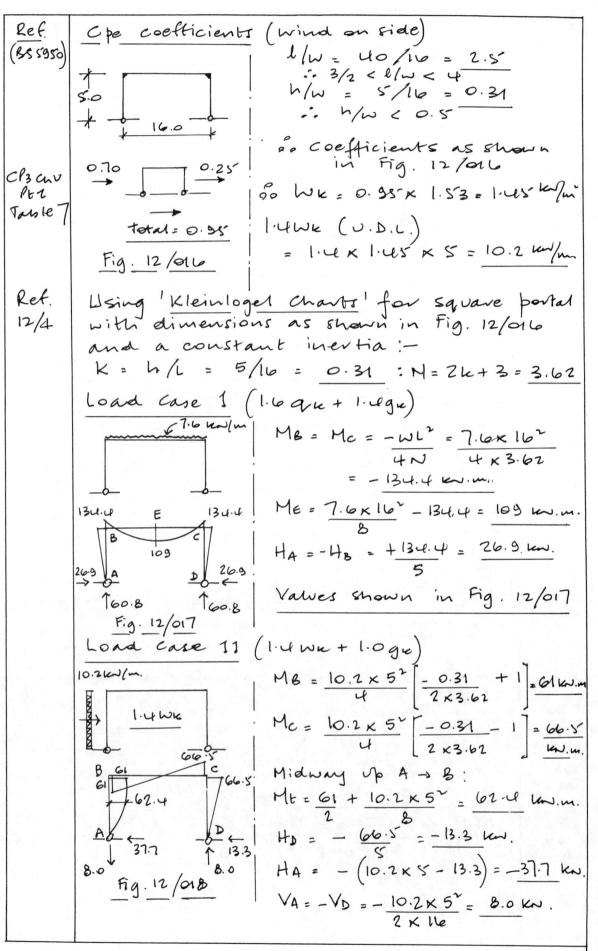

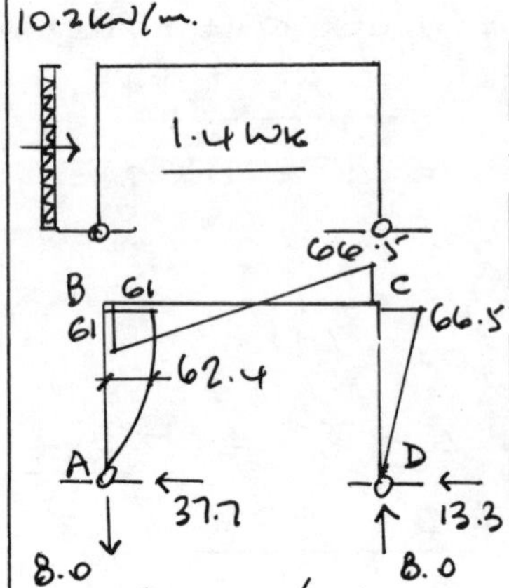

Fig. 12/018

$M_B = \dfrac{10.2 \times 5^2}{4}\left[\dfrac{-0.31}{2 \times 3.62} + 1\right] = 61 \; kN.m$

$M_C = \dfrac{10.2 \times 5^2}{4}\left[\dfrac{-0.31}{2 \times 3.62} - 1\right] = 66.5 \; kN.m.$

Midway up $A \rightarrow B$:

$M_t = \dfrac{61}{2} + \dfrac{10.2 \times 5^2}{8} = 62.4 \; kN.m.$

$H_D = -\dfrac{66.5}{5} = -13.3 \; kN.$

$H_A = -(10.2 \times 5 - 13.3) = -37.7 \; kN.$

$V_A = -V_D = -\dfrac{10.2 \times 5^2}{2 \times 16} = 8.0 \; kN.$

Ref. (BS5950)		
	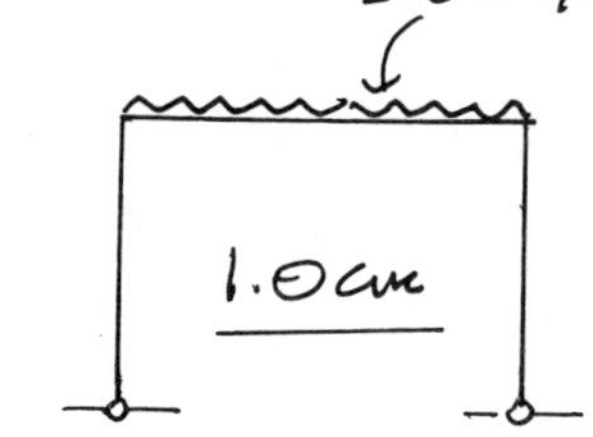	1.0qk : U.D.L. = 1.0 × 0.4 × 5 = 2.0 kN/m. $M_B = M_C = \frac{2 \times 16^2}{4 \times 3.62} = 35.4$ kN.m. $M_E = \frac{2 \times 16^2}{8} - 35.4 = 28.6$ kN.m
	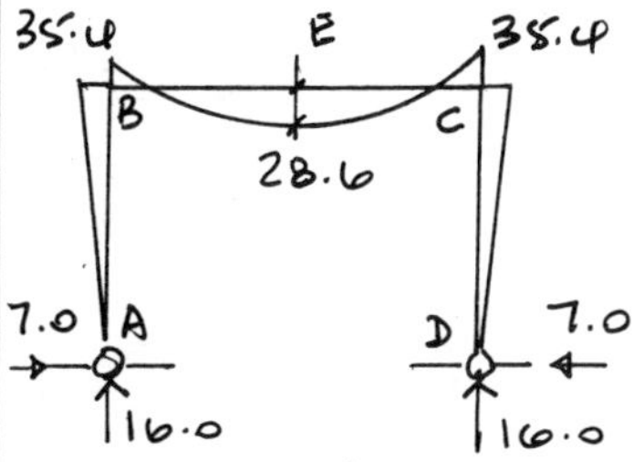Fig. 12/019	$H_A = -H_B = \frac{35.4}{5} = 7.0$ kN. $V_A = V_D = \frac{2 \times 16}{2} = 16$ kN. 1.4Wk + 1.0qk : $M_B = +61 - 35.5 = +35.5$ kN.m $M_C = -66.5 - 35.5 = -102$ kN.m $M_E = +28.6 - 2.8 = +25.8$ kN.m
	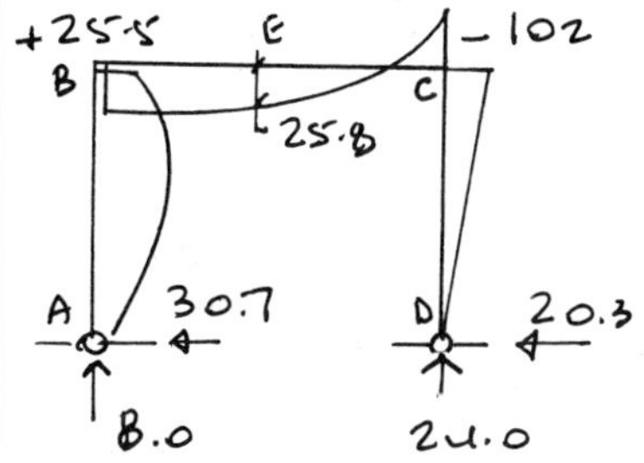Fig. 12/020 II 1.4Wk + 1.0qk (reversible)	**Member checks:** A: stanchions Design to L.C.1 (Fig 12/017) & check to member C−D (Fig. 12/020)
Ref. 12/2 Table 6 BS5950	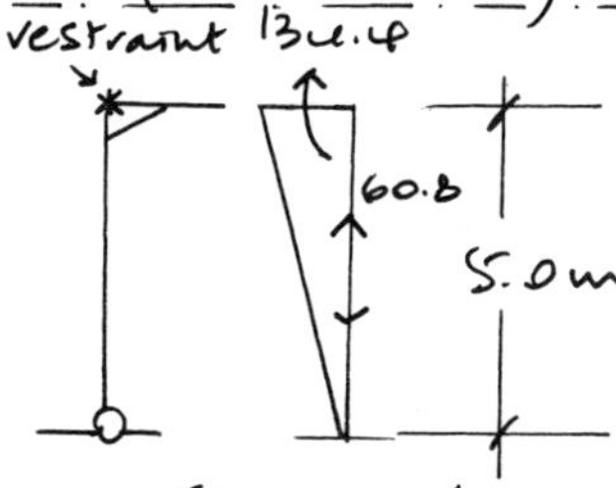Fig. 12/021	Try a 305 × 165 × 46 U.B. $r_{yy} = 3.9$ cm. ; $A = 58.9$ cm² ; $S_{xx} = 723$ cm³ ; $u = 0.890$; $x = 27.2$ $T = 11.8$ mm < 16 ; $p_y = 275$ N/mm². $l_{ey} = 1.0 \times 5000 = 5000$ mm.
Table 27(b) BS5950.	$\therefore \lambda = \frac{5000}{39} = 128.2$: $p_y = 97$ N/mm²	
	$\therefore P_{cy} = \frac{97 \times 5890}{10^3} = 571$ kN.	
Cl. 4.3.7.3	M_b : $\frac{\lambda}{x} = \frac{128.2}{27.2} = 4.71$; $v = 0.83$; $m = 0.57$	
	$\therefore \lambda_{LT} = 1.0 \times 0.890 \times 0.83 \times 128.2 = 94.7$ $p_b = 134$ N/mm²	
Cl. 4.8.3.3.1	$\therefore M_b = 134 \times 723/10^3 = 97$ kN.m.	
	$\therefore \frac{60.8}{571} + \frac{0.57 \times 134.4}{97} = 0.896 < 1.0$	
	$\therefore$ 305 × 165 × 46 UB stanchion o.k. for L.C.1	

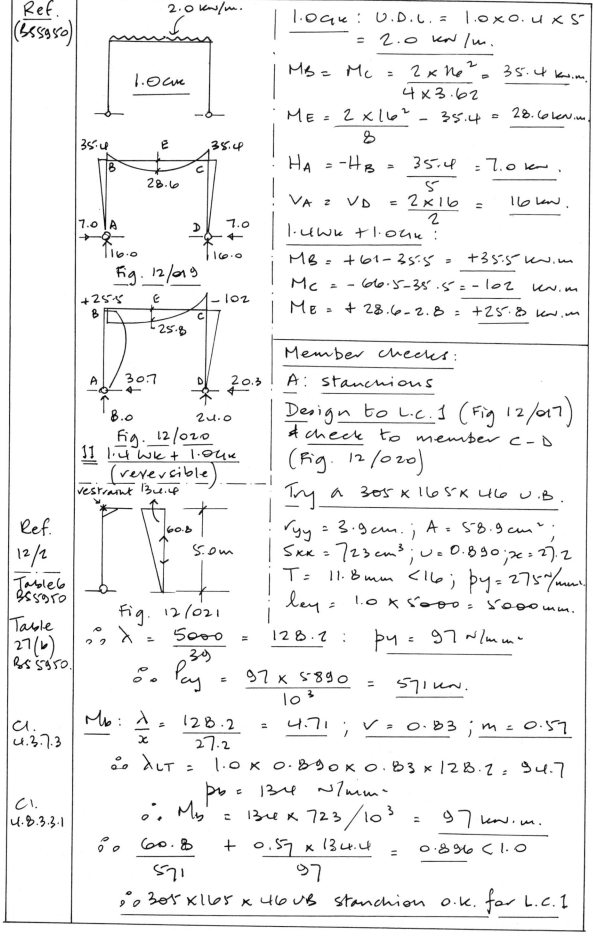

Ref. (BS 5950)

Check column for L.C. II

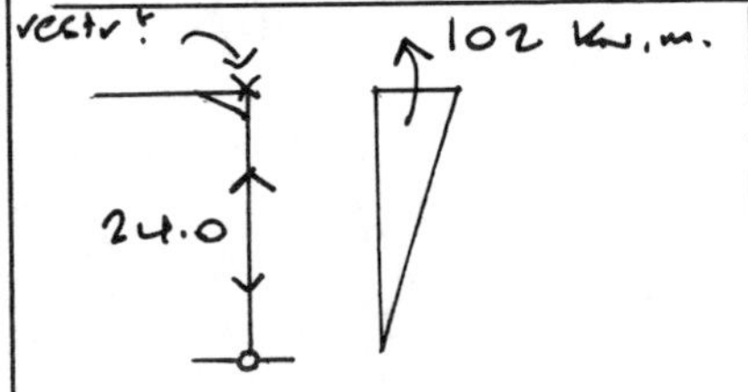

Fig. 12/022

By inspection, both moment & axial load are less than L.C. 1

∴ 305 × 165 × 46 U.B.'s O.K. for stanchions.

B: Rafters – assume purlins @ 1.6m. crs.

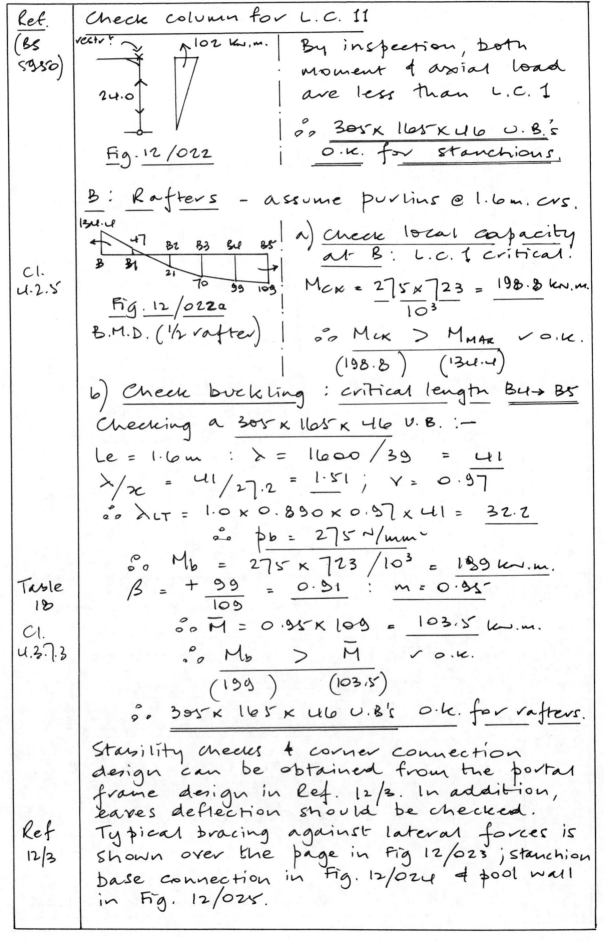

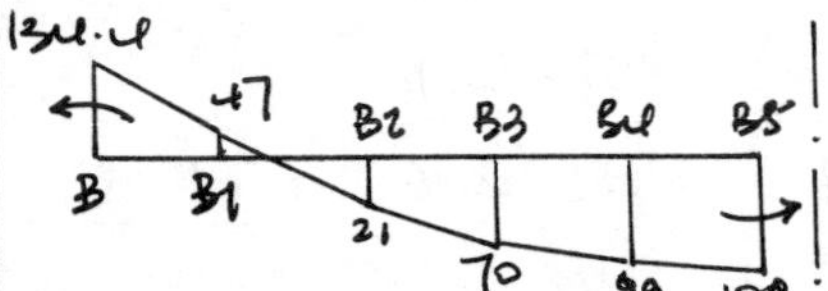

Fig. 12/022a
B.M.D. (½ rafter)

a) Check local capacity at B: L.C. 1 critical.

Cl. 4.2.5

$M_{cx} = \frac{275 \times 723}{10^3} = 198.8$ kN.m.

∴ $M_{cx} > M_{MAX}$ ✓ O.K.
(198.8) (134.4)

b) Check buckling: critical length B4 → B5

Checking a 305 × 165 × 46 U.B. :–

$L_e = 1.6m$: $\lambda = 1600/39 = 41$

$\lambda/x = 41/27.2 = 1.51$; $v = 0.97$

∴ $\lambda_{LT} = 1.0 \times 0.890 \times 0.97 \times 41 = 32.2$

∴ $p_b = 275$ N/mm²

∴ $M_b = 275 \times 723 / 10^3 = 199$ kN.m.

Table 18 — $\beta = +\frac{99}{109} = 0.91$: $m = 0.95$

Cl. 4.3.7.3 — ∴ $\bar{M} = 0.95 \times 109 = 103.5$ kN.m.

∴ $M_b > \bar{M}$ ✓ O.K.
(199) (103.5)

∴ 305 × 165 × 46 U.B.'s O.K. for rafters.

Stability checks & corner connection design can be obtained from the portal frame design in Ref. 12/3. In addition, eaves deflection should be checked. Typical bracing against lateral forces is shown over the page in Fig 12/023; stanchion base connection in Fig. 12/024 & pool wall in Fig. 12/025.

Ref 12/3

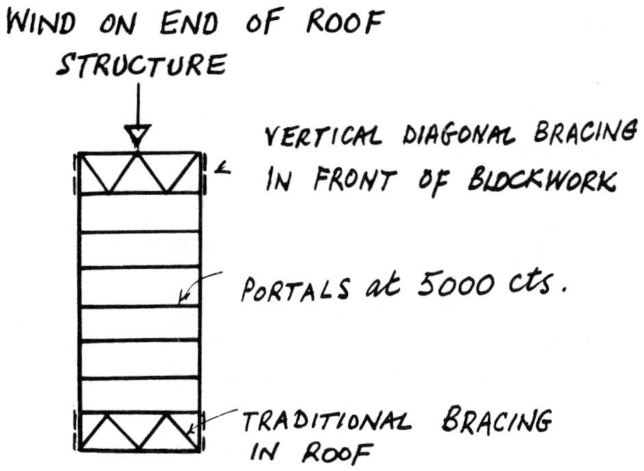
WIND ON END OF ROOF STRUCTURE
VERTICAL DIAGONAL BRACING IN FRONT OF BLOCKWORK
PORTALS at 5000 cts.
TRADITIONAL BRACING IN ROOF

Fig.12/023

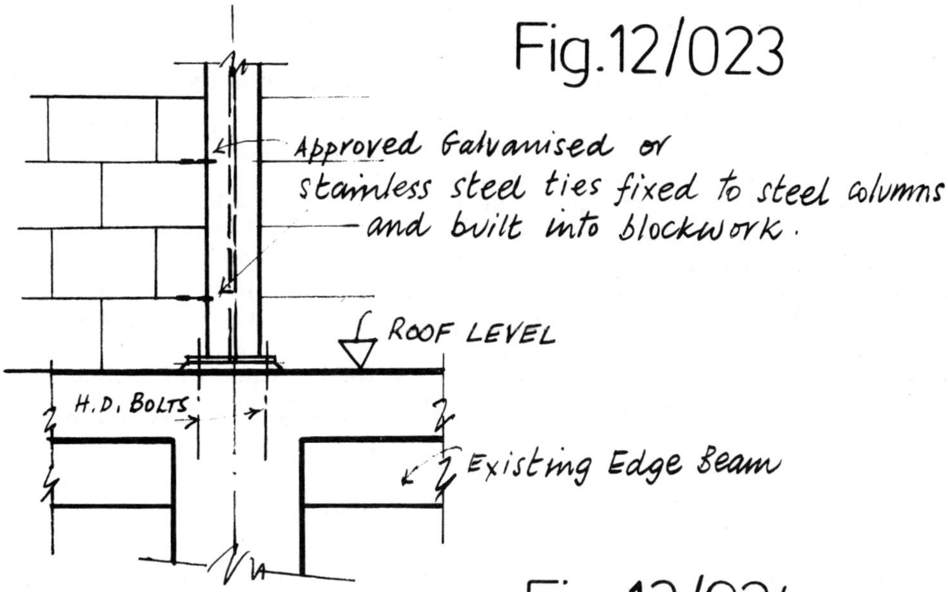
Approved Galvanised or stainless steel ties fixed to steel columns and built into blockwork.
ROOF LEVEL
H.D. BOLTS
Existing Edge Beam

Fig.12/024

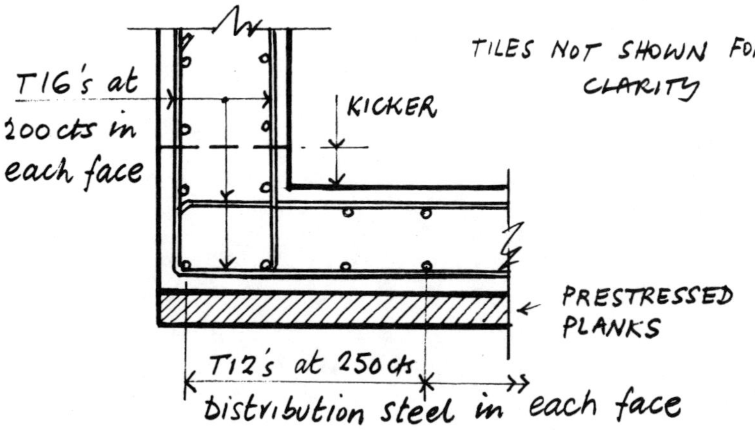
TILES NOT SHOWN FOR CLARITY
T16's at 200 cts in each face
KICKER
PRESTRESSED PLANKS
T12's at 250 cts distribution steel in each face

Fig.12/025

Chapter 13 Multi-storey car-park with integral car parking

Offices are required for a developer, who requires integral car-parking on two levels. The building is L-shaped, requiring the excavation of part of the site for a basement and sub-basement. The scheme is shown in plan and section in Figure 13/001. Excavation is required because the land level to the West of line J is +10.0m., whilst to the East of line A it is level with the sub-basement floor. Between A and J, the site slopes uniformly. In the scheme chosen, the car-parking spaces, 2.4 m. wide x 4.8 m. deep can be accommodated in pairs between columns at 5.0 m. centres (minimum). Across the building there is room for a central 1-lane 4.8 m. wide traffic aisle. Therefore, up to ground floor level, there are four rows of columns – 2 external and two internal. Above the ground floor level, the Architect requires only one line of internal columns up to roof level. At roof level, the false Mansard structures hide plant, which occupies the space between the Mansards. Access to the basement car-park level is by external 1-way ramps. The car-park levels are partly-clad, externally, by dwarf walls (apart from the retaining wall elevations), whilst above ground level, the external walls are clad in cavity walls. The cavity walls comprise facing brick outer skin and blockwork inner skin. Certain key elements of the structure have been designed. Namely, a two-way spanning basement floor slab with supporting beam, a one-way spanning office floor slab, a ground floor transfer beam together with supporting column with base, and a comment on the overall stability of the structure. The base design is limited to a check on safe bearing stresses.

Loading – qk

Office levels – 5.0 kN/m^2
Car-parking – 2.5 kN/m^2
Plant level – 7.5 kN/m^2
Roof level – 0.75 kN/m^2

Loading – gk

False ceiling & services in offices – 1.0 kN/m^2
Mansard roof – 3.3 kN/m^2

Site conditions

A thorough site investigation revealed the following information:-

(a) Ground to the East of line A – 2.5 m. of loose gravel (with water seepage) overlying stiff clay (c = 150 kN/m^2) to a substantial depth.

(b) Ground to the West of line J – the same stiff clay, to a substantial depth.

Design Code of Practice

The building is to be designed to BS 8110 (Structural Concrete).

Materials

Concrete – fcu = 40 N/mm^2
Reinforcement – fy = 460 N/mm^2

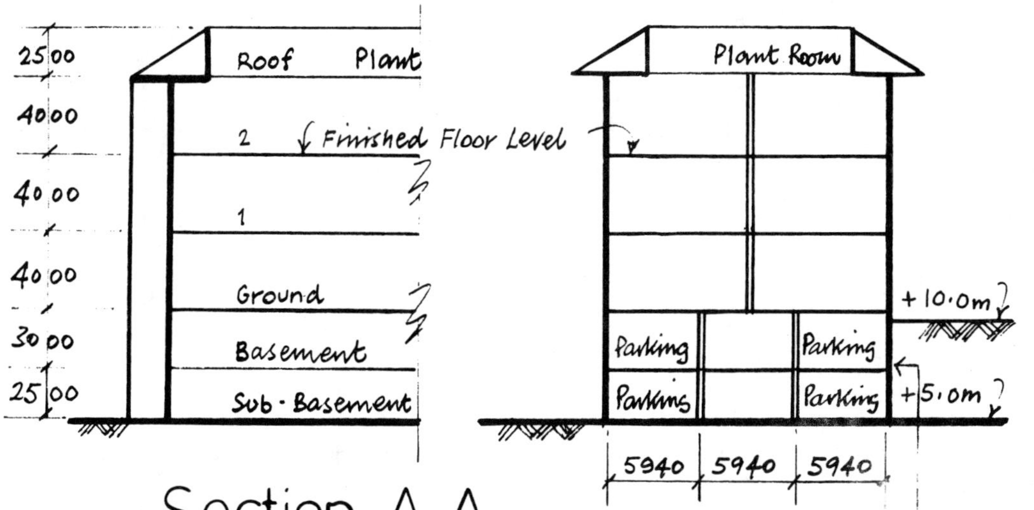

Section A.A.

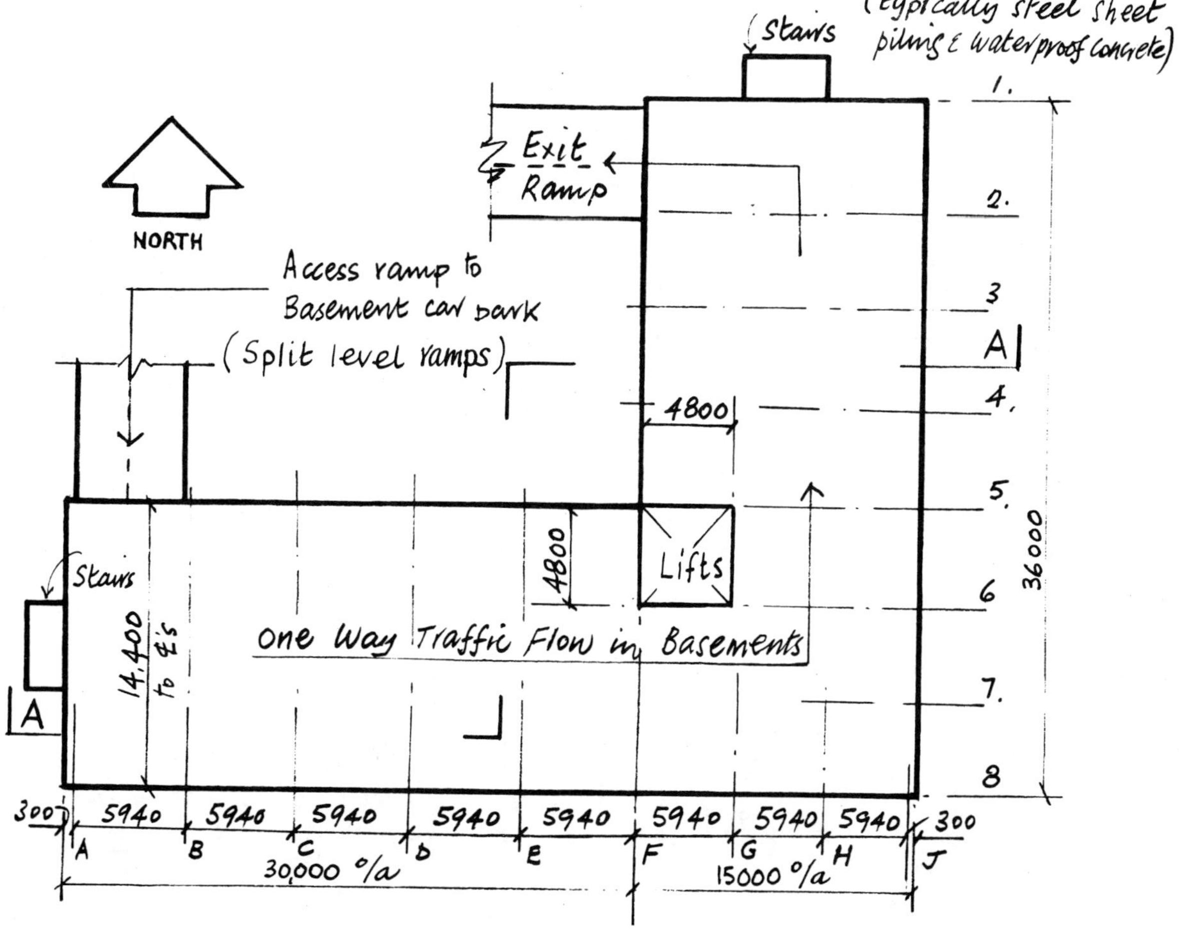

Floor Plan

Fig.13/001

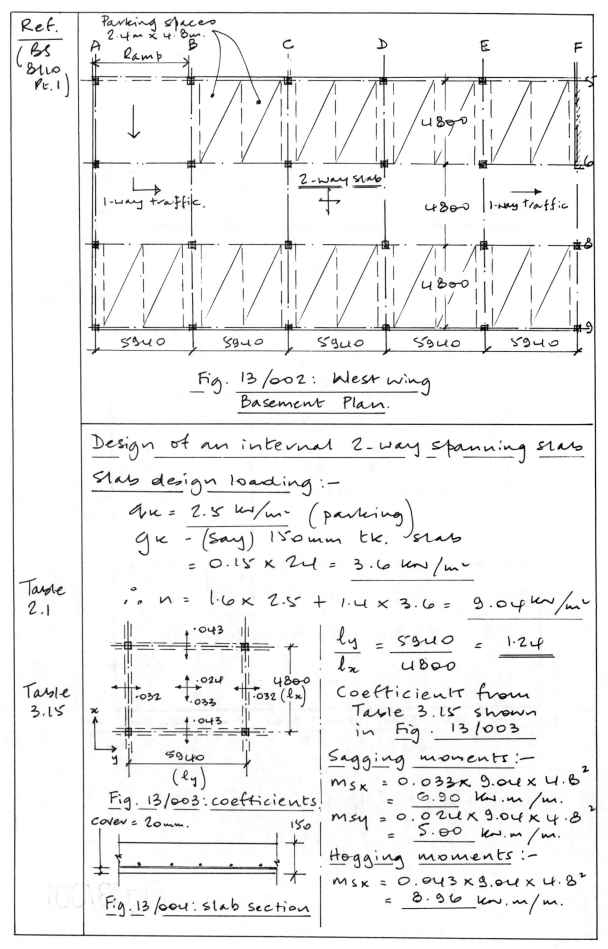

Ref.
(BS 8110 Pt.1)

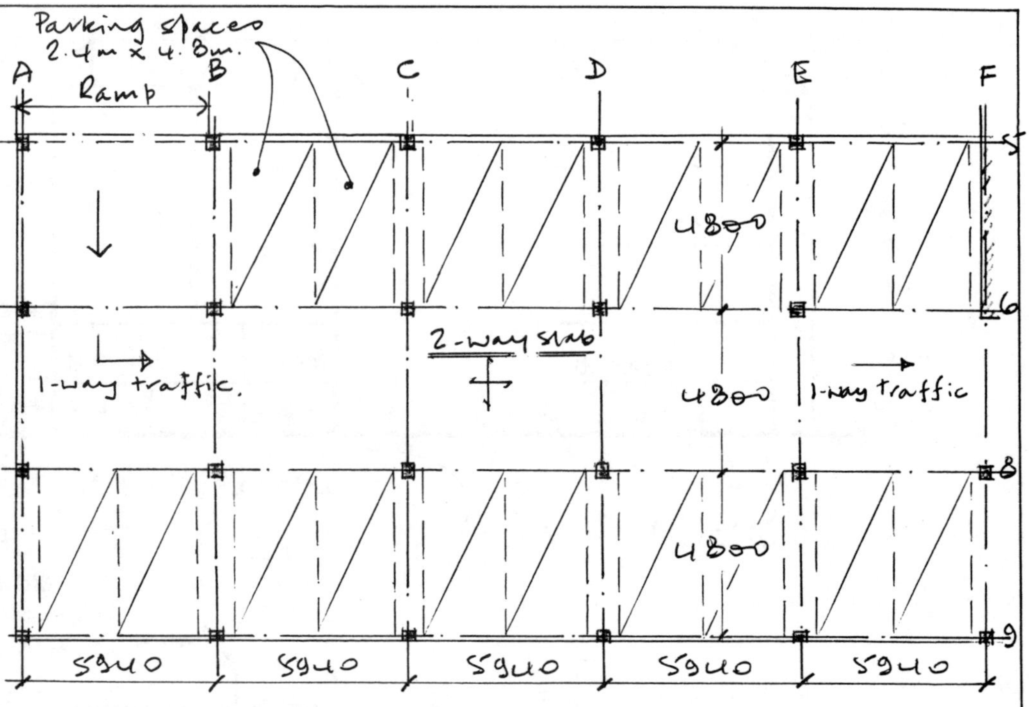

Fig. 13/002: West wing Basement Plan.

Design of an internal 2-way spanning slab

Slab design loading :-

$q_k = 2.5 \text{ kN/m}^2$ (parking)

g_k - (say) 150mm tk. slab

$= 0.15 \times 24 = 3.6 \text{ kN/m}^2$

Table 2.1

$\therefore n = 1.6 \times 2.5 + 1.4 \times 3.6 = 9.04 \text{ kN/m}^2$

Table 3.15

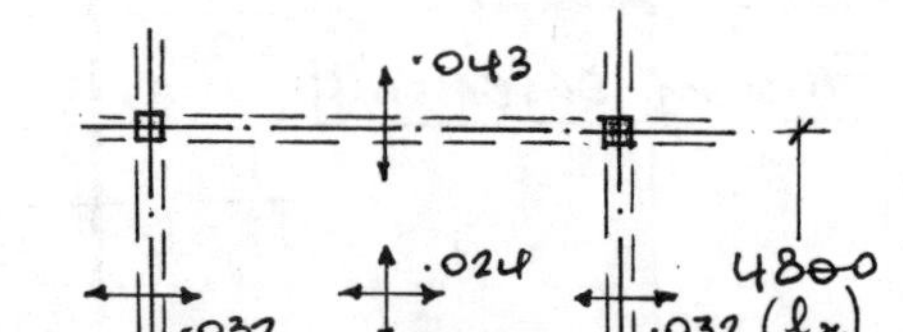

Fig. 13/003: coefficients

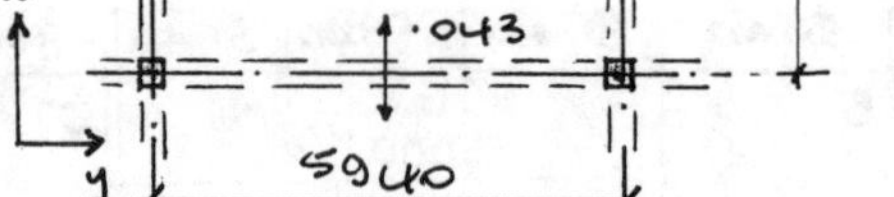

Fig. 13/004: slab section

$\dfrac{l_y}{l_x} = \dfrac{5940}{4800} = 1.24$

Coefficients from Table 3.15 shown in Fig. 13/003

Sagging moments :-

$m_{sx} = 0.033 \times 9.04 \times 4.8^2$
$= 6.90 \text{ kN.m/m.}$

$m_{sy} = 0.024 \times 9.04 \times 4.8^2$
$= 5.00 \text{ kN.m/m.}$

Hogging moments :-

$m_{sx} = 0.043 \times 9.04 \times 4.8^2$
$= 8.96 \text{ kN.m/m.}$

Ref. (BS8110 Pt.1)	
	Finally, $m_{sy} = 0.032 \times 9.04 \times 4.8^2 = 6.70$ kN.m/m
	As moments similar in both directions – design same steel in both directions for critical moment.
	Sagging : $m_{MAX} = 6.70$ kN.m/m.
	Effective depth (see Fig 13/004) – 2 layers of T10 bars : $d_{av} = 150 - 20 - 10 = 120$ mm
This book – App. A1	$\frac{M}{f_{cu}bd^2} = \frac{6.70 \times 10^6}{40 \times 10^3 \times 120^2} = 0.012$: $\frac{z}{d} = 0.95$
	∴ $z = 114$ mm.
	A_s reqd. $= \frac{6.70 \times 10^6}{0.87 \times 460 \times 114} = 147$ mm²/m.
Table 3.27	Min. steel $= \frac{0.13}{100} \times 10^3 \times 150 = 195$ mm²/m.
	∴ Min. steel critical : T10's @ 300 (262 mm²/m)
	Hogging : $m_{MAX} = 8.96$ kN.m/m.
	Effective depth (see Fig 13/004) – 1 layer of T10 bars : $d_{av} = 150 - 20 - 10 = 120$ mm.
This book – App A1	$\frac{M}{f_{cu}bd^2} = \frac{8.96 \times 10^6}{40 \times 10^3 \times 120^2} = 0.016$: $\frac{z}{d} = 0.95$
	∴ $z = 114$ mm.
	A_s reqd $= \frac{8.96 \times 10^6}{0.87 \times 460 \times 11} = 196$ mm²/m.
Table 3.27	∴ Min. steel critical : T10's @ 300 (262 mm²/m)
Cl.3.4.6 & Cl.3.5.7	Deflection check (based on shorter span)
	Allowable span/d = 26 × m.f.
	A_s reqd : shorter span
	$\frac{M}{f_{cu}bd^2} = \frac{6.70 \times 10^6}{40 \times 10^3 \times 120^2} = 0.012$: $\frac{z}{d} = 0.95$
	∴ $z = 114$ mm.
	A_s reqd $= \frac{6.70 \times 10^6}{0.87 \times 460 \times 114} = 147$ mm²/m
Table 3.11	∴ $f_s = \frac{5}{8} \times 460 \times \frac{147}{262} = 161$: m.f. = 2.0
	Allowable = 2 × 26 = 52 } o.k.
	Actual = 4800/120 = 40 }

Ref. (BS8110 Pt.1)

Check shear in slab:

Table 3.16

$V_{sx} = \beta_{vx} . n . l_x = 0.40 \times 9.06 \times 4.8 = \underline{17.40 kN/m}$

$V_{sy} = \beta_{vy} . n . l_x = 0.33 \times 9.04 \times 4.8 = \underline{14.32 kN/m}$

$\therefore V_{sx}$ is critical: $v = \dfrac{17.40 \times 10^3}{10^3 \times 125} = \underline{0.14 N/mm^2}$

Table 3.9

$\therefore \dfrac{100 A_s}{bd} = \dfrac{100 \times 262}{10^3 \times 120} = \underline{0.22}$: $\underline{v_c = 0.52 N/mm^2}$

$\therefore$ No shear reinf. required

Design of supporting beams in N-S direction (see Fig. 13/001) - grids B, C, D & E (Fig. 13/002)

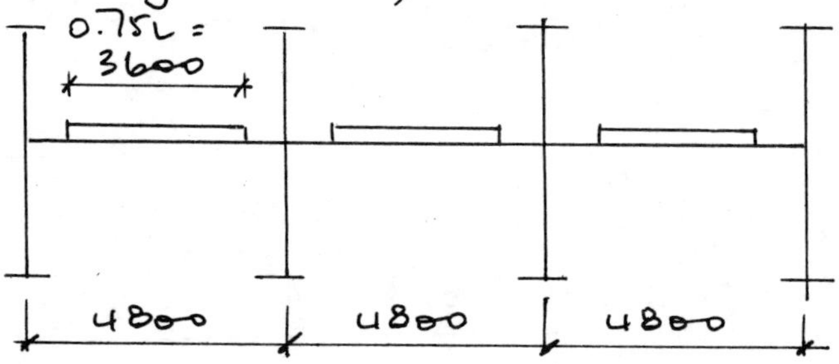

Beam idealised into 3 span continuous beam

Fig. 13/005

From above, shear from slabs spanning in y-direction = $\underline{14.32}$ kN/m.

2 slabs spanning onto main frame (see Fig 13/001) : total shear = 2 × 14.32 = $\underline{28.64}$ kN/m.

Total shear on central 0.75L of beam (see Fig. 13/005) = 28.64 × 3.6 = $\underline{103.1}$ kN.

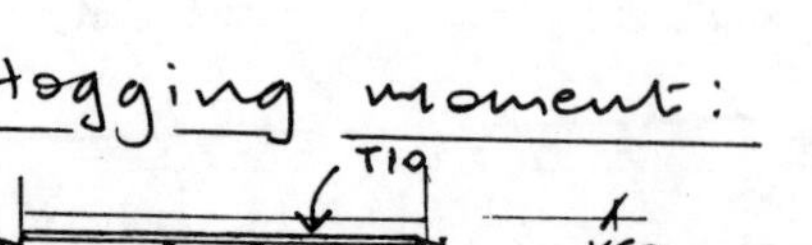

Fig. 13/006 – F.E.M coeffts – beams loaded middle 3/4

From Fig. 13/006; Hogging moment
= 0.10156 × 103.1 × 4.8
= $\underline{50.26}$ kN.m.

Sagging moment
= 0.05444 × 103.1 × 4.8
= $\underline{28.94}$ kN.m.

Hogging moment: try a 400 deep × 300 beam

Fig. 13/007

For hogging steel, assuming T16 bars :-

d = 400 – 20 – 10 – 8
= $\underline{362 mm}$

(see Fig. 13/007)

Ref.	
(BS8110 Pt.1) This book – App. A1	$\therefore \dfrac{M}{f_{cu}bd^2} = \dfrac{50.26\times10^6}{40\times300\times362^2} = \underline{0.032}$: $\dfrac{z}{d} = \underline{0.95}$
	$\therefore z = \underline{344\ mm.}$
	$\therefore As\ req^d. = \dfrac{50.26\times10^6}{0.87\times460\times344} = \underline{365\ mm^2}$
	Use 2 no. T16's (402 mm^2)
Table 3.27	Min. steel $= \dfrac{0.13}{100}\times300\times400 = \underline{156\ mm^2}$ ✓ ok.
	Sagging moment:
	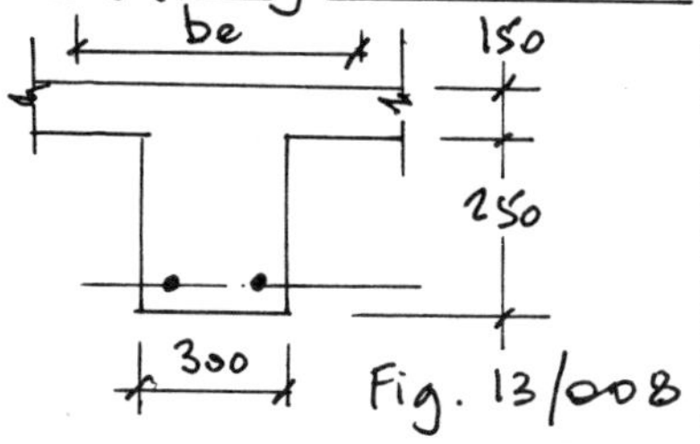
	Fig. 13/008
	For sagging steel, assuming T16 bars & T10 links
	$d = 400 - 20 - 10 - 8 = \underline{362mm.}$
	(see Fig. 13/008)
	$be = \left(\dfrac{0.7\times4800}{5}\right) + 300 = \underline{972mm}$
This book – App. A1	$\therefore \dfrac{M}{f_{cu}bd^2} = \dfrac{26.94\times10^6}{40\times972\times362^2} = \underline{0.006}$: $\dfrac{z}{d} = 0.95$
	$\therefore z = \underline{344\ mm.}$
	$\therefore As\ req^d. = \dfrac{26.94\times10^6}{0.87\times460\times344} = \underline{195\ mm^2}$
	Use 2No. T16's (402 mm^2)
	Check shear at support:
Cl. 3.4.5.2	$V = \dfrac{103.1}{2} = \underline{51.6\ kN.}$
	$\therefore v = \dfrac{51.6\times10^3}{300\times362} = \underline{0.48}\ N/mm^2$
Table 3.9	$\dfrac{100As}{bd} = \dfrac{100\times402}{300\times362} = \underline{0.37}$: $v_c = \underline{0.48}\ N/mm^2$
	$\therefore \underset{(0.24)}{0.5v_c} < \underset{(0.48)}{v} < \underset{(0.88)}{(v_c+0.4)}$
	$\therefore$ Minimum links req^d.
	For T10 links: $Asv = 157\ mm^2$ (2 legs)
	$\therefore Sv = \dfrac{157\times0.87\times460}{0.4\times300} = \underline{523\ mm.}$
	Minimum spacing $= 0.75d = \underline{271.5mm}$
	$\therefore$ Use T10's @ 250 crs.

Ref.

(BS 8110 Pt. 1)

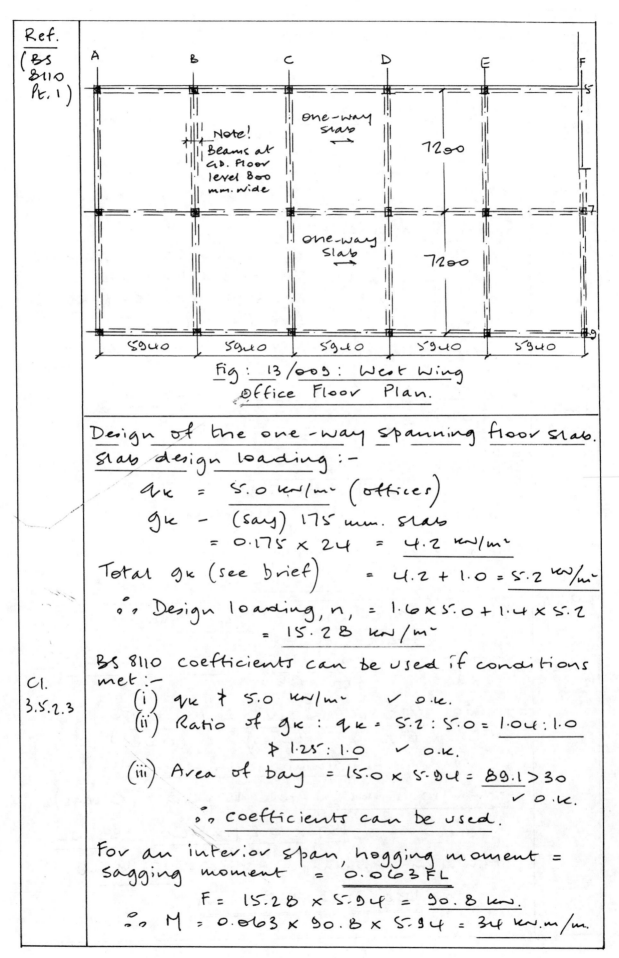

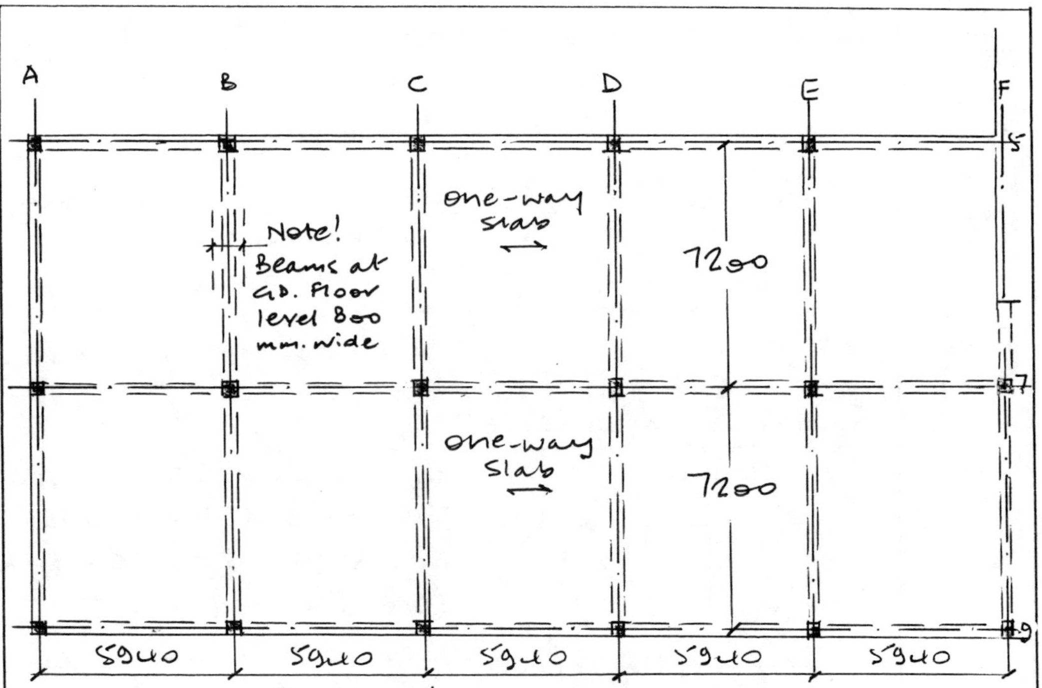

Fig: 13/009: West Wing Office Floor Plan.

Design of the one-way spanning floor slab.

Slab design loading:-

q_k = 5.0 kN/m² (offices)

g_k - (say) 175 mm. slab

= 0.175 × 24 = 4.2 kN/m²

Total g_k (see brief) = 4.2 + 1.0 = 5.2 kN/m²

∴ Design loading, n, = 1.6 × 5.0 + 1.4 × 5.2

= 15.28 kN/m²

BS 8110 coefficients can be used if conditions met:-

Cl. 3.5.2.3

(i) q_k ≯ 5.0 kN/m² ✓ o.k.

(ii) Ratio of $g_k : q_k$ = 5.2 : 5.0 = 1.04 : 1.0

≯ 1.25 : 1.0 ✓ o.k.

(iii) Area of bay = 15.0 × 5.94 = 89.1 > 30

✓ o.k.

∴ coefficients can be used.

For an interior span, hogging moment = sagging moment = 0.063 FL

F = 15.28 × 5.94 = 90.8 kN.

∴ M = 0.063 × 90.8 × 5.94 = 34 kN.m/m.

Ref.	
(BS8110 Pt. 1)	Effective depth : allow T12 bars for top & bottom steel : $d = 175 - 20 - 6 = 149$ mm.
	$\frac{M}{f_{cu}bd^2} = \frac{34 \times 10^6}{40 \times 10^3 \times 149^2} = 0.04$
This book App. A1	$z/d = 0.95$: $z = 141.5$ mm.
	A_s req^d $= \frac{34 \times 10^6}{0.87 \times 460 \times 141.5} = 600$ mm²/m
	Use : T12's @ 125 crs (904.3 mm²/m)
Table 3.27	Minimum steel $= \frac{0.13}{100} \times 175 \times 10^3 = 228$ mm²/m ✓ o.k.
	Check shear in slab over support :
Cl. 3.4.5.2	Shear/m. $= 0.5F = 0.5 \times 90.8 = 45.4$ kN.
	$v = \frac{45.4 \times 10^3}{10^3 \times 141.5} = 0.32$ N/mm²
Table 3.9	$\frac{100A_s}{bd} = \frac{100 \times 904.3}{10^3 \times 149} = 0.61$: $v_c = 0.79$ N/mm²
	∴ $v < 0.5v_c$: no reinf. required
	(0.32) (0.40)
	Check deflection :
	Actual span/d ratio $= 5940/149 = 40$
	Allowable span/d $=$ m.f. $\times 26$.
	Modification factor :–
	$\frac{M}{bd^2} = \frac{34 \times 10^6}{10^3 \times 149^2} = 1.53$
Table 3.11	$f_s = \frac{5}{8} \times 460 \times \frac{600}{904.3} = 190.7$ N/mm²
	m.f. $= 0.55 + \frac{(477 - 190.7)}{120(0.9 + 1.53)} = 1.532 \not> 2.0$
	∴ m.f. $= 1.532$.
	∴ Allowable $= 1.532 \times 26 = 40 = 40$ ✓ o.k.
	Design of supporting beams in N-S direction
	Floors 1 & 2 (Section A-A, Fig 13/001), grids B, C, D & E (Figs 13/001 & 13/009)
	Beams idealised into 2 span continuous beams.

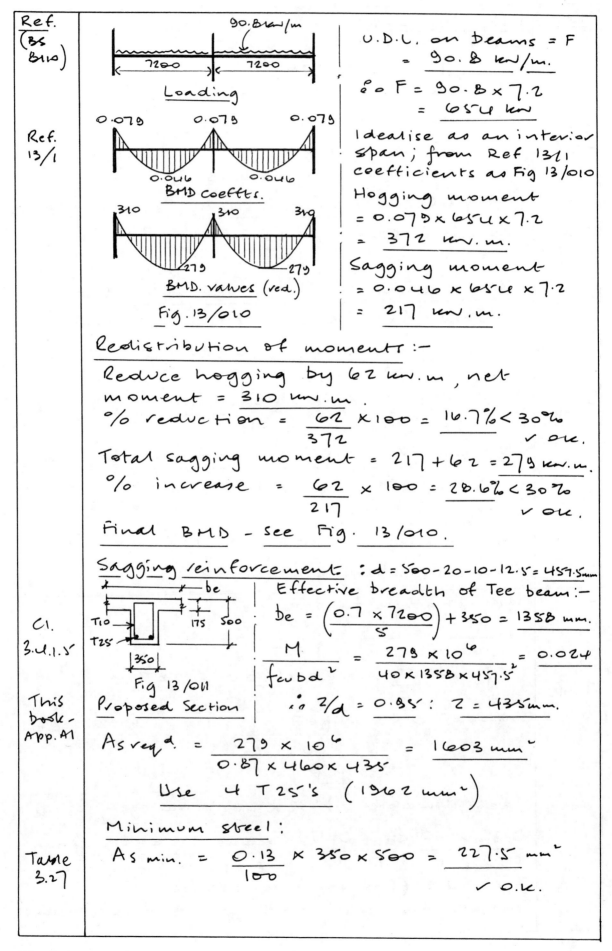

Ref.	
(BS 8110)	

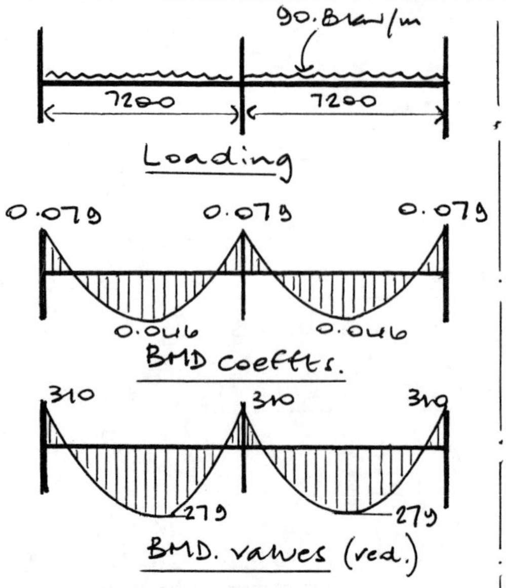

Fig. 13/010

U.D.L. on beams = F = 90.8 kN/m.

∴ F = 90.8 × 7.2 = 654 kN

Ref. 13/1 — Idealise as an interior span; from Ref 13/1 coefficients as Fig 13/010

Hogging moment = 0.079 × 654 × 7.2 = 372 kN.m.

Sagging moment = 0.046 × 654 × 7.2 = 217 kN.m.

Redistribution of moments:–

Reduce hogging by 62 kN.m, net moment = 310 kN.m.

% reduction = $\frac{62}{372} \times 100 = 16.7\% < 30\%$ ✓ ok.

Total sagging moment = 217 + 62 = 279 kN.m.

% increase = $\frac{62}{217} \times 100 = 28.6\% < 30\%$ ✓ ok.

Final BMD – see Fig. 13/010.

Sagging reinforcement : d = 500 − 20 − 10 − 12.5 = 457.5 mm

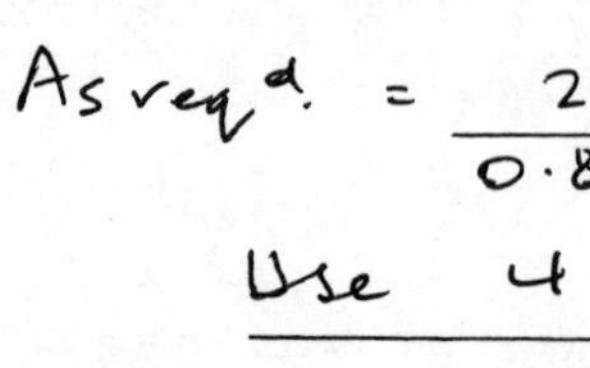

Fig 13/011
Proposed Section

Cl. 3.4.1.5 — Effective breadth of Tee beam:–

$b_e = \left(\frac{0.7 \times 7200}{5}\right) + 350 = 1358$ mm.

$\frac{M}{f_{cu} b d^2} = \frac{279 \times 10^6}{40 \times 1358 \times 457.5^2} = 0.024$

This book – App. A1 — ∴ $z/d = 0.95$: $z = 435$ mm.

$A_{s\,req^d} = \frac{279 \times 10^6}{0.87 \times 460 \times 435} = 1603$ mm²

Use 4 T25's (1962 mm²)

Minimum steel:

Table 3.27 — $A_{s\,min.} = \frac{0.13}{100} \times 350 \times 500 = 227.5$ mm² ✓ O.K.

Ref.	
(BS8110 Pt.1)	**Hogging reinforcement** : $d = 500 - 20 - 12 - 25 - 5 = 438$ mm.
	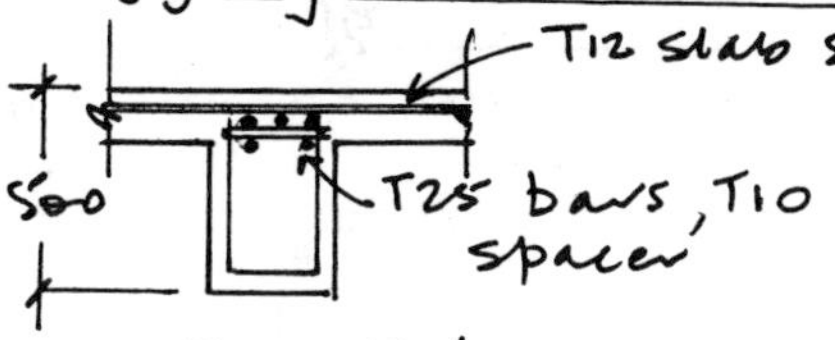 Fig. 13/012
	$\frac{M}{f_{cu} b d^2} = \frac{310 \times 10^6}{40 \times 350 \times 438^2} = 0.114$
This book - App. A1	$\therefore z/d = 0.85$: $z = 372.3$ mm.
	$As_{req^d} = \frac{310 \times 10^6}{0.87 \times 460 \times 372.3} = 2081$ mm.
	Use 5 T25's ($2453\ mm^2$)
	(min steel = $227.5\ mm^2$ ✓ ok).
	Check shear: F = 654 kN.
	V = F/2 = 327 kN.
Cl. 3.4.5.2	$\therefore v = \frac{327 \times 10^3}{350 \times 438} = 2.13\ N/mm^2$
Table 3.9	$\frac{100 As}{bd} = \frac{100 \times 2453}{350 \times 438} = 1.60$: $v_c = 0.86\ N/mm^2$
Table 3.8	$\therefore (v_c + 0.4) < v < 0.8\sqrt{f_{cu}}$
	(1.26) (2.13) (5.0)
	$\therefore$ Design links required : T10 - $A_{sv} = 157\ mm^2$
	$\therefore S_v = \frac{157 \times 0.87 \times 460}{350\ (2.13 - 0.86)} = 141$ mm.
	Use T10 links @ 125 crs.
	Design of ground floor collector beam (grids B, C, D & E) - see Fig 13/013
	Calculation of axial load on central column ('N' in Fig. 13/013)
	Office floors 1 & 2
	Load / floor = 2 × V (from office floor calcs)
	= 2 × 327 = 654 kN.
	Roof $q_k = 0.75\ kN/m^2$: $g_k = 1.5\ kN/m^2$ (brief)
	$\therefore$ Design loading = $1.6 \times 0.75 + 1.4 \times 1.5$
	= $3.3\ kN/m^2$

Ref.

(BS 8110 Pt. 1)

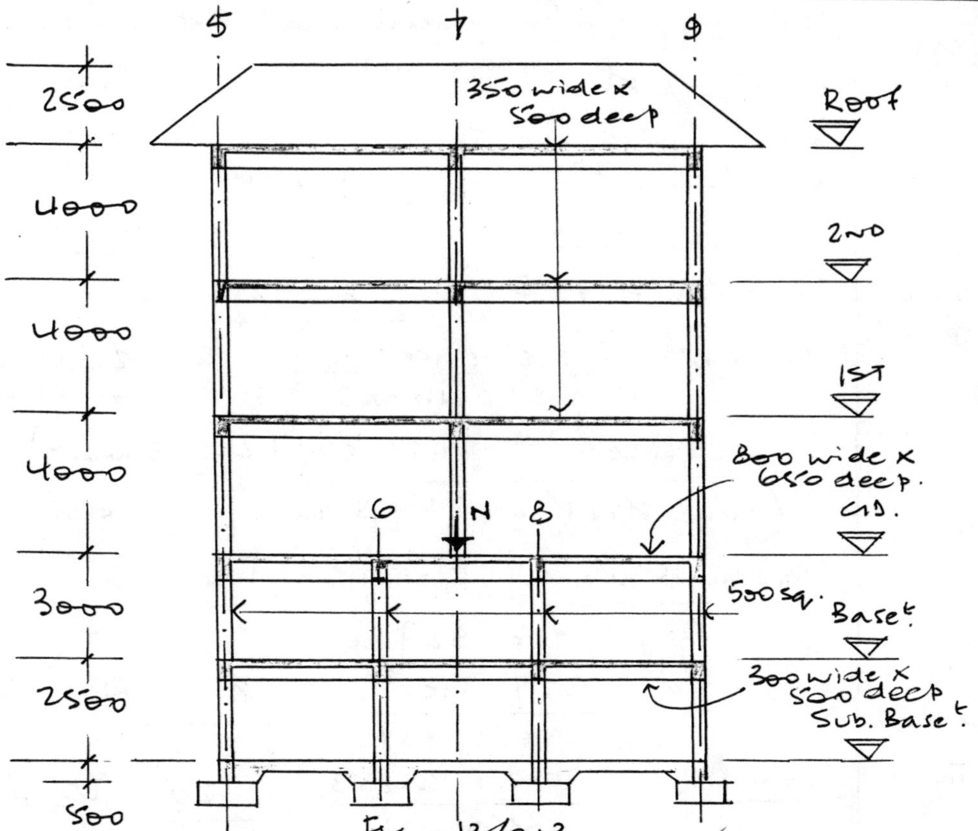

Fig. 13/013

Plant room. Assume a 200 mm tk. slab

$g_k = 0.2 \times 24 = 4.8\ \text{kN/m}^2$ + ceiling

Total = $5.8\ \text{kN/m}^2$

∴ Design loading = $1.6 \times 7.5 + 1.4 \times 5.8 = 20.12\ \text{kN/m}^2$

Area carried by central column (Fig 13/009)

$= 7.2 \times 5.94 = 42.8\ \text{m}^2$

From brief, roof = $1.6 \times 0.75 + 1.4 \times 3.3 = 5.8\ \text{kN/m}^2$

∴ Estimate of load on central column from roof & plant room

$= (5.8 + 20.12) \times 42.8 = 1109\ \text{kN}$.

∴ at ground floor level:

$N = 2 \times 654 + 1109 = 2417\ \text{kN}$.

From office floor calcs., U.D.L. on ground floor beam from slab = $90.8\ \text{kN/m}$.

Assuming a beam 650 mm deep x 800 mm. wide, $I_{xx} = 80 \times 65^3/12 = 1830833\ \text{cm}^4$

$A = 80 \times 65 = 5200\ \text{cm}^2$

Ref.	
(BS8110 Pt.1)	Results of plane frame program – see Fig. 13/014

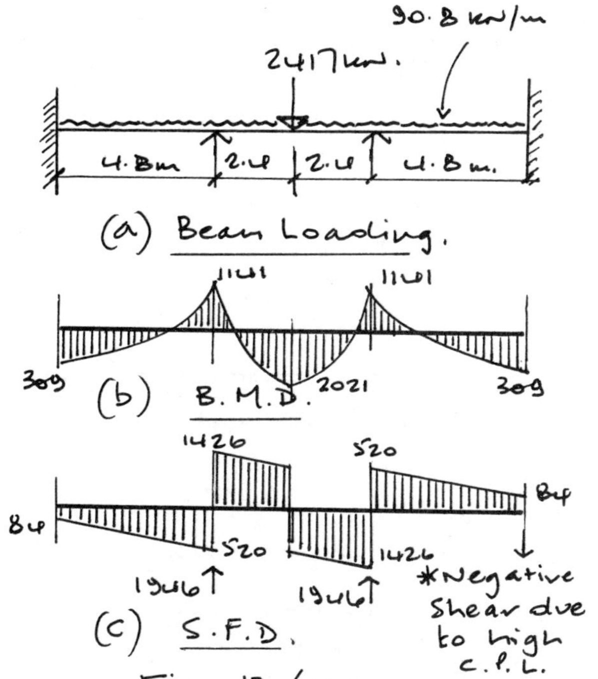

(a) Beam Loading.

(b) B.M.D.

(c) S.F.D.

Fig. 13/014

Beam effective depths :–

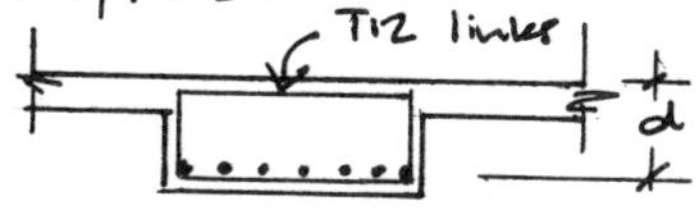

$d = 650 - 20 - 12 - 16$
$= 602$ mm

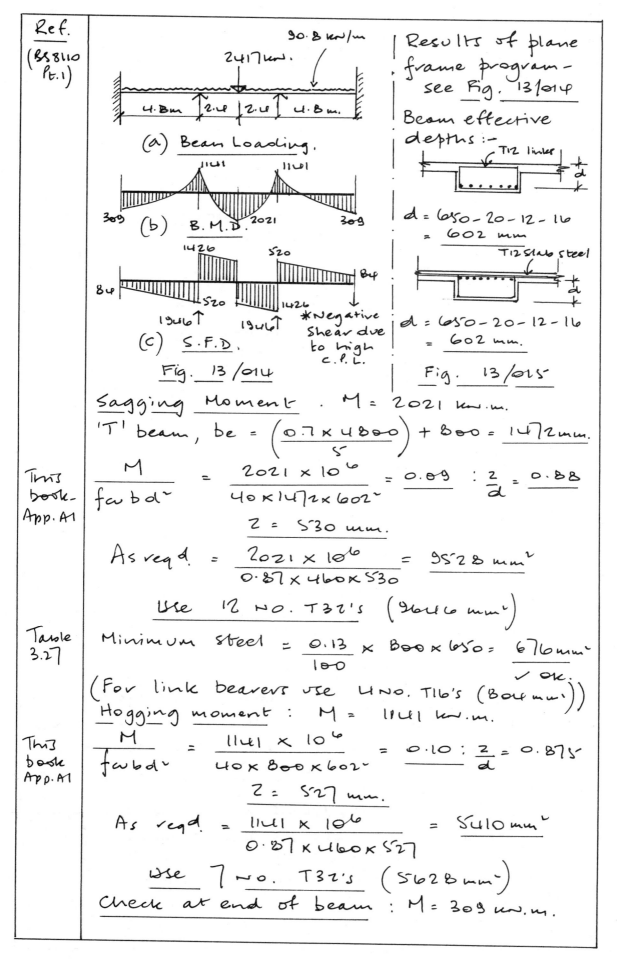

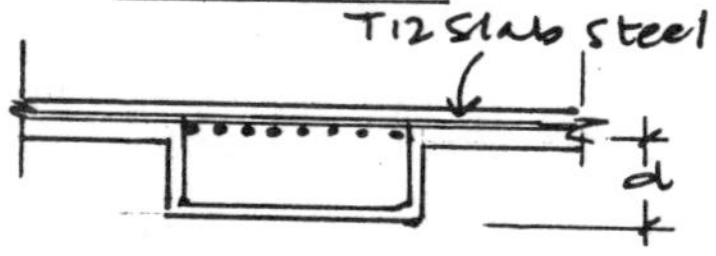

$d = 650 - 20 - 12 - 16$
$= 602$ mm.

Fig. 13/015

Sagging Moment . $M = 2021$ kN.m.

'T' beam, $b_e = \left(\frac{0.7 \times 4800}{5}\right) + 800 = 1472$ mm.

This book – App. A1: $\frac{M}{f_{cu}bd^2} = \frac{2021 \times 10^6}{40 \times 1472 \times 602^2} = 0.09$: $\frac{z}{d} = 0.88$

$z = 530$ mm.

A_s reqd. $= \frac{2021 \times 10^6}{0.87 \times 460 \times 530} = 9528$ mm^2

Use 12 No. T32's (9646 mm^2)

Table 3.27: Minimum steel $= \frac{0.13}{100} \times 800 \times 650 = 676$ mm^2 ✓ ok.

(For link bearers use 4 No. T16's (804 mm^2))

Hogging moment : $M = 1141$ kN.m.

This book App. A1: $\frac{M}{f_{cu}bd^2} = \frac{1141 \times 10^6}{40 \times 800 \times 602^2} = 0.10$: $\frac{z}{d} = 0.875$

$z = 527$ mm.

A_s reqd. $= \frac{1141 \times 10^6}{0.87 \times 460 \times 527} = 5410$ mm^2

Use 7 No. T32's (5628 mm^2)

Check at end of beam : $M = 309$ kN.m.

Ref.	
(BS 8110 Pt 1) This ~~book~~- App. A1	$\frac{M}{f_{cu}bd^2} = \frac{309 \times 10^6}{40 \times 800 \times 602^2} = 0.027 : \frac{z}{d} = 0.95$ $z = 572$ mm. A_s req^d $= \frac{308 \times 10^6}{0.87 \times 460 \times 572} = 1350$ mm^2 Use 3 No. T25's (1472 mm^2)

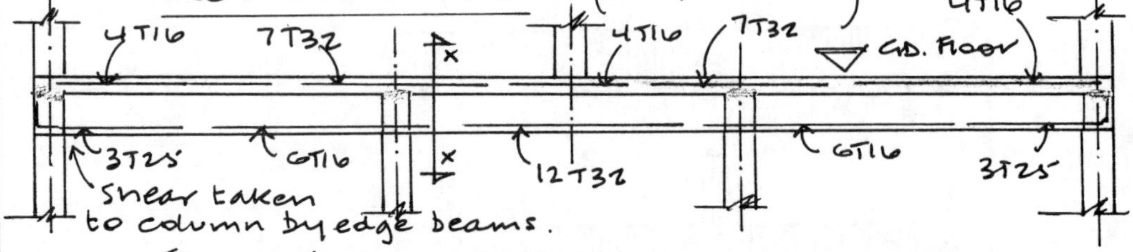

Fig. 13/016 - Designed main steel.

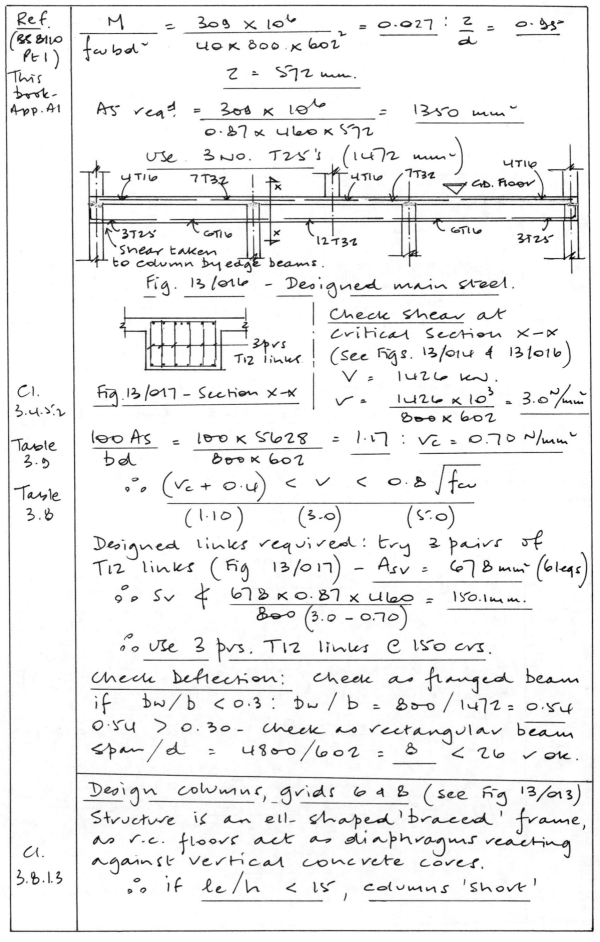

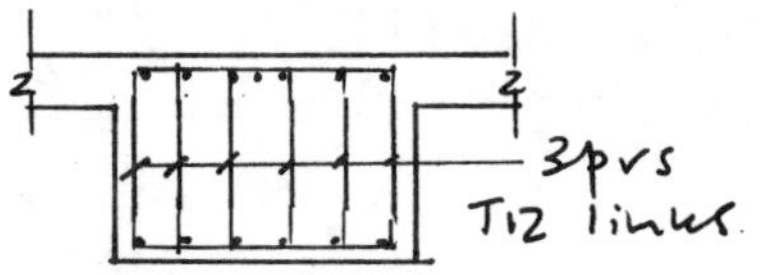

Fig. 13/017 - Section x-x

	Check shear at critical section x-x (see Figs. 13/014 & 13/016) $V = 1426$ kN.
Cl. 3.4.5.2	$v = \frac{1426 \times 10^3}{800 \times 602} = 3.0$ N/mm^2
Table 3.9	$\frac{100 A_s}{bd} = \frac{100 \times 5628}{800 \times 602} = 1.17 : v_c = 0.70$ N/mm^2
Table 3.8	$\therefore \underline{(v_c + 0.4) < v < 0.8\sqrt{f_{cu}}}$ $(1.10) \quad (3.0) \quad (5.0)$ Designed links required: try 3 pairs of T12 links (Fig 13/017) - $A_{sv} = 678$ mm^2 (6 legs) $\therefore s_v \ngtr \frac{678 \times 0.87 \times 460}{800(3.0 - 0.70)} = 150.1$ mm. $\therefore$ Use 3 prs. T12 links @ 150 crs. Check Deflection: check as flanged beam if $b_w/b < 0.3$: $b_w/b = 800/1472 = 0.54$ $0.54 > 0.30$ - check as rectangular beam span/d $= 4800/602 = 8 < 26$ ✓ ok.
	Design columns, grids 6 & 8 (see Fig 13/013) Structure is an ell-shaped 'braced' frame, as r.c. floors act as diaphragms reacting against vertical concrete cores.
Cl. 3.8.1.3	$\therefore$ if $l_e/h < 15$, columns 'short'

Ref. (BS 8110 Pt.1)

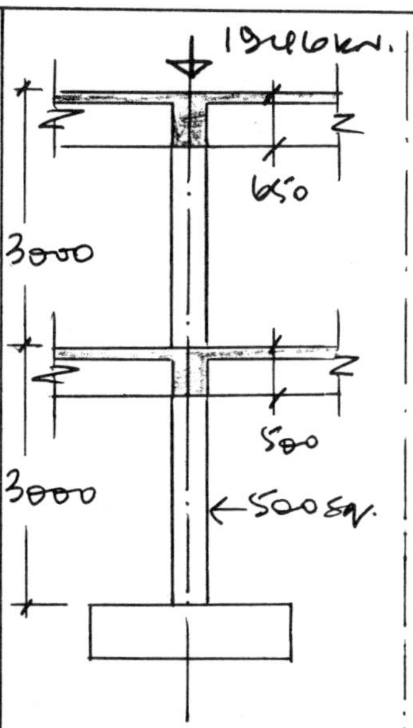

Fig. 13/018

Top length: from Fig 13/018,

$l_o = 3000 - 650 = \underline{2350 \text{ mm}}$.

Both ends - condition ①

$\therefore \beta = 0.75$

$l_e = 0.75 \times 2350 = 1762 \text{ mm}$

$l_e/h = 1762/400 = \underline{4.4} < 15$ ✓ o.k.

'Short column'

Cl. 3.8.2.4

Minimum moment:

$e_{min} = 0.05 \times 400 = \underline{20 \text{ mm}}$

$M_{min} = \dfrac{1946 \times 20}{10^3} = \underline{38.9 \text{ kN.m}}$

Chart 39

$\therefore \dfrac{M}{bh^2} = \dfrac{38.9 \times 10^6}{400^3} = \underline{0.61}$

$\dfrac{N}{bh} = \dfrac{1946 \times 10^3}{400^2} = \underline{12.20}$

$\left.\right\} \dfrac{100A_{sc}}{bh} = \underline{0.4}$ (minimum)

$A_{sc} = \dfrac{0.4 \times 400^2}{100} = \underline{640 \text{ mm}^2}$

Use 4 T16's (804 mm^2)

Bottom length: from Fig 13/018

$l_o = 3000 - 500 = \underline{2500 \text{ mm}}$.

Moment base: both ends condition ①

$\therefore \beta = 0.75$

$l_e = 0.75 \times 2500 = \underline{1875} \text{ mm}$

$l_e/h = 1875/400 = \underline{4.7} < 15$ ✓ o.k.

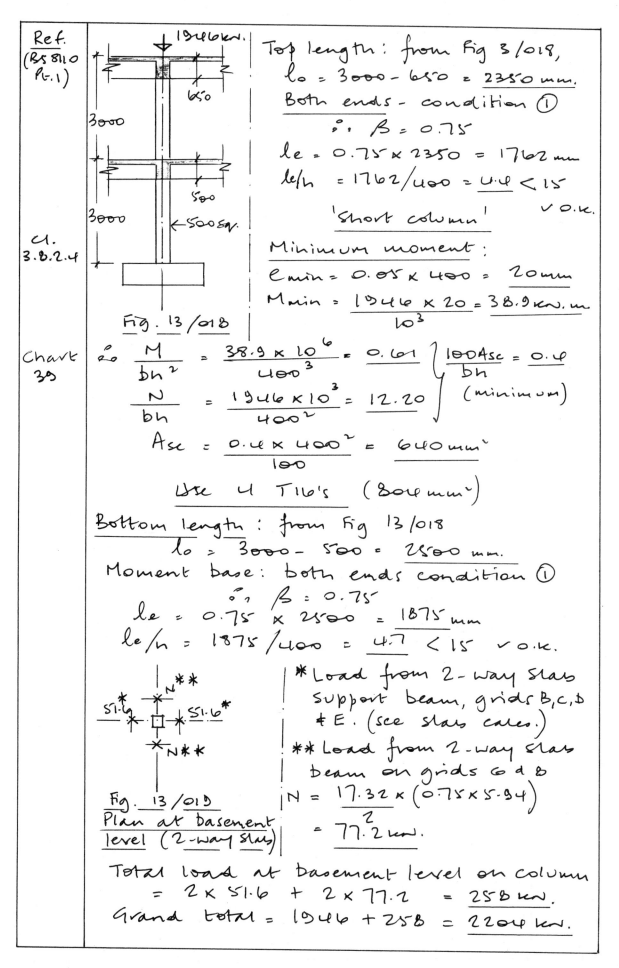

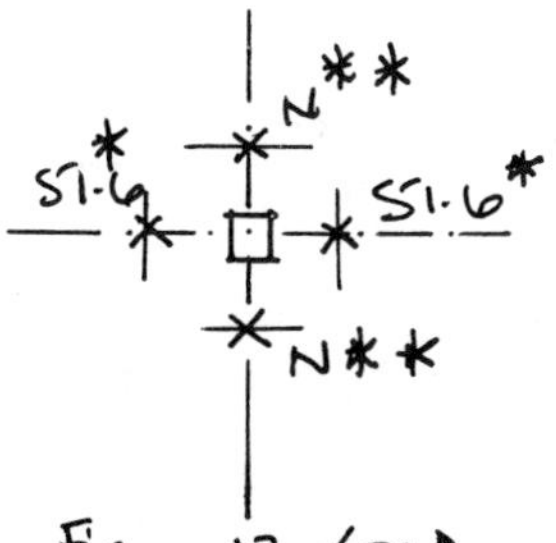

Fig. 13/019
Plan at basement level (2-way slab)

* Load from 2-way slab support beam, grids B, C, D & E. (see slab calcs.)

** Load from 2-way slab beam on grids 6 & 8

$N = \dfrac{17.32 \times (0.75 \times 5.94)}{2}$

$= \underline{77.2 \text{ kN}}$.

Total load at basement level on column

$= 2 \times 51.6 + 2 \times 77.2 = \underline{258 \text{ kN}}$.

Grand total $= 1946 + 258 = \underline{2204 \text{ kN}}$.

Ref. (BS8110 Pt.1)

$M_{min} = \dfrac{2204 \times 20}{10^3} = 44.1$ kN.m.

$\dfrac{M}{bh^2} = \dfrac{44.1 \times 10^6}{400^3} = 0.69$

Chart 39

$\dfrac{N}{bh} = \dfrac{2204 \times 10^3}{400^2} = 13.8$

$\dfrac{100 A_s}{bh} = 0.4$

$\therefore A_{sc} = 640$ mm² (as upper length)

Use 4 T16's (804 mm²)

Pad foundation base size

From brief, bearing soil is stiff clay, c = 150 kN/m²

Safe bearing ≃ 2.0c = 2.0 × 150 = 300 kN/m²

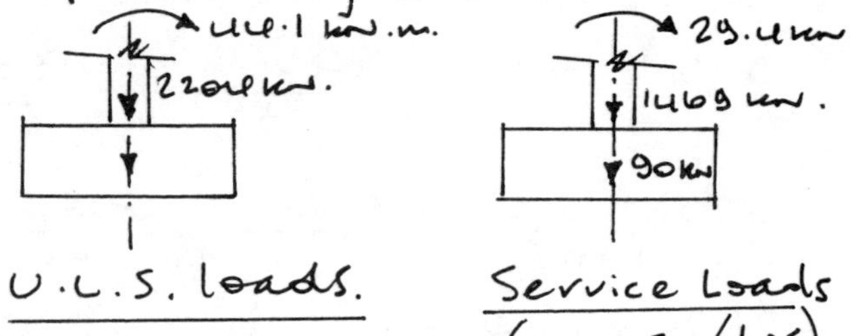

Try base 2.5 m² × 0.6 m deep.
Swt (service) = $2.5^2 \times 0.6 \times 24$ = 90 kN.
(see Fig 13/020)

Fig. 13/020

From Fig. 13/020, total downward load = 1469 + 90 = 1559 kN.

'Z' of base $= \dfrac{bd^2}{6} = \dfrac{2.5^3}{6} = 2.60$ m³

Area of base $= bd = 2.5^2 = 6.25$ m²

Max. pressure $= \dfrac{1559}{6.25} + \dfrac{29.4}{2.60} = +261$ (< 300 kN/m²)

Min. pressure $= \dfrac{1559}{6.25} - \dfrac{29.4}{2.60} = +238$ (< 300 kN/m²)

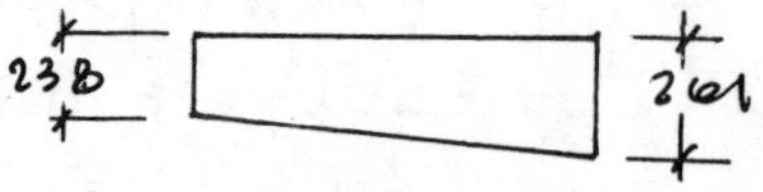

Fig. 13/021

Note! Tie stability checks also required on this type of structure.

Chapter 14 Residential accommodation in load-bearing masonry and reinforced concrete

Due to expansion, a University requires three bedroom blocks, linked by bridges to a central access tower. A typical wing contains 96 bedrooms on 8 floors, with openings in the gable walls leading to the central tower, and , at the other end, to an external emergency staircase. The scheme, shown in Figure 14/001, consists of pre-cast concrete plank floors carried on load-bearing masonry down to a concrete transfer structure at first floor. The reason for the transfer structure is that no walls are to be allowed between ground and first floor, with space being allowed for support columns only. The plank floors and hollow block walls allow the necessary tie forces for the 'robustness' to be accommodated in preventing progressive collapse. The hollow block walls can also accommodate tension reinforcement required at the junction between the gable walls and the transfer structure slab due to overall stability overturning. The key design elements concentrated on are the load-bearing walls, load-bearing gable walls, wall panels subject to lateral loading, overall stability check and accidental damage check.

Loading – qk

Bedroom floors	– 2.0 kN/m^2
Corridors	– 4.0 kN/m^2
Roof – no access	– 0.75 kN/m^2 (snow)

Loading – gk

Services	– 0.2 kN/m^2
Ceilings and finishes	– 1.5 kN/m^2

Loading – wk

Basic wind speed – 40 m/sec. : location – city centre site.

Site conditions

A thorough ground investigation revealed the following information:-
2.5 m. of sand and gravel (N=15) overlying 10.0 m. of stiff clay (C = 135 kN/m^2), overlying granite bedrock. Ground water encountered at 1.5 m. down from ground level.

Design Codes of Practice

Building to be designed to BS 5628 (structural Masonry) and BS 8110 (Structural Concrete).

Materials

Structural blockwork – 7.0 N/mm^2 : facing bricks – 7% – 12% water absorption in mortar designation (iii)
Concrete – fcu = 40 N/mm^2
Reinforcement (in the blockwork) – fy = 460 N/mm^2

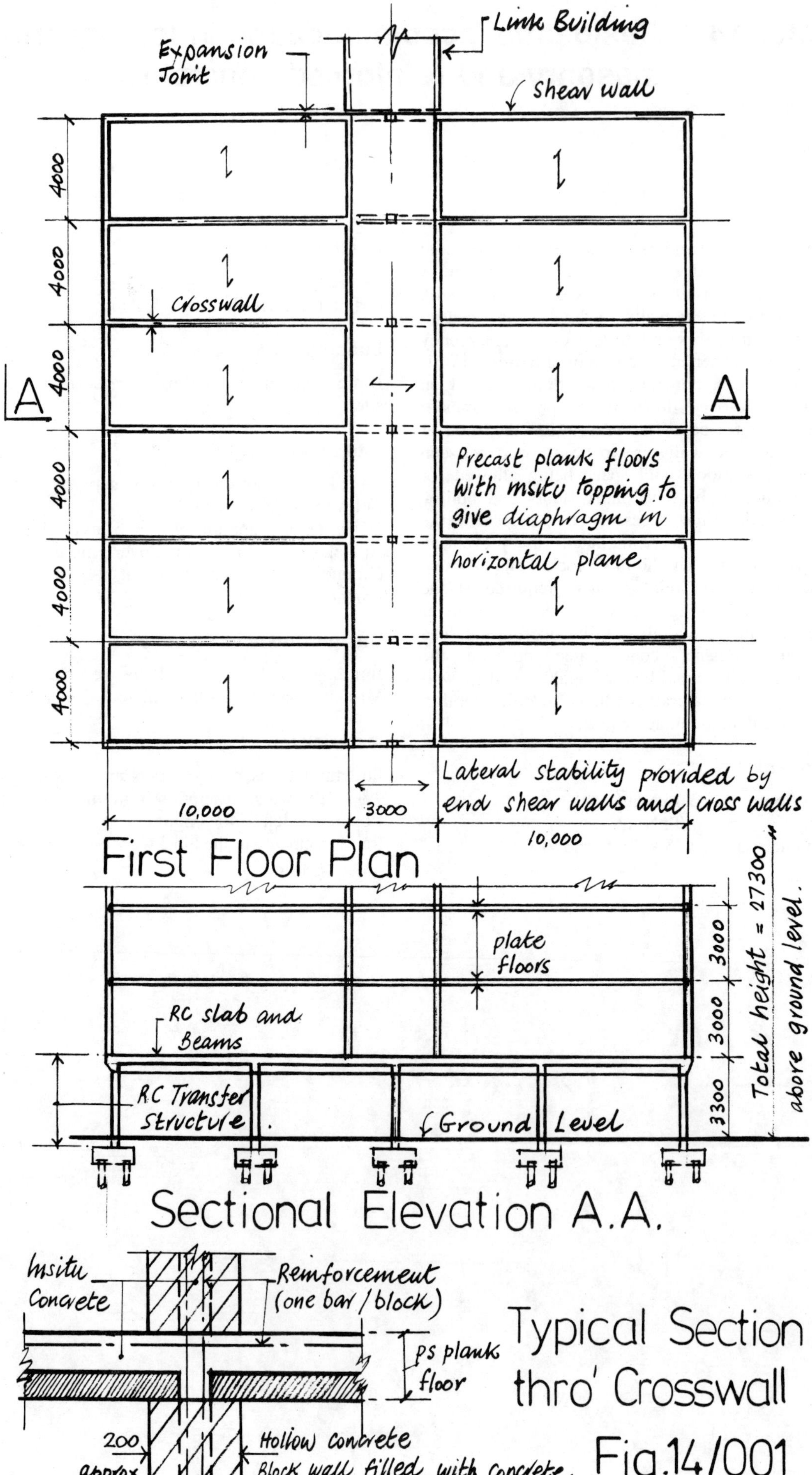
Link Building
Expansion Joint
Shear wall
4000
4000
4000
4000
4000
4000
Crosswall
A
A
Precast plank floors with insitu topping to give diaphragm in horizontal plane
Lateral stability provided by end shear walls and cross walls
10,000
3000
10,000
First Floor Plan
plate floors
3000
3000
3300
Total height = 27,300 above ground level.
RC slab and Beams
RC Transfer Structure
Ground Level
Sectional Elevation A.A.
Insitu Concrete
Reinforcement (one bar / block)
PS plank floor
200 approx
Hollow concrete Block wall filled with concrete.
Typical Section thro' Crosswall

Fig.14/001

Ref. (BS. 5628)

O/all geometry of building (see Figs 14/001 + 002)

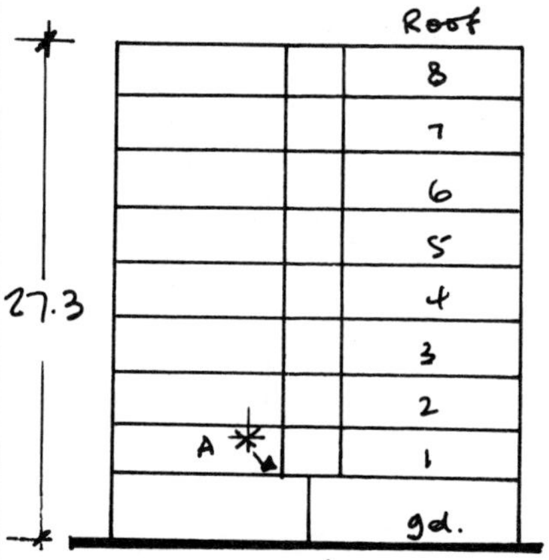

Fig. 14/002
(Typical cross-section)

Overall height = 27.3 m.
— length = 24.3 m.
— width = 23.3 m.
Flr./flr. height = 3.0 m.
Span of r.c. slabs (plank + in-situ) = 4.0 m.
Take $\gamma_{masonry}$ = 19 kN/m³
$\gamma_{concrete}$ = 24 kN/m³.
γ_m = 2.5 (close testing & supervision controls)

Characteristic (unfactored) loads :-

Roof : dead loads, g_k

Plank + in-situ slab = 0.15 × 24 = 3.6 kN/m²
Screed = 1.0 kN/m²
Total = 4.6 kN/m²

Live loads, q_k, no access = 0.75 kN/m²

Floors : dead loads, g_k (power-floated slab)

Plank + in-situ slab = 0.15 × 24 = 3.6 kN/m²
Services = 0.2 kN/m²
Finishes, inc. ceilings = 1.5 kN/m²
Total = 5.3 kN/m²

Loadings on crosswalls (see Fig 14/001)

Table 14/1	g_k loads (Floors & roof)	kN/m	q_k loads (Floors* & roof) * q_k = 2.0 kN/m²		q_k less BS6399 reduction	kN/m
Roof	4 × 4.6 =	18.4	4 × 0.75	=	3.0 – 0% =	3.0
8th	18.4 + (4×5.3) =	39.6	3 + (4×2)	=	11.0 – 10% =	9.9
7th	39.6 + 21.2 =	60.8	11 + 8	=	19.0 – 20% =	15.2
6th	60.8 + 21.2 =	82.0	19 + 8	=	27.0 – 30% =	18.9
5th	82.0 + 21.2 =	103.2	27 + 8	=	35.0 – 40% =	21.0
4th	103.2 + 21.2 =	124.4	35 + 8	=	43.0 – 40% =	25.8
3rd	124.4 + 21.2 =	145.6	43 + 8	=	51.0 – 40% =	30.6
2nd	145.6 + 21.2 =	166.8	51 + 8	=	59.0 – 40% =	35.4
1st	166.8 + 21.2 =	188.0	59 + 8	=	67.0 – 40% =	40.2

Ref. (B.S. 5628)	Loadings on spine walls (see Fig. 14/001) Corridor slab spans 3.0 m onto spine walls : q_k = 4.0 kN/m² for corridor.

Table 14/2	g_k loads (Floors & roof)	kN/m	q_k loads (corridors* & roof) * q_k = 4.0 kN/m²	q_k less BS 6399 reduction	kN/m.
Roof	3/2 × 4.6 =	6.9	3/2 × 0.75 =	1.125 − 0% =	1.125
8th	6.9 + (3/2 × 5.3) =	14.85	1.125 + 3/2 × 4 =	7.125 − 10% =	6.41
7th	14.85 + 7.95	22.8	7.125 + 6 =	13.125 − 20% =	10.5
6th	22.8 + 7.95	30.75	13.125 + 6 =	19.125 − 30% =	13.4
5th	30.75 + 7.95	38.7	19.125 + 6 =	25.125 − 40% =	15.1
4th	38.7 + 7.95	46.65	25.125 + 6 =	31.125 − 40% =	18.7
3rd	46.65 + 7.95	54.6	31.125 + 6 =	37.125 − 40% =	22.3
2nd	54.6 + 7.95	62.55	37.125 + 6 =	43.125 − 40% =	25.9
1st	62.55 + 7.95	70.5	43.125 + 6 =	49.125 − 40% =	29.5

Both crosswalls & corridor walls are of same construction (200 mm hollow blocks with concrete infill) : crosswalls critical.

Check that blocks with f_k = 7 N/mm² 200 mm thick, reinforced, are satisfactory at position 'A' – Fig. 14/002.

Dead load, floors & roof = 166.8 kN/m

Factored = 1.4 × 166.8 = 233.52 kN/m.

Live load, floors & roof = 35.4 kN/m

Factored = 1.6 × 35.4 = 56.64 kN/m.

Dead load, masonry (γ = 19 kN/m³)

Factored = 1.4 × 0.2 × 24.0 × 19 = 127.68 kN/m

Total n_w = 417.84 kN/m.

Ref. 14/2 — Vertical capacity of a reinforced block wall, $N_d = \beta\left(\frac{f_k \cdot A}{\gamma_{mm}} + \frac{0.8 \times f_y \times A_s}{\gamma_{ms}}\right)$

Table 7 — where β is from Table 7 & a factor of slenderness & load eccentricity.

t_{ef} = 200 mm : h_{ef} = 0.75 × height (in-situ slabs provide enhanced support.

∴ S.r. = (0.75 × 3000)/200 = 11.25

Ref. (B.S. 5628)

$e = 0.05t$: ∴ $\beta = 0.956$ (Table 7)

f_k for blocks $= 7.0$ N/mm²;

A_s = area of steel /m. of wall.

Try 1 No. T12 bar / block (440 mm. long blocks: $A_s = \frac{\pi \times 12^2}{4} \times \frac{1000}{440} = 257$ mm²/m

Ref 14/2

$\gamma_{ms} = 1.15$: $\gamma_{mm} = 2.5$ (see brief)

$$\therefore N_d = 0.956\left(\frac{7 \times 200 \times 10^3}{2.5 \times 10^3} + \frac{0.8 \times 460 \times 257}{1.15 \times 10^3}\right)$$

$N_d = 614$ kN/m.

$$\therefore \underset{(614)}{N_d} > \underset{(417.84)}{N_w}$$

Check gable wall: only 50% of load, but load eccentric at floor level.

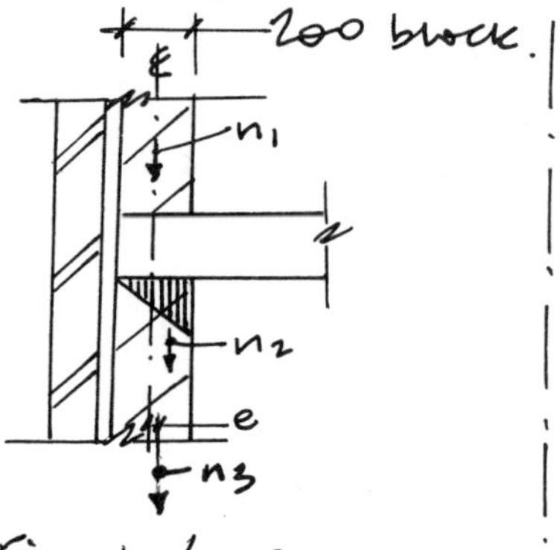

Fig. 14/003

From Table 14/1, n_1 = loads from 3rd floor.

Cl. 22

$g_k = 145.6 \times 1.4$	=	203.8
$q_k = 30.6 \times 1.6$	=	49.0
wall $= 1.4 \times 0.2 \times 21 \times 19$	=	111.72
n_1	=	364.52

On gable wall, reaction is half floor load + wall

∴ $n_1 = (252.8 + 49.0)/2 + 111.72 = 238.12$ kN/m

n_2 is load from 2nd floor only

$= (2 \times 5.3) \times 1.4 + (2 \times 2) \times 1.6 = 21.24$ kN/m.

Eccentricity of n_2 is $t/6$ due to triangular distribution of stress.

Taking moments about centreline:–

$$(238.12 + 21.24) \times e = 21.24 \times t/6$$

Table 7

$$\therefore e = 0.0136t < 0.05t$$

∴ β is same as previous calc.: load is ≃ 50% and by inspection, gable wall O.K. under vertical loading

Ref.	
(BS 5628)	Check lateral loading on a panel on side wall: Location – 8TH floor. Walls subject to high lateral loading, and low compressive load must be checked for failure due to flexural tensile stresses.
	Wind loading:
	From o/all geometry of building at beginning of calcs:–
	$l/w = 24.3/23.3 \simeq 1.0 \therefore 1 < l/w \leq 3/2$
	$h/w = 27.3/23.3 = 1.2 \therefore 1/2 < h/w \leq 3/2$
	V = 40 m/sec (see brief)
Table 3 CP3 ChV Pt 2	From Table 3 (4) Class B; $S_2 = 0.83$ for $h = 27.3$ m.; $V_s = 0.83 \times 40 = 33.2$ m/sec.
	$q = 0.613 \cdot V_s^2 = \dfrac{0.613 \times 33.2^2}{10^3} = 0.676$ kN/m²
Table 7 CP3 ChV Pt. 2	From Table 7, $C_{pe} = +0.7$ on windward wall. $C_{pi} = -0.3$.
	$\therefore W_k = (C_{pe} - C_{pi}) \times q = (0.7 - (-0.3)) \times 0.676$
	$= 0.676$ kN/m²
	Typical section through side wall is shown in Fig. 14/001. Panel is simply-supported, top & bottom, by ties onto slabs & continuous past cross walls.
Cl. 36.3(b)(2)	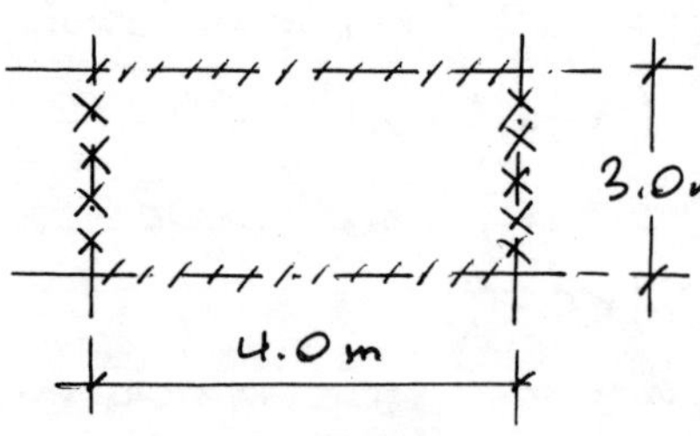 Fig. 14/001 wall panel.
	Check panel dimensions. From Cl. 36.3 (b) (2) $h \times L \ngtr 2025\, t_{ef}^2$
	From Fig. 14/001 & Cl. 28.3.1.1, $t_{ef} = \frac{2}{3}(100 + 200) = 200$ mm.
	$h \times L = 3000 \times 4000 = 12 \times 10^6$ mm²
	$2025\, t_{ef}^2 = 2025 \times 200^2 = 81 \times 10^6$ mm²
	$\therefore h \times L \ngtr 2025\, t_{ef}^2$ ✓ o.k.
Cl. 36.3	Also, $50 t_{ef}$ > greatest dimension.
	$50 t_{ef} = 50 \times 200 = 10{,}000 > 4000$ ✓ ok.

Ref	
(BS 5628) Cl. 36.4.1.	Design moment on panel, $M = \alpha . W_k . \gamma_f . L^2$
Table 9	For $\mu = 0.35$ (say), $\alpha = 0.033$ from section 4, Table 9
	$W_k = 0.676$ kN/m² from previous calcs.
Cl. 22	$\gamma_f = 1.2$ (Cl. 22) – panel failure will not cause structure collapse, as the floor slabs span onto the cross walls
	$L = 4.0$ m : $M = 0.033 \times 0.676 \times 1.2 \times 4.0^2$
	$M = 0.43$ kN.m. /m.
Cl. 36.4.3	From Cl. 36.4.3, design moment of resistance, $M_R = \frac{f_{kx} . Z}{\gamma_m}$ (sum of both leaves): strength in mortar (ii) & (iii) =
	Bricks 7% → 12% absorption : $f_{kx} = 1.1$ N/mm²
	Blocks 7 N/mm² : $f_{kx} = 0.35$ N/mm²
	$Z_{bricks} = \frac{10^3 \times 100^2}{6} = 1.67 \times 10^6$ mm³
	$Z_{blocks} = \frac{10^3 \times 200^2}{6} = 6.67 \times 10^6$ mm³
	Total flexural strength $= \frac{1.1 \times 1.67 \times 10^6}{2.5 \times 10^6} + \frac{0.35 \times 6.67 \times 10^6}{2.5 \times 10^6}$
	$= 1.66$ kN.m/m.
	$\therefore M_R > M$ ✓ panel o.k. (1.66) (0.43)
	Check lateral stability of block with wind on side – ($1.4W_k + 0.9G_k$)
	Value of wind pressure, q, (calculated for $h = 27.3$ m in panel design) will be calculated for the S_2 value in each storey in the following Table H4/3.
Table 10 CP3 ChV Pt 2.	The overall force coefficient for the building geometry, $C_f = 0.96$.
	Finally, $1.4W_k \times$ Area is calculated, per storey, per 4.0 m width, as design is done per crosswall.

Ref. (B.S. 5628)	Table 14/3 H (m.)	S2 Value	Vs (m/sec)	q = 0.613 × Vs² (kN/m²)	Wk = q × Cf (kN/m²)	1.4 Wk × area (kN)	Lateral forces on side of bldg.
	27.3	0.83	33.2	0.676	0.649	10.9	10.9 →
	24.3	0.80	32.0	0.628	0.603	10.13	10.13 →
	21.3	0.76	30.4	0.566	0.543	9.12	9.12 →
	18.3	0.73	29.2	0.523	0.502	8.43	8.43 →
	15.3	0.69	27.6	0.467	0.448	7.53	7.53 →
	12.3	0.66	26.4	0.427	0.410	6.89	6.89 →
	9.3	0.61	24.4	0.365	0.350	5.88	5.88 →
	6.3	0.56	22.4	0.308	0.296	5.0	5.0 →
	3.3	0.52	20.8	0.265	0.254	4.3	4.3 →

Overturning moment / (pair of crosswalls*)
(* each side of corridor)

$= [10.9 \times 22.5 + 10.13 \times 19.5 + 9.12 \times 16.5 + 8.43 \times 13.5 + 7.53 \times 10.5 + 6.89 \times 7.5 + 5.88 \times 4.5 + 5 \times 1.5]$

$= \underline{872 \text{ kN.m}}$

OTM / crosswall = 872 / 2 = $\underline{436 \text{ kN.m.}}$

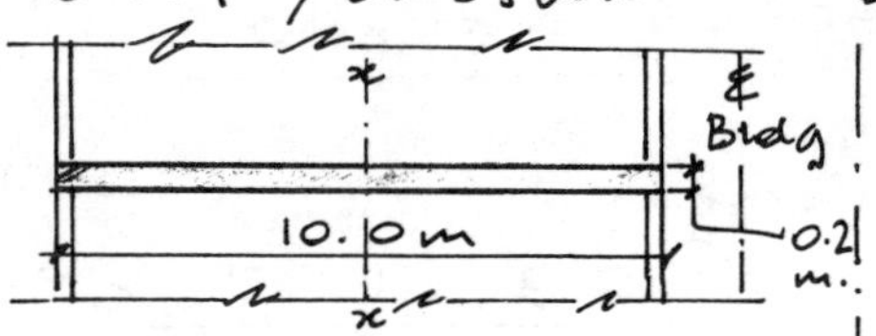

Fig. 14/005 - Plan on crosswall.

From fig 14/005, regarding crosswall as a cantilever from transfer structure slab level:–

$Z_{xx} = \dfrac{0.2 \times 10.0^2}{6} = 3.33 \text{m}^3$

$\therefore f = \pm \dfrac{M}{Z} = \pm \dfrac{436}{3.33} = \underline{\pm 131 \text{ kN/m}^2}$

From Table 14/1, at 1st Floor level,

$\underline{1.0 G_k = 188 \text{ kN/m.}}$

Area of crosswall $= 0.2 \times 10 = \underline{2} \text{ m}^2$

$\therefore \dfrac{0.9 G_k}{A} = \dfrac{0.9 \times 188 \times 10}{2} = \underline{846 \text{ kN/m}^2}$

$\therefore$ For 1.4 Wk + 0.9 Gk :–

Max. pressure = 846 + 131 = $\underline{977 \text{ kN/m}^2}$

Min. pressure = 846 − 131 = $\underline{715 \text{ kN/m}^2}$

$\therefore$ No tension: but vertical bars will be anchored to slab, in any case.

$\therefore$ O.T.M. is O.K.

Ref	
Ref (BS 5628) section 5 Cl. 37	This building is 'robust', according to the Code, Section 5, on "accidental damage", with ties both vertical & horizontal. However, being over 5 storeys in height, it is classed as a Category 2 building.
Table 12	Table 12 gives 3 options for designing and detailing such a structure for accidental damage. Option 1., in which vertical and
Ref. 14/1	horizontal elements (unless 'protected'*) are to be proved removable, one at a time, without causing collapse, is considered the most appropriate for this type of building. [Note * – a 'protected' element is one that will stand a lateral load of 34 kN/m²].

i) Crosswall panel removed by explosion.

Fig. 14/006

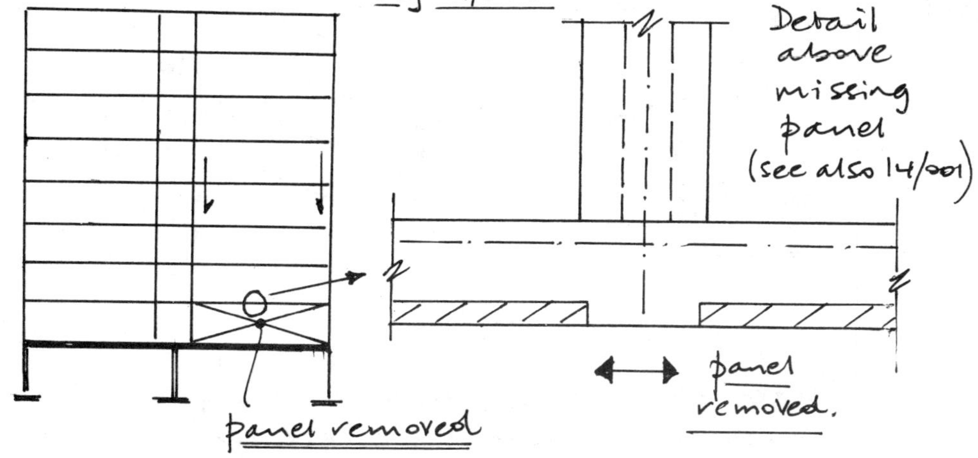

The floor slabs above the removed panel, except the immediate slab, can be considered to be supported as normal by the cross walls (reinforced), which arch and react, through bonding and shear forces onto the corridor walls & side walls (Fig 14/006). This leaves the 2 slab spans suspended by embedded T12 bars (Fig 14/006).

Factored reaction $= 2 \times \frac{4}{2}\left[1.4 \times 5.3 + 1.6 \times 2.0\right]$

$= \underline{42.5}$ kN/m.

Ref (BS 5628)

Tension capacity of bars:

1 T12 / block (see previous calcs.)

$\therefore A_s = 257 \text{ mm}^2/\text{m}$.

$$T = \frac{0.87 \times 460 \times 257}{10^3} = 102.8 \text{ kN/m}.$$

$\therefore$ T > Load ✓ ok.
(102.8) (42.5)

ii) Side wall panel removed.

No structural implications as panel is non-load bearing vertically.

iii) Gable wall panel removed.

Ref 14/1

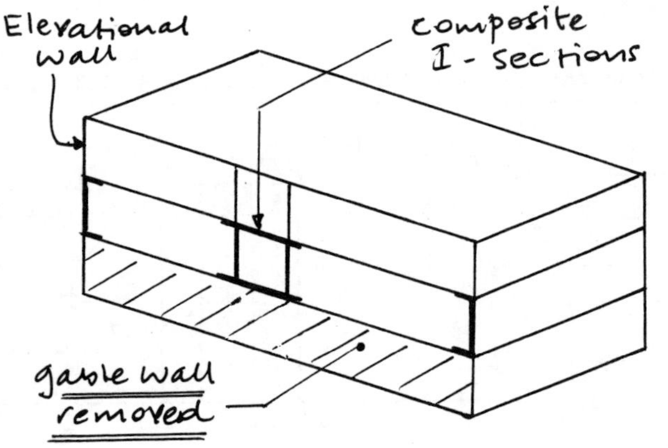

Fig. 14/007

Ref 14/1 suggests that, in this instance, the designer might consider 2 adjacent floor slabs acting as the flanges of deep I beams with the corridor and elevational walls between them acting as the webs of the same beam.

These composite sections may be used to cantilever from the last cross wall & support, at the end of the cantilever, a similar I-shaped composite beam utilising the gable wall as the web. Thus, a framework of composite beams is provided, and reinforced accordingly, to support the structure over (see Fig. 14/007).

Note!! The design of the r.c. transfer structure would have to accomodate all the various loadings from the load bearing masonry structure.

Chapter 15 National Park restaurant and gift shop

The Peak District National Park Board require a Restaurant and Gift Shop at a viewpoint of natural beauty, overlooking the cliff of an old sandstone quarry. The scheme, shown in plan in Figs. 15/001 & 15/002, and in section in Fig. 15/003, comprises of a gift shop of timber trussed rafter roof on brick and block cavity walling with external timber cladding. These are, in turn, supported by a concrete floor on composite steel beams. The lower floor is also of concrete slabs acting compositely with steel beams anchored down to the sandstone. The load bearing external Restaurant walls are of reconstituted stone block outer skin 150 mm. thick and 215 mm. block inner skin. The shop walls are of 100 mm. brick clad with timber, along with an inner skin of 215 mm. block. The following key elements of the structure have been designed : timber trussed rafters, composite and non-composite steel beams, masonry panels and padstones.

Loading – qk

Pitched roof of shop (snow) – 0.75 kN/m^2
Both floors – 4.0 kN/m^2

Loading – gk

Floors are of proprietary precast prestressed planks with structural screed. For the spans in the scheme, and live loading in the brief, a total **gk** from the manufacturer's catalogue is 5.0 kN/m^2.
Shop roof, including timber and insulation – 1.0 kN/m^2

Loading – wk

Basic wind speed – 50 m/sec.

Site conditions

0.5 m. of soft sandstone overlying an indeterminate depth of excavatable sandstone, safe bearing 1000 kN/m ^{2}, and capable of holding rock anchorages.

Design Codes of Practice

BS 5950 (Part 3) and Ref 15/1 for the Composite Steel beams.
BS 5268 for the Structural Timber.
BS 5628 for the Structural Masonry.
BS 5950 for the general Structural Steelwork.

Materials

Timber – SC3 with one hour fire resistance required.
Steel – To BS EN10025 – S275.
Precast planks and structural screed – fcu = 30 N/mm^2
Masonry blocks – fk = 7.0 N/mm ^{2}

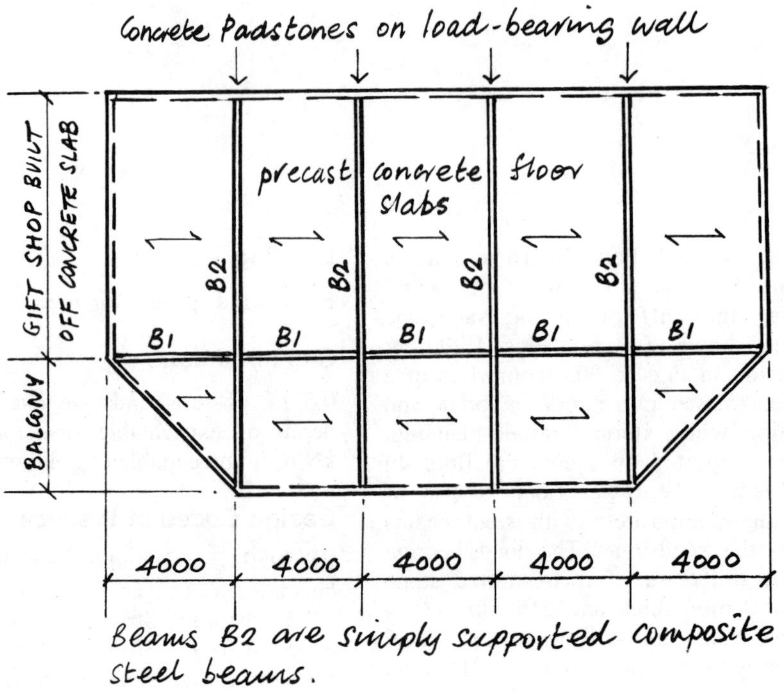

First Floor Plan

Fig.15/001

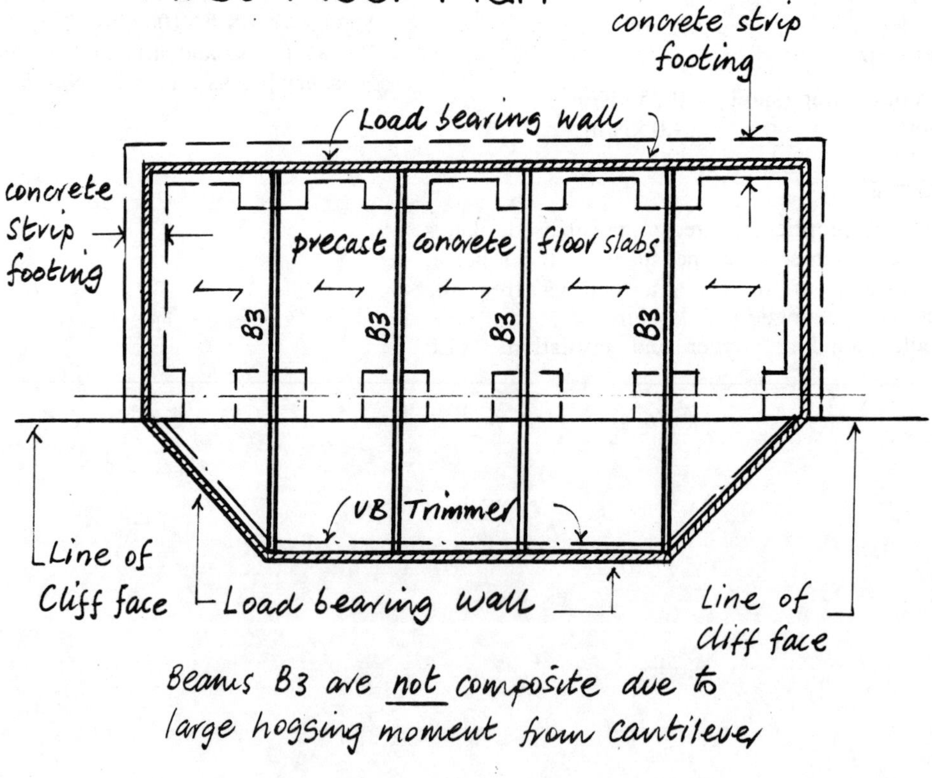

Ground Floor Plan

Fig.15/002

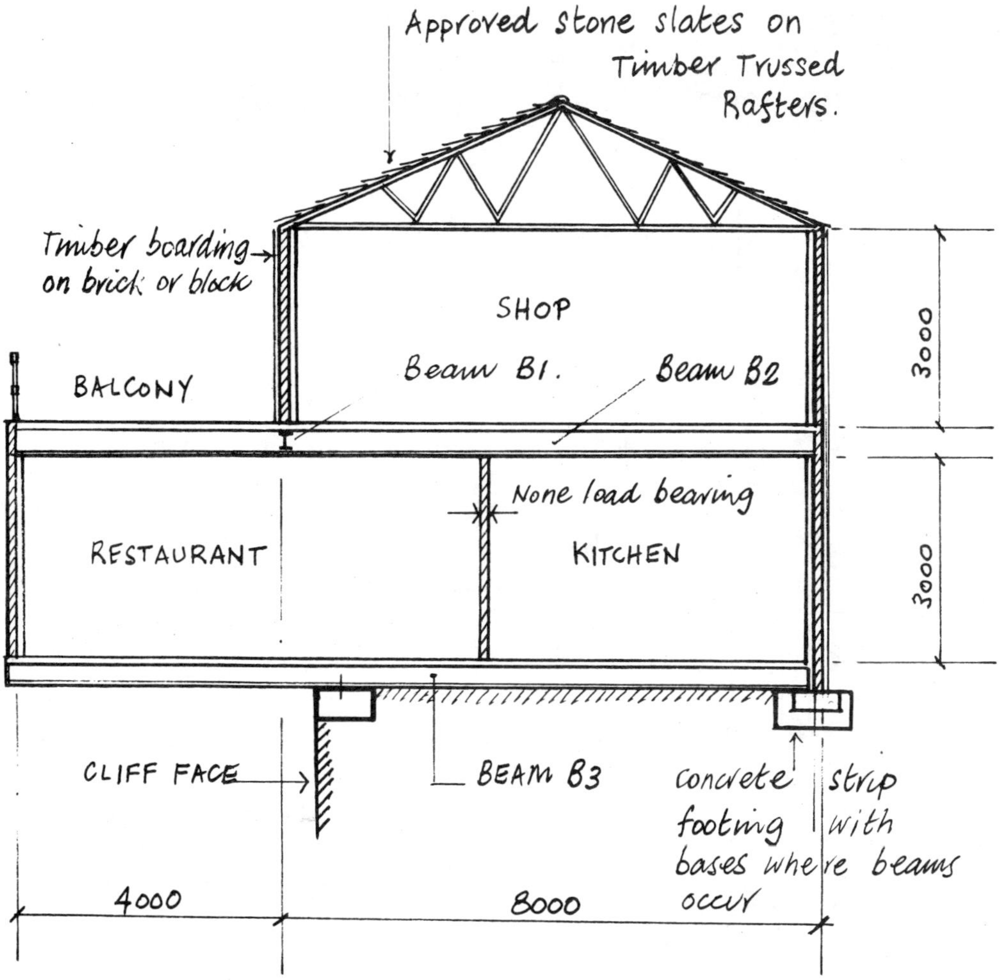
Approved stone slates on Timber Trussed Rafters.
Timber boarding on brick or block
SHOP
BALCONY
Beam B1.
Beam B2
3000
None load bearing
RESTAURANT
KITCHEN
3000
CLIFF FACE
BEAM B3
concrete strip footing with bases where beams occur
4000
8000

Typical Section

Fig.15/003

Ref.	
	Design loadings
	Shop roof: for timber design – code works on 'service' loading, design load = 0.75 + 1.0 = 1.75 kN/m²
	For calculations to BS 5950 or BS 5628, Design load = 0.75 × 1.6 + 1.0 × 1.4 = 2.6 kN/m²
	Floor loading: from brief: gk = 5.0 kN/m²
	Design load = 1.6 × 4.0 + 1.4 × 5.0 = 13.4 kN/m²
	Shop walls: boarding on 102.5 mm brick, 215 mm block inner skin.
	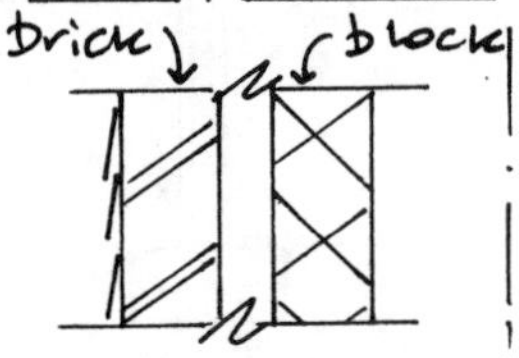 Fig. 15/004
	gk = 0.1 × 18 + 0.215 × 14 = 4.8 kN/m²
	Design load = 1.4 × 4.8 = 6.74 kN/m²
	Lower external walls.: 150 mm reconstituted stone, 215 mm block inner skin.
	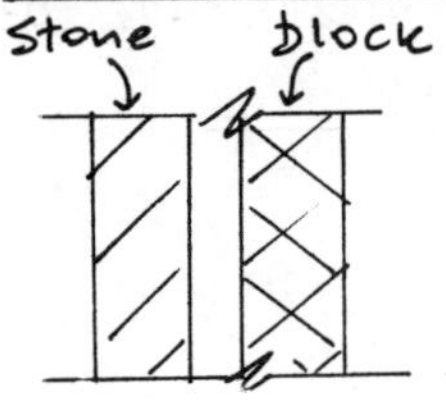 Fig. 15/005
	gk = 0.15 × 18 + 0.215 × 14 = 5.7 kN/m²
	Design load = 1.4 × 5.7 = 8.0 kN/m²
	Steel floor beams: (see Fig. 15/001)
	A: Beams 'B1' – upper floor, non-composite, unrestrained: span 4.0 m.
	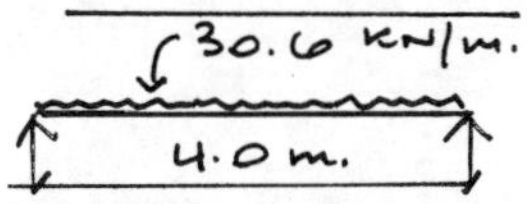 Fig. 15/006
	Loading: U.D.L. reaction from roof = 2.6 × 8/2 = 10.4 kN/m.
	U.D.L. from wall = 6.74 × 3 = 20.2 kN/m.
	Total = 30.6 kN/m.
	$M_{MAX} = \frac{wL^2}{8} = \frac{30.6 \times 4^2}{8} = 61.2$ kN.m.
	$F_v = \frac{wL}{2} = \frac{30.6 \times 4}{2} = 61.2$ kN.
Ref 15/2	Try a 254 × 146 × 37 U.B.
	$S_{xx} = 485$ cm³; $I_{xx} = 5560$ cm⁴; $r_{yy} = 34.7$ mm
	$U = 0.889$; $x = 24.3$; $T = 10.9$ mm.

Ref.	
	Classification: $T = 10.9\,mm < 16\,mm$
	$\therefore$ $p_y = 275\ N/mm^2$
	$\varepsilon = \sqrt{\frac{275}{275}} = 1.0$
Ref. 15/2	**Flange:** $b/T = 6.72 < 8.5\varepsilon$ – 'plastic'
	Web: $d/t = 34.2 < 79\varepsilon$ – 'plastic'
	$\therefore$ Section 'plastic'
	Calculation of M_b: $L_e = 4.0\,m$
	Beam loaded between restraints: $m = 1.0$, 'n' varies
BS5950 Table 16	For $\beta = 0$ & $V = 0$; $n = 0.94$
	$\lambda = \frac{4000}{34.7} = 115.3$; $\frac{\lambda}{x} = \frac{115.3}{24.3} = 4.74$
BS5950 Table 14	$\therefore$ $v = 0.83$
	$\lambda_{LT} = nuv\lambda$
	$\therefore$ $\lambda_{LT} = 0.94 \times 0.889 \times 0.83 \times 115.3 = 77$
BS5950 Table 11 & cl. 4.3.7.3	$\therefore$ $p_b = 171.6\ N\,mm^2$
	$\therefore$ $M_b = \frac{171.6 \times 485}{10^3} = 83.2\ kN.m.$
	$\therefore$ $M_b > M_{MAX}$ ✓ o.k.
	(83.2) (61.2)
Ref 15/2	**Check Shear:** $t = 6.4\,mm.$: $D = 256\,mm.$
	$\therefore$ $P_v = \frac{0.6 \times 275 \times 6.4 \times 256}{10^3} = 270\ kN.$
	$\therefore$ $P_v > F_v$ ✓ o.k.
	(270) (61.2)
	Check Deflection: $q_k = 0.75 \times \frac{8}{2} = 3\ kN/m.$
	$\therefore$ $\Delta = \frac{5}{384} \times \frac{3 \times (4000)^4}{205 \times 10^3 \times 5560 \times 10^4}$
	$= 8.8\,mm$ *
BS5950 Table 5	Allowable $= 4000/360 = 11.1\,mm.$
	Allowable > Actual ✓ o.k.
	$\therefore$ For 'B1' use a 254 × 146 × 37 U.B.
	* Dead load deflection can be accomodated by pre-cambering the beam.

Ref.	
	B : Beams 'B2' – upper floor, composite

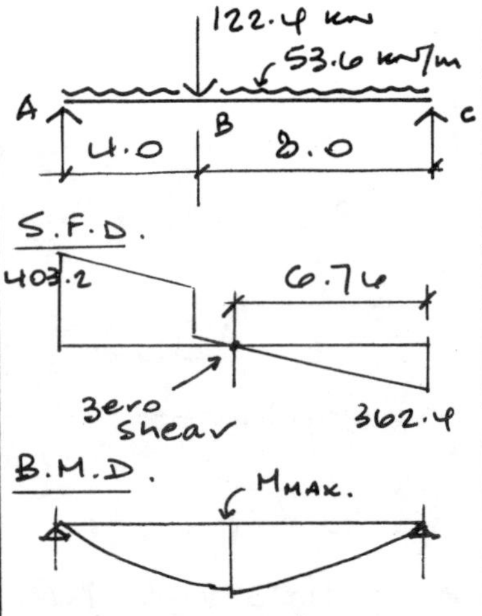

Fig. 15/007

span 12.0 m : 4.0 m. crs. : propped.

U.D.L. from slab

$= 13.4 \times 4 = \underline{53.6 \text{ kN/m}}$.

2 Reactions from beam 'B1'

$= 2 \times 61.2 = \underline{122.4 \text{ kN}}$. (P.L.)

$R_A \times 12 = \dfrac{53.6 \times 12^2}{2} + 122.4 \times 8.0$

$\underline{R_A = 403.2 \text{ kN}}$

$\underline{R_C = 362.4 \text{ kN}}$.

$\therefore M_{MAX} = 362.4 \times 6.76 - 53.6 \times \dfrac{6.76^2}{2}$

$\underline{M_{MAX} = 1225 \text{ kN.m}}$.

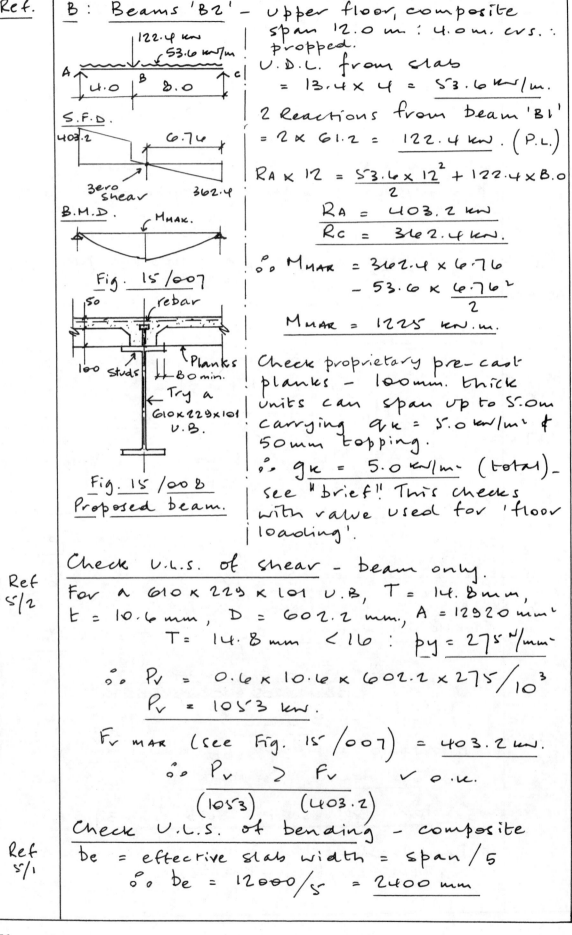

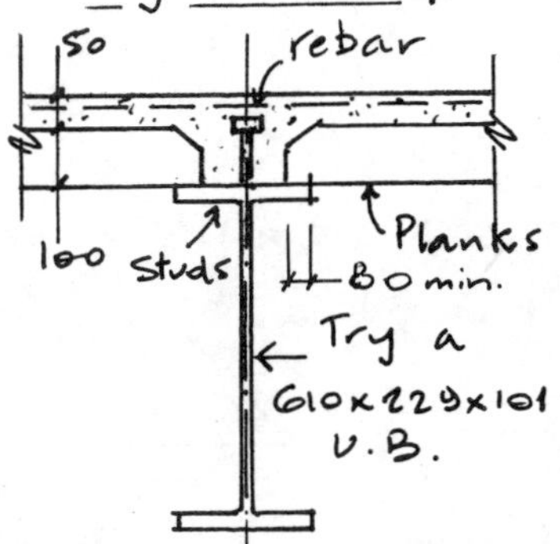

Fig. 15/008
Proposed beam.

Check proprietary pre-cast planks – 100mm. thick units can span up to 5.0m carrying $q_k = 5.0 \text{ kN/m}^2$ & 50mm topping.

$\therefore q_k = \underline{5.0 \text{ kN/m}^2}$ (total) – see "brief"! This checks with value used for 'floor loading'.

Check U.L.S. of shear – beam only.

Ref 5/2

For a 610 × 229 × 101 U.B, T = 14.8 mm, t = 10.6 mm, D = 602.2 mm, A = 12920 mm²

T = 14.8 mm < 16 : $\underline{p_y = 275 \text{ N/mm}^2}$

$\therefore P_v = 0.6 \times 10.6 \times 602.2 \times 275/10^3$

$\underline{P_v = 1053 \text{ kN}}$.

$F_{v\,MAX}$ (see Fig. 15/007) $= \underline{403.2 \text{ kN}}$.

$\therefore \underset{(1053)}{P_v} > \underset{(403.2)}{F_v}$ ✓ o.k.

Check U.L.S. of bending – composite

Ref 5/1

b_e = effective slab width = span / 5

$\therefore b_e = 12000/5 = \underline{2400 \text{ mm}}$

Ref.	
Ref 5/1	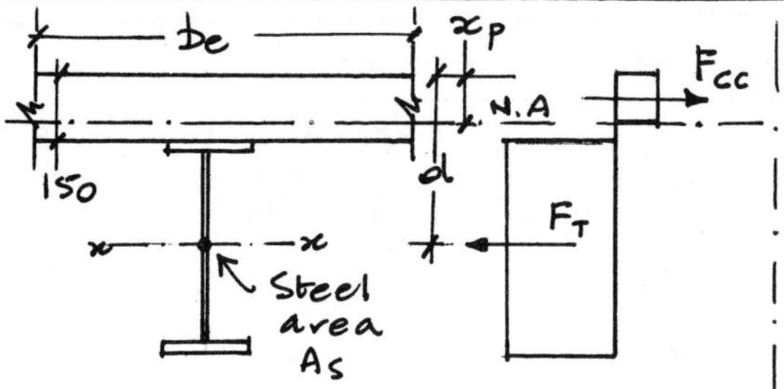

Fig. 15/009 - Moment Capacity - n.a. in slab

In Fig. 15/009 :-

$F_{cc} = 0.4 f_{cu} . b_e . x_p$

$F_T = A_s . p_y$

Equating forces :-

$$x_p = \frac{A_s . p_y}{0.4 f_{cu} . b_e}$$

$$x_p = \frac{12900 \times 275}{0.4 \times 30 \times 2400}$$

$x_p = 123$ mm.

$\therefore x_p <$ slab thickness (150 mm.)

$\therefore$ N.A. in slab.

Taking moments of forces :- (Ref 5/1)

$$M_c = A_s . p_y (d - x_p/2)$$

$d = 602.2/2 + 150 = 451$ mm.

$$\therefore M_c = \frac{12920 \times 275}{10^6} (451 - 123/2)$$

$M_c = 1384$ kN.m.

$\therefore M_c > M_{MAX}$ ✓ O.K.

(1384) (1225)

Shear connectors.

Shear connector force, $F_{cc} = 0.4 f_{cu} . b_e . x_p$ (Ref 5/1)

$\therefore F_{cc} = 0.4 \times 30 \times 2400 \times 123 / 10^3$

$F_{cc} = 3542$ kN.

Try 25 mm ϕ studs, 100 mm long in concrete, $f_{cu} = 30$ N/mm^2 : $P_k = 154$ kN. (Ref. 5/1)

Number of connectors, $N_c = \dfrac{F_{cc}}{0.75 P_k}$.

$$\therefore N_c = \frac{3542}{0.75 \times 154} = 31$$

These are spaced over each half of span

$$\therefore \text{Spacing} = \frac{12000/2}{31} = 193 \text{ mm.}$$

$\therefore$ use 25ϕ, 100 mm studs @ 175 crs.

Ref.	
	Check local concrete failure.
Ref 5/1	$q = F_{cc} / (span/2) = \frac{3542 \times 10^3}{6000} = 590\ N/mm$
	'q' must be less than both of :-
	1) $0.15 L_s . f_{cu}$ & 2) $0.9 L_s + 0.7 A_e . p_y$,
	where :-
	L_s = 1.5 × (stud dia.) + 2 × height of stud
	= 1.5 × 25 + 2 × 100 = 237.5 mm.
	(But, $L_s \ngtr$ 2 × slab depth = 300 mm ✓ O.K.)
	A_e = area of reinforcement in slab.
	Say, T20's @ 250 c/c = 1256 mm^2/m
	= 1.26 mm^2/mm.
	∴ 1) = 0.15 × 237.5 × 30 = 1069 N/mm ✓ O.K.
	2) = 0.9 × 237.5 + 0.7 × 1.26 × 460 = 619 N/mm ✓ O.K.
	Check deflection: elastically, neutral axis is in the steel.
Ref 5/1	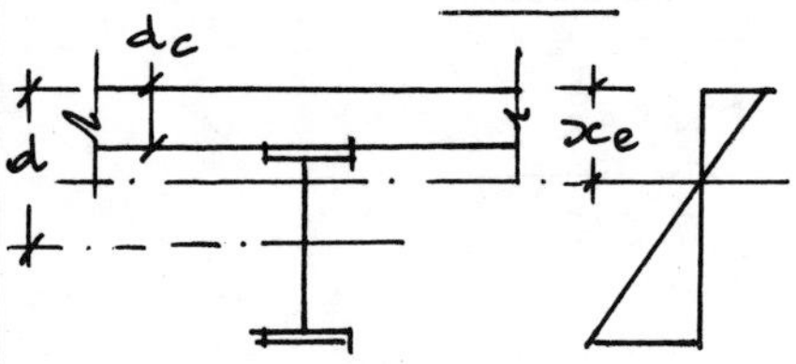 Fig. 15/010 Transformed section
	Take modular ratio, $m = 14.6$
	$x_e = \frac{(m . A_s . d + b_e d_c^2/2)}{(m . A_s + b_e . d_c)}$
	$= \frac{(14.6 \times 12920 \times 451 + \frac{2400 \times 150^2}{2})}{(14.6 \times 12920 + 2400 \times 150)}$
	∴ x_e = 204 mm (> 150 mm)
	$I_{es} = \frac{b_e d_c^3}{12m} + b_e d_c (x_e - \frac{d_e}{2})^2 / m + I_s + A_s (d - x_e)^2$
	$= \frac{2400 \times 150^3}{12 \times 14.6} + \frac{2400 \times 150 (204 - 150/2)^2}{14.6}$
	$+ 75700 \times 10^4 + 12900 (451 - 204)^2$
	$I_{es} = 2 \times 10^9\ mm^4$
	Unfactored live load U.D.L. = 4 × 4 = 16 kN/m.
	Unfactored point load (take as a central point load) = $0.75 \times \frac{8}{2} \times 4$ = 12 kN.

Ref.	
	$\therefore \Delta = \frac{5}{384} \times \frac{16 \times (12000)^4}{205 \times 10^3 \times 2 \times 10^9} + \frac{1}{48} \times \frac{12 \times 10^3 (12000)^3}{205 \times 10^3 \times 2 \times 10^9}$
	$= 10.5 + 1.0 = \underline{11.5 \text{ mm.}}$
BS5950 Table 5	Allowable $= \frac{12000}{360} = \underline{33.3 \text{ mm.}}$ ✓ o.k.
	$\therefore$ For '<u>B2</u>' use a <u>610 × 229 × 101 UB + 100 mm precast slab + 50 mm structural topping + 25φ studs × 100 mm. long @ 175 crs.</u>

<u>C</u>: <u>Beams 'B3'</u> - due to long cantilever, beams are in constant hogging - composite construction not appropriate.

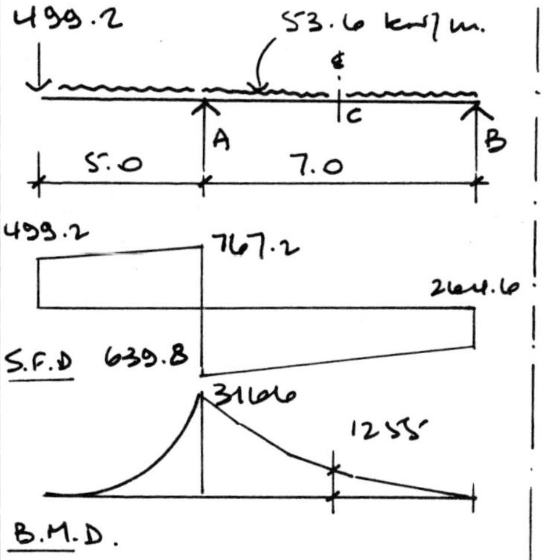

<u>Fig. 15/011</u>

<u>Loading</u>:

U.D.L. from slab
$= \underline{53.6 \text{ kN/m}}$ (as 'B2')

Point load :-

a) Reaction from 'B2' above $= \underline{403.2 \text{ kN}}$.

b) Reaction from trimmer beam carrying wall
$= 8.0 \times 3.0 \times 4.0 = \underline{96 \text{ kN}}$
(see 'lower external wall loading' earlier)

$\therefore$ Total point load $= 403.2 + 96 = \underline{499.2 \text{ kN}}$

$R_A \times 7 = \frac{499.2 \times 12 + 53.6 \times 12^2}{2} = \underline{1407 \text{ kN}}.$

$\therefore \underline{R_B = -2644.6}$ (<u>uplift reaction</u>)

<u>Moments</u>: (see also Fig 15/011)

Cantilever moments
$= 499.2 \times 5 + \frac{53.6 \times 5^2}{2} = \underline{3166 \text{ kN.m.}}$

Moment at c
$= 499.2 \times 8.5 + \frac{53.6 \times 8.5^2}{2} - 1407 \times 3.5$
$= \underline{1255} \text{ kN.m.}$

<u>Shear forces</u> as per Fig. 15/011.

Ref.	
	Try a 914 × 419 × 343 U.B.
Ref 15/2	$r_{yy} = 94.6$ mm : $S_{xx} = 15500\ cm^3$:
	$x = 30.1$: $U = 0.883$: $T = 32.0$ mm. :
	$t = 19.4$ mm : $D = 911.4$ mm.
BS 5950 Table 10	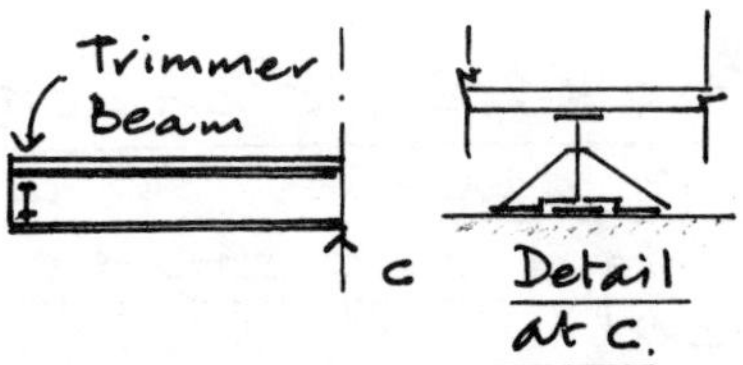 Fig. 15/012
	Cantilever: Restraints – at tip, 'torsional restraint'; at support – 'continuous with lateral + torsional restraint' – see Fig 15/012 ∴ $L_e = 0.8L$.
	Lateral torsional buckling:
	$l_e = 0.8 \times 5000 = 4000$ mm.
	∴ $\lambda = 4000/94.6 = 42.3$
	$\lambda/x = 42.3/30.1 = 1.4$: $v = 0.97$
	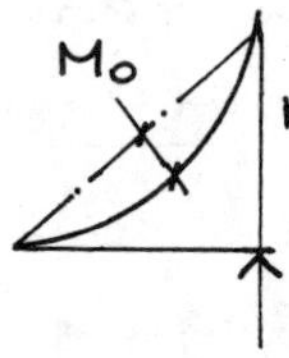 Fig. 15/013
	Calculation of 'n'
	$M_o = \frac{53.6 \times 5^2}{8} = 167.5$ kN.m
	∴ $\gamma = \frac{M}{M_o} = \frac{3166}{167.5} = 18.9$
	$\beta = 0$: $n = 0.80$
BS5950 Cl. 4.3.7.5	∴ $\lambda_{LT} = 0.80 \times 0.883 \times 0.97 \times 42.3 = 20.5$
	For $T = 32.0$ mm : $p_y = 265\ N/mm^2$
BS5950 Table 11	∴ $p_b = 265\ N/mm^2$ (Table 11)
	∴ $M_b = 265 \times 15500/10^3$
	$= 4108$ kN. m.
	∴ $M_b > M_{MAX}$ ✓ o.k.
	(4108) (3166)
	Check shear: $F_v = 499.2 + 5 \times 53.6 = 767.2$ kN
	$P_v = 0.6 \times 19.4 \times 911.4 \times 265/10^3$
	$= 2811$ kN.

Ref.	
	$\therefore$ $\underline{P_v > F_v}$ ✓ o.k. (2811) (767.2)

Check deflection:

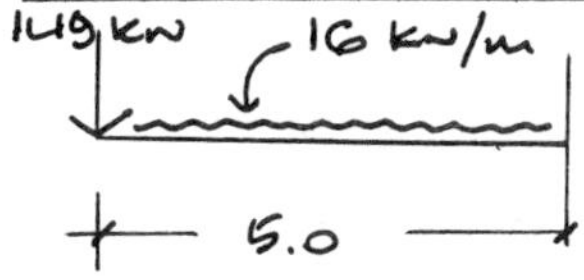

Fig. 15/014

Unfactored live loads:-

By proportion, point load $= \frac{4}{13.4} \times 499.2 = \underline{149\text{ kN}}$.

Live load, U.D.L., $= 4 \times 4 = \underline{16\text{ kN/m}}$.

$$\therefore \Delta = \frac{1}{8} \times \frac{16 \times (5000)^4}{205 \times 10^3 \times 625 \times 10^7} + \frac{1}{3} \times \frac{149 \times 10^3 \times (5000)^3}{205 \times 10^3 \times 625 \times 10^7}$$

$\underline{\Delta = 5.9\text{ mm}}$.

BS5950 Table 5 — Allowable $= 5000/180 = \underline{27.7\text{ mm}}$. ✓ o.k.

For 7.0m span; $\lambda_{LT} = 57.12$; $p_b = 213$; $M_b = 3317$ kN.m.

$\therefore$ $\underline{M_b > M_{MAX}}$ ✓ o.k. (3317) (3166)

End anchorage:

Either use a rock anchor system, or a gravity / friction pad footing.

Consider a load case with cantilever fully loaded, and 1.0 g_k on 7m. span:-

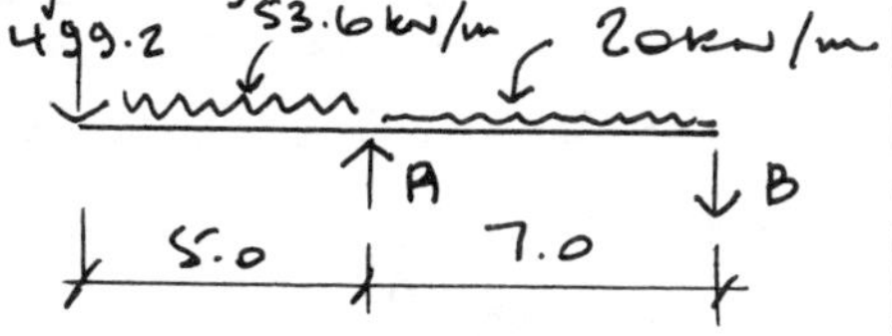

Fig. 15/015

By proportion, $1.0 g_k = \frac{5}{13.4} \times 53.6 = \underline{20\text{ kN/m}}$

$499.2 \times 12 + 53.6 \times 5 \times 9.5 + \frac{20 \times 7^2}{2} = R_A \times 7.0$

$\therefore$ $\underline{R_A = 1289\text{ kN}}$ & $\underline{R_B = -381.8\text{ kN}}$. *

* critical, when compared with previous load case $= \underline{-264.6\text{ kN}}$.

Try a base 1.5m. deep × 2.75m² plan.

Self weight $= 1.5 \times 2.75^2 \times 24 = \underline{272.5\text{ kN}}$,

'Service' uplift $\simeq \frac{381.8}{1.5} = \underline{254.5\text{ kN}}$. * ✓ o.k.

* If s.w. of rear wall included, f.o.s. is higher.

Ref.	Masonry checks :

BS5628 Cl.34

A: Concentrated loads :-

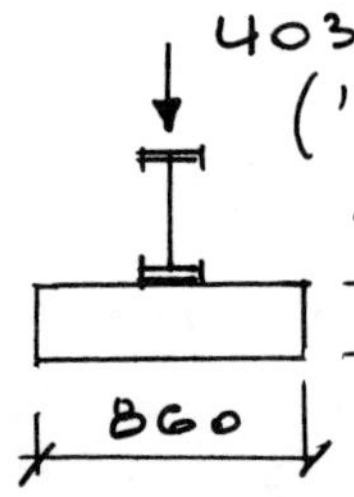

Fig. 15/016

Check padstone supporting beam 'B2'

Wall thickness, 215 mm, $f_k = 7\ N/mm^2$ (block).

Take $\gamma_m = 3.5$

Bearing Type 2 :

design resistance $= 1.5 f_k \times (\text{area}) / \gamma_m$

$$= \frac{1.5 \times 7 \times 860 \times 215}{10^3 \times 3.5} = \underline{555\ kN}.$$

Resistance > U.L.S. load
(555) (403.2)

B: Check wall under padstone :-

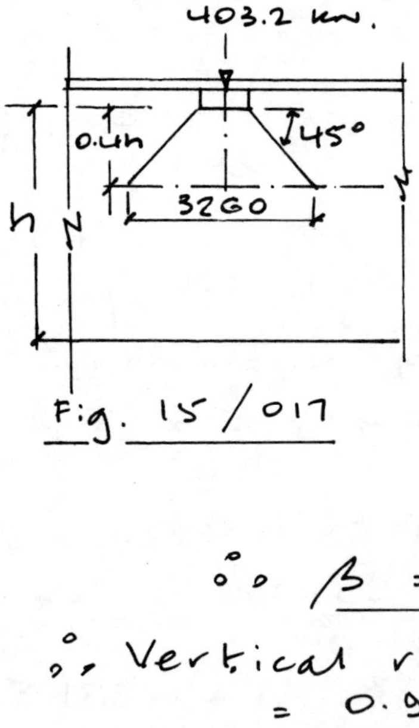

Fig. 15/017

t_{ef} for wall

$$= \frac{2}{3}(150 + 215) = \underline{243\ mm}.$$

$h = h_{ef} = \underline{3000}\ mm.$

$\therefore$ S.r. $= \dfrac{3000}{243} = \underline{12.3}\ (< 27)$

For load-bearing skin, eccentricity = zero (padstone)

$\therefore \beta = \underline{0.92}$

BS 5628 Cl. 32.2.1

$\therefore$ Vertical resistance

$$= \frac{0.92 \times 215 \times 7}{3.5} = \underline{396}\ kN/m.$$

U.L.S. load at a level 0.4h below padstone $= 403.2/3.26 = \underline{123.6}\ kN/m.$

$\therefore$ Resistance > U.L.S. load ✓ O.K.
(396) (123.6)

Ref.	
	C : Lateral wind loads:
	Overall stability assured, for shop, as 'box' shape ensures stiffness, when combined with diaphragm of braced trussed rafters & concrete slab.
	Check shop gable wall panel:–

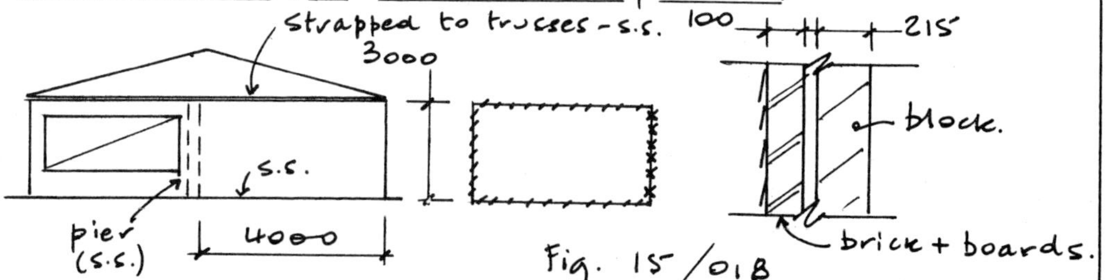

Fig. 15/01B

Ref.	
	Calculation of V_s & q :–
	$V = 50$ m/sec. (see brief)
	$V_s = S_1 . S_2 . S_3 . V$,
	where, $S_1 = 1.36$ (max.) – edge of cliff
CP3 ch V Pt 2:	$S_2 = 0.87$
	$S_3 = 1.0$.
Cl. 5.1	∴ $V_s = 1.36 \times 0.87 \times 1.0 \times 50 = 59.2$ m/sec.
	∴ $q = \frac{0.613 \times 59.2^2}{10^3} = 2.15$ kN/m²
	Calculation of W_k. (shop): $h_{av} = 5.0$m.
	$h/w = 5/8 = 0.62$: $1/2 < h/w < 3/2$
	$l/w = 20/8 = 2.5$: $3/2 < l/w < 4$
	∴ C_{pe} on gable $= +0.7$: $C_{pi} = -0.3$
	∴ $C_f = 0.7 + 0.3 = 1.0$.
	∴ $W_k = 1.0 \times 2.15 = 2.15$ kN/m²
	Check on panel dimensions (see Fig. 15/01B)
BS 5628 Cl. 3.6.3 (a)	$t_{ef} = \frac{2}{3}(100 + 215) = 210$mm. or 215 mm.
	Height × length ≯ $2025 t_{ef}^2$
	$h \times L = 3000 \times 4000 = 12 \times 10^6$ mm²
	$2025 t_{ef}^2 = 2025 \times 215^2 = 93.6 \times 10^6$ mm² ✓ o.k.
	& $50 t_{ef} = 50 \times 215 = 10750 > 4000$ ✓ o.k.
	∴ Panel dimns. o.k.

Ref.	
	Check bending strength of panel.
	Design moment $= \alpha . W_k . \gamma_f . L^2$
BS5628 Table 9F	For $h/L = 0.75$, $\mu = 0.35$: $\alpha = 0.041$
	$\therefore M = 0.041 \times 2.15 \times 1.20 \times 4^2 = 1.69$ kN.m/m
	Moment of resistance:
	Brick skin: $Z = \dfrac{10^3 \times 100^2}{6} = 1.67 \times 10^6\ mm^3$
	$f_{kx} = 0.9\ N/mm^2$ (common brick)
	$\therefore M_R = \dfrac{0.9 \times 1.67 \times 10^6}{3.5 \times 10^6} = 0.43$ kN.m/m.
	Block skin: $Z = \dfrac{10^3 \times 215^2}{6} = 7.7 \times 10^6\ mm^3$
	$f_{kx} = 0.6\ N/mm^2$ (7 Newton block)
	$\therefore M_R = \dfrac{0.6 \times 7.7 \times 10^6}{3.5 \times 10^6} = 1.32$ kN.m/m.
	Total, both skins $= 0.43 + 1.32 = 1.75$ kN.m/m
	$\therefore M_R > M$ ✓ o.k.
	(1.75) (1.69)

Trussed rafters to shop (8m. span)

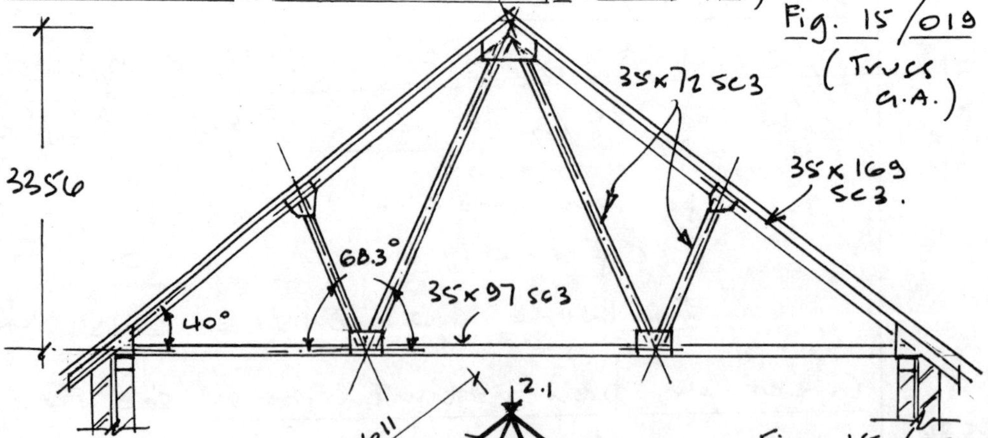

Fig. 15/019 (Truss G.A.)

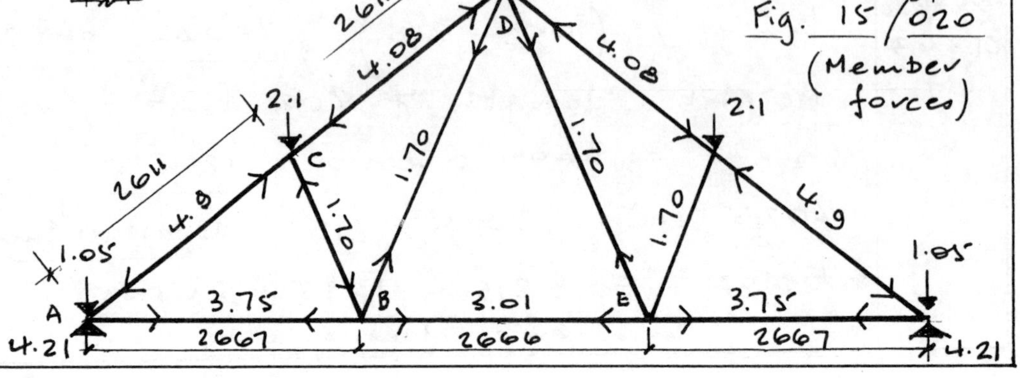

Fig. 15/020 (Member forces)

Ref.	
	Trusses (see Fig. 15/019) @ 600 crs. in S.C.3 timber.
	Roof loading (timber code) = $1.75\ kN/m^2$ (see earlier).
	∴ U.D.L. on truss = $0.6 \times 1.75 = 1.05\ kN/m$.
	Point load (idealized) value = $1.05 \times 2 = 2.1\ kN$ (see Fig 15/020).
	Reaction = $4.2\ kN$. – member forces as Fig 15/020.
	Check at Node 'D'
	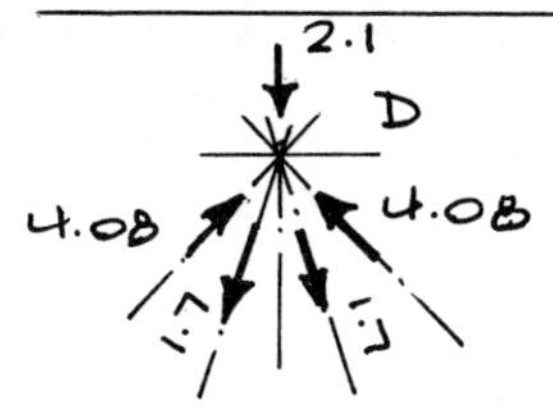 Fig. 15/021
	↑ $4.08 \times 2 \times \sin 40^\circ = 5.24\ kN$.
	↓ $2.1 + 2 \times 1.7 \sin 68.3^\circ = 5.26\ kN$.
	✓ checks o.k.
	A: Check top boom (35 × 169 SC3)
	Area = $35 \times 169 = 5915\ mm^2$
	$Z = 35 \times 169^2/6 = 167 \times 10^3\ mm^3$
	$I_{xx} = 35 \times 169^3/12 = 14.1 \times 10^6\ mm^4$
	(I_{yy} not required, as tile battens restrain member on y–y axis).
	∴ $r_{xx} = \sqrt{\frac{I_{xx}}{A}} = \sqrt{\frac{14.1 \times 10^6}{5915}} = 50\ mm$
	Member A–C is in axial compression + bending
	Bending ≃ $0.10\,wL^2 = 0.10 \times 1.05 \times 2.61^2 = 0.72\ kN.m.$
	∴ $\sigma_{m,a\parallel} = \frac{0.72 \times 10^6}{167 \times 10^3} = 4.3\ N/mm^2$
	Axial load = $4.9\ kN$ (see Fig. 15/020)
	∴ $\sigma_{c,a\parallel} = \frac{4.9 \times 10^3}{5915} = 0.83\ N/mm^2$
BS5268 Cl. 15.6	$\frac{\sigma_{m,a\parallel}}{\sigma_{m,adm,\parallel}\left(1 - \frac{1.5\,\sigma_{c,a\parallel}}{\sigma_e} . K_{12}\right)} + \frac{\sigma_{c,a\parallel}}{\sigma_{c,adm\parallel}} < 1.0$

Ref.	
	$\sigma_{m,adm\parallel} = \sigma_{m,grade} \times K_3 \times K_7 \times K_8$
	$K_7 = \left(\frac{300}{169}\right)^{0.11} = 1.05$
	$\therefore \sigma_{m,adm\parallel} = 5.3 \times 1.25 \times 1.05 \times 1.1$
	$= 7.65\ N/mm^2$
	$\sigma_{c,adm} = \sigma_{c,grade} \times K_3 \times K_8 \times K_{12}$
	K_{12} : $l_{ex} = 0.85 \times 2611 = 2219\ mm.$
	$(l_e/r)_x = 2219/49 = 45$
BS5268 Table 20	$E_m/\sigma_c = 5800/6.8 = 853$: $K_{12} = 0.762$
	$\therefore \sigma_{c,adm} = 6.8 \times 1.25 \times 1.1 \times 0.762$
	$= 7.13\ N/mm^2$
	$\sigma_e = \frac{\pi^2 \times 5800}{45^2} = 28.3\ N/mm^2$
	$\therefore \frac{4.3}{7.65\left(1 - \frac{1.5 \times 0.8 \times 0.762}{28.3}\right)} + \frac{0.83}{7.13}$
	$= 0.96 < 1.0$ ✓ o.k.
	$\therefore$ 35 × 197 SC3. satisfactory.
BS5268 Cl. 16	Tension boom – member A – B. (35 × 97 SC3)
	$\sigma_{t,a} = \frac{3.75 \times 10^3}{35 \times 97} = 1.1\ N/mm^2$
	$\sigma_{t,adm} = \sigma_{t,grade} \times K_3 \times K_{14}$
	$\therefore \sigma_{t,adm.} = 3.2 \times 1.25 \times 1.17 = 4.7\ N/mm^2$
	$\therefore \sigma_{t,adm} > \sigma_{t,a}$ ✓ o.k.
	(4.7) (1.1)
	Internal tension member B–D (35 × 72 SC3)
	$\sigma_{t,a} = \frac{1.70 \times 10^3}{35 \times 72} = 0.674\ N/mm^2$
	$\therefore \sigma_{t,adm.} > \sigma_{t,a}$ ✓ o.k.
	(4.7) (0.674)

Ref.	
	Internal compression member B-C (35×72 SC3)
	Length = 1805 mm : Eff. length = 1.0 × 1805
	$I_{yy} = \frac{35^3 \times 72}{12} = 257 \times 10^3 \text{ mm}^4$: $A = 2520 \text{ mm}^2$
	$r_{yy} = \sqrt{\frac{257 \times 10^3}{2520}} = 10.1 \text{ mm.}$
	$l_e / r_{yy} = \frac{1.0 \times 1805}{10.1} = 178$
BS 5268 Table 20	For $E / \sigma_{c\parallel} = 853 \text{ N/mm}^2$: $K_{12} = 0.146$
	$\therefore \sigma_{c,adm.} = 6.8 \times 1.25 \times 1.1 \times 0.146 = 1.365 \text{ N/mm}^2$
	$\sigma_{c,all} = \frac{1.7 \times 10^3}{2520} = 0.674 \text{ N/mm}^2$
	$\therefore \sigma_{c,adm} > \sigma_{c,all}$
	(1.365) (0.674)
	$\therefore$ 35 × 72 SC3 satisfactory – internals.
	Check truss deflection:
Ref 15/4	A simple empirical formula for deflection of symmetrical flat & pitched trusses:–
	$\Delta = \frac{L^2}{4290\,H}\left[\frac{L}{26.6} + 1\right]$
	Where L = span of truss : H = height at midspan.
	$\therefore \Delta = \frac{8^2}{4290 \times 3.356}\left[\frac{8}{26.6} + 1\right]$
	$= 0.006 \text{ m.} = 6 \text{ mm.}$
	Allowable $= 0.003 \times 8000 = 24 \text{ mm}$ ✓ o.k.
	For connection design – see Ref. 15/4.

Chapter 16 Refurbishment and conversion of a fire-damaged domestic property into offices

A National Insurance Firm require a small branch office, and have elected to convert and re-furbish a fire-damaged listed domestic property. The Architect's survey of the original building, plus the proposed scheme, are shown in Fig. 16/001. The scheme includes provision for a 3-storey office block, with storage in a new basement. The existing timber floors, roof and brick staircase walls are damaged beyond any further use, and must be replaced. The Architect stipulates that no internal load-bearing walls or columns are to be used, other than the existing cross-walls. Also, a minimum headroom of 2.4 m. must be provided in a watertight basement, with a suggested underpinning detail, as shown in Fig. 16/002. The Structural Engineer's scheme is also shown in Figure 16/003, comprising timber floors spanning from the inner skins of the external walls onto Universal Beams. The survey reveals that the roof is damaged beyond repair, and a traditional rafter and steel purlin arrangement is proposed.

Loading – qk

All floors, including basement	= 5.0 kN/m^2 (inc. an allowance for partitions).
Roof (snow)	= 0.75 kN/m^2

Loading – gk

Timber floors (including ceiling)	= 1.0 kN/m^2
Roof – blue slate (inc. insulation and timbers)	= 1.2 kN/m^2

Soil safe bearing

The existing corbelled brickwork footings, shown on Fig. 16/001, are founded on clay, safe bearing = 100 kN/m^2.

Design Codes of Practice

BS 5268 – Timber: BS 5628 – Structural Masonry: BS 5950 – Structural Steelwork

Materials

Timber elements – SC3
Steelwork – BS EN 10025 S275, 1-hour minimum fire resistance
Existing brickwork unit compressive strength = 20 N/mm^2. Mortar used was 1:1:6 lime mortar. Stresses from point loads to be restricted to 0.5 N/mm^2.
New blockwork strength to be 7 N/mm^2.
Concrete – underpinning and basement slab – fcu = 35 N/mm^2: cover = 40 mm.
Reinforcement in basement slab – fy = 460 N/mm^2

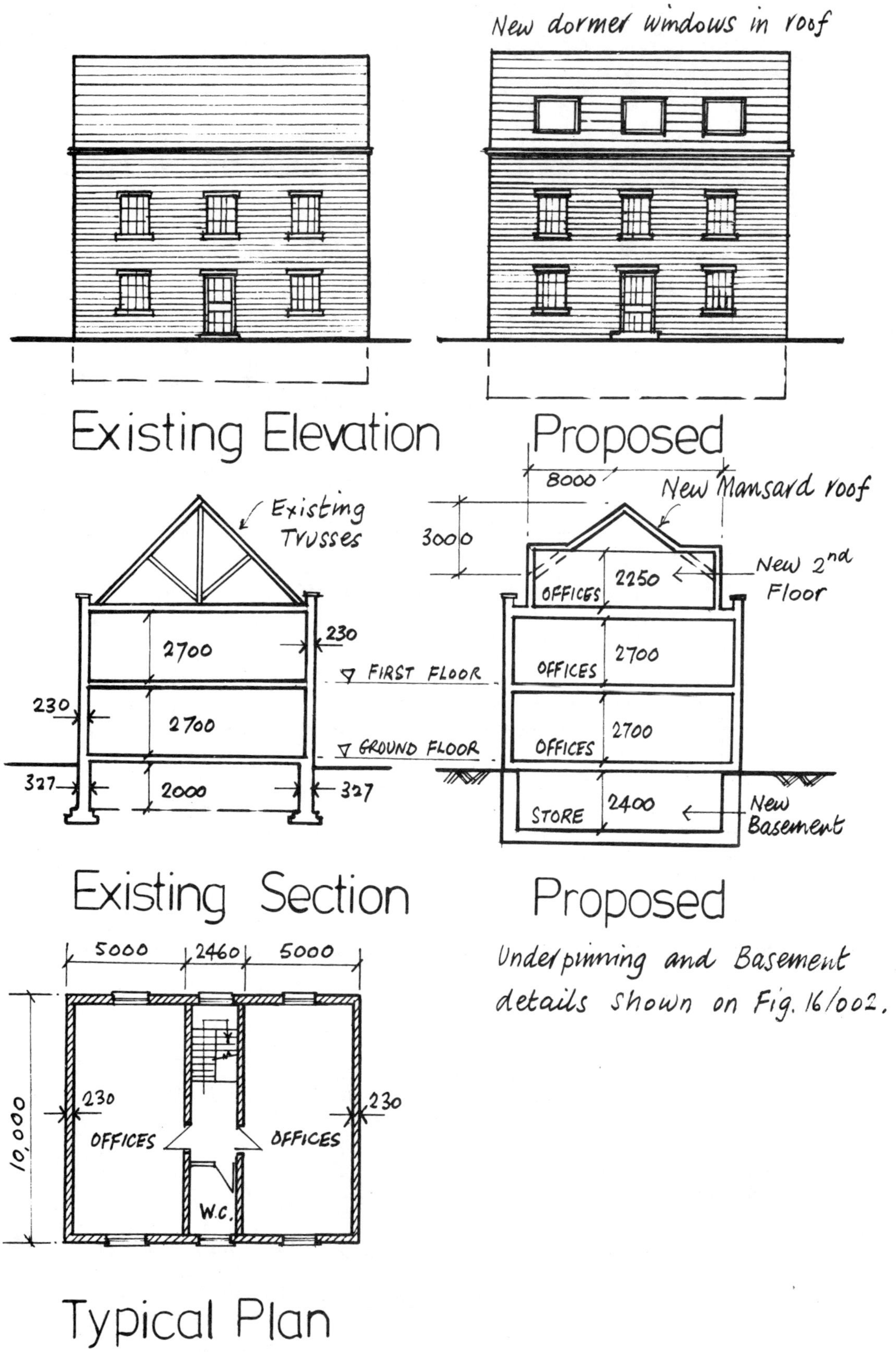
New dormer windows in roof
Existing Elevation
Proposed
8000
New Mansard roof
Existing Trusses
3000
2250
OFFICES
New 2nd Floor
230
2700
FIRST FLOOR
OFFICES
2700
230
2700
GROUND FLOOR
OFFICES
2700
327
2000
327
STORE
2400
New Basement
Existing Section
Proposed
Underpinning and Basement details shown on Fig. 16/002.
5000
2460
5000
10,000
230
230
OFFICES
OFFICES
W.C.
Typical Plan

Fig.16/001

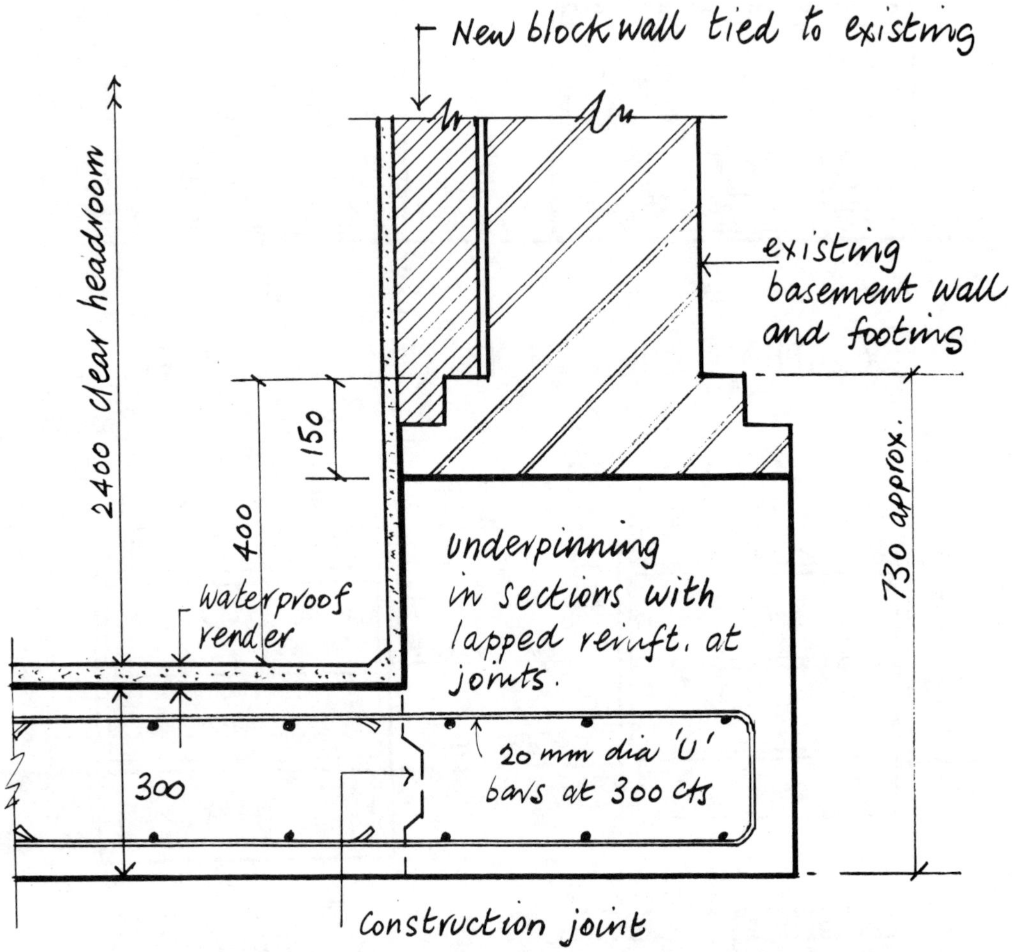

Concrete underpinning tied into new basement slab.

Proposed Underpinning and New Basement Slab

Alternative is asphalt tanking with floor screed and block walls for waterproofing Basement.

Fig.16/002

Ref.

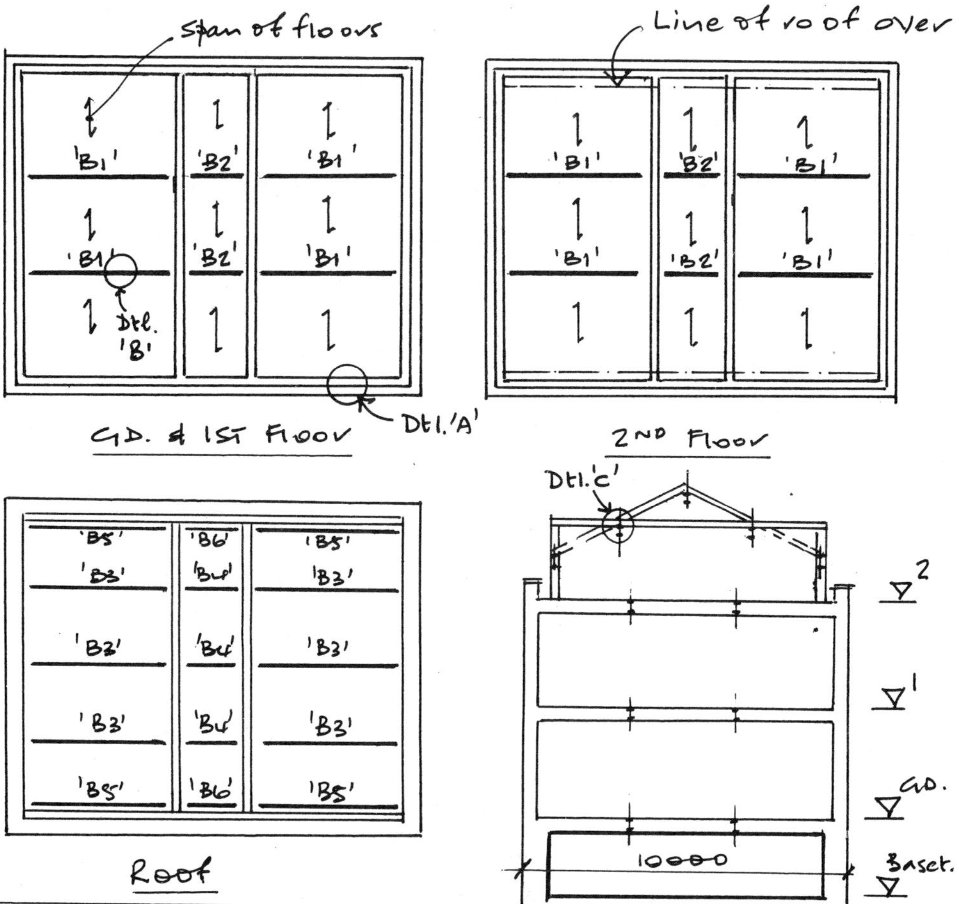

For dimensions - see Fig 16/001

Fig. 16/003 (Details Fig. 16/004 & 16/002)

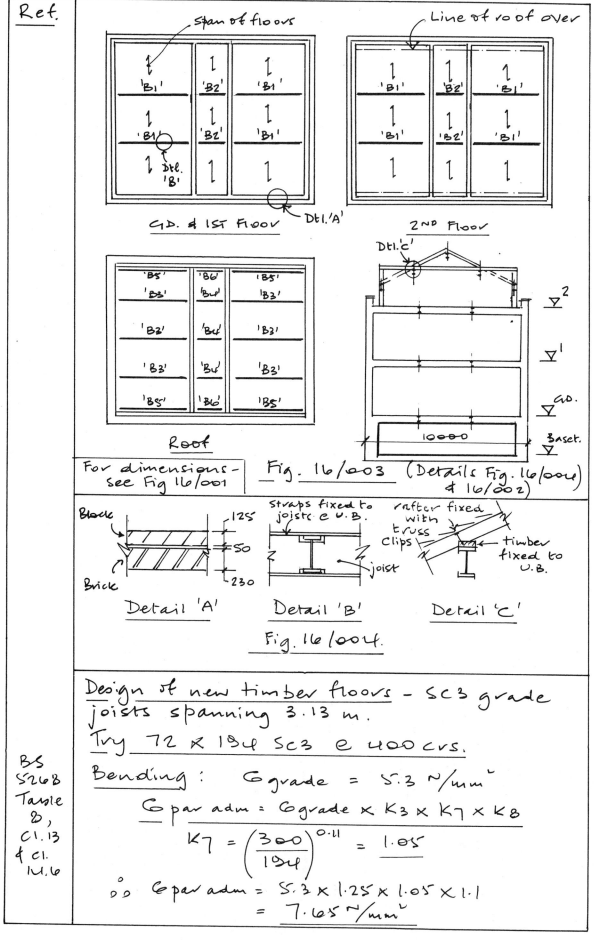

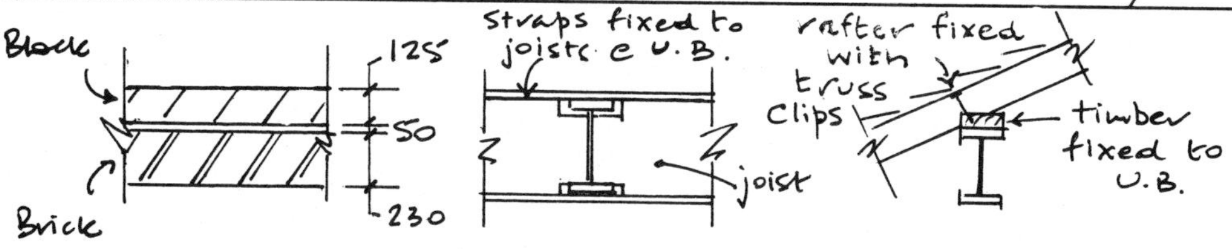

Detail 'A' Detail 'B' Detail 'C'

Fig. 16/004.

Design of new timber floors - SC3 grade joists spanning 3.13 m.

Try 72 × 194 SC3 @ 400 crs.

BS 5268 Table 8, Cl. 13 & Cl. 14.6

Bending: $\sigma_{grade} = 5.3\ N/mm^2$

$\sigma_{par\ adm} = \sigma_{grade} \times K_3 \times K_7 \times K_8$

$K_7 = \left(\frac{300}{194}\right)^{0.11} = \underline{1.05}$

$\therefore\ \sigma_{par\ adm} = 5.3 \times 1.25 \times 1.05 \times 1.1$

$= \underline{7.65\ N/mm^2}$

Ref.	
	For a 72×194 joist, 'Z' = $452 \times 10^3 mm^3$
	∴ $M_R = fz = \frac{7.65 \times 452 \times 10^3}{10^6} = 3.46$ kN.m.
	U.D.L. per joist = 0.4 (5.0 + 1.0) = 2.4 kN/m.
	∴ $M_{MAX} = \frac{2.4 \times 3.13^2}{8} = 2.94$ kN.m.
	$M_R > M_{MAX}$ ✓ o.k.
BS 5268 Cl.4.7	Deflection: Limit = 0.003L = 0.003 × 3130 = 9.4 mm.
	For a 72 × 194 joist, $I_{XX} = 43.8 \times 10^6 mm^4$
	E_{mean} for SC3 = 8800 N/mm^2
	$\Delta = \frac{5}{384} \times \frac{2.4 \times (3130)^4}{8800 \times 43.8 \times 10^6} = 7.8$ mm
	∴ Actual (7.8) < Limit (9.4) ✓ o.k
	Shear: $F_v = \frac{2.4 \times 3.13}{2} = 3.76$ kN./joist
	∴ $V_{actual} = 1.5 \left(\frac{3.76 \times 10^3}{72 \times 194}\right) = 0.4 N/mm^2$
BS 5268 Table 8	$V_{adm\ par} = V_{par\ grade} \times K_3 \times K_8$
	= 0.67 × 1.25 × 1.1
	= 0.92 N/mm^2
	∴ $V_{adm\ par}$ (0.92) > V_{actual} (0.4) ✓ o.k.
	∴ 72 × 194 SC3 grade joists @ 400 crs. O.K.
	Steel Floor Beams:
	A: Beams 'B1' – 5.0 m. span
	Gd. & 1st floor design loading = 1.6 × 5.0 + 1.4 × 1.0 = 9.4 kN/m^2
	Timber floors span 3.13 m onto beams:
	U.D.L. = 9.4 × 3.13 = 29.4 kN/m.
	$M_{MAX} = \frac{29.4 \times 5^2}{8} = 91.9$ kN.m.
	$F_v = \frac{29.4 \times 5}{2} = 73.5$ kN.

Ref.	
	Try a 305 × 102 × 28 U.B. (restrained – see Dtl. 'B')
Ref 16/1	$D = 308.9$ mm.; $t = 6.1$ mm; $T = 8.9$ mm;
BS5950 Table 6,	$p_y = 275$ N/mm²; $b/T = 5.72$; $d/t = 45.2$;
Table 7	$S_{xx} = 408$ cm³; $I_{xx} = 5439$ cm⁴.
	Classification: ($\varepsilon = 1.0$)
	flange : $b/T = 5.72 < 8.5\varepsilon$: 'plastic'
	web : $d/t = 45.2 < 79\varepsilon$: 'plastic'
	Check Bending: (section 'plastic')
BS5950 Cl. 4.2.5	$M_{cx} = \dfrac{275 \times 408}{10^3} = 112.2$ kN.m.
	$\therefore$ $M_{cx} > M_{MAX}$ ✓ o.k. (112.2) (91.9)
	Check Shear:
BS5950 Cl. 4.2.3	$P_v = 0.6 \times 275 \times 308.9 \times 6.1 / 10^3$
	$P_v = 311$ kN.
	$\therefore$ $P_v > F_v$ ✓ o.k. (311) (73.5)
	Check Deflection: $q_k = 5 \times 3.13 = 15.65$ kN/m.
	$\Delta = \dfrac{5}{384} \times \dfrac{15.65 \times (5000)^4}{205 \times 10^3 \times 5439 \times 10^4} = 11.42$ mm.
BS5950 Table 5	Allowable $= 5000/360 = 13.9$ mm ✓ o.k.
	'B1' : 305 × 102 × 28 U.B.
	B: Beam 'B2' – span 2.5m
	Design U.D.L. as 'B1' = 29.4 kN/m.
	$M_{MAX} = \dfrac{29.4 \times 2.5^2}{8} = 23$ kN.m.
	$F_v = \dfrac{29.4 \times 2.5}{2} = 36.8$ kN.
	Try a 152 × 89 × 16 U.B. (restrained – Dtl. 'B')
Ref 16/1	$D = 152.4$ mm.; $t = 4.6$ mm; $T = 7.7$ mm.;
BS5950 Table 6	$p_y = 275$ N/mm²; $b/T = 5.77$; $d/t = 26.5$;
	$S_{xx} = 124$ cm³; $I_{xx} = 838$ cm⁴

Ref.	
	Classification : ($\varepsilon = 1.0$)
BS5950 Table 7	Flange : $b/T = 5.77 < 8.5\varepsilon$: 'plastic'
	Web : $d/t = 26.5 < 79\varepsilon$: 'plastic'
	∴ Section 'plastic'
	Check bending :
BS5950 Cl. 4.2.5	$M_{cx} = \frac{275 \times 124}{10^3} = 34.1$ kN.m.
	∴ $M_{cx} > M_{max}$ (34.1) (23) ✓ o.k.
	Check shear :
BS5950 Cl. 4.2.3.	$P_v = 0.6 \times 275 \times 152.4 \times 4.6 / 10^3$
	$P_v = 116$ kN.
	∴ $P_v > F_v$ (116) (36.7) ✓ o.k.
	Check deflection: $q_k = 15.65$ kN/m (as 'B1')
	$\Delta = \frac{5}{384} \times \frac{15.65 \times (2500)^4}{205 \times 10^3 \times 838 \times 10^4} = 4.6$ mm
BS5950 Table 5	Allowable $= 2500/360 = 6.9$ mm. ✓ o.k.
	'B2' : 152 x 89 x 16 U.B.
	Steel Roof Beams.
	C : Beam 'B3' - span 5.0 m.
	Roof design loading $= 1.6 \times 0.75 + 1.4 \times 1.2$ $= 2.9$ kN/m²
	Roof slope $= \sqrt{4.0^2 + 3.0^2} = 5.0$ m.
	Purlins take a width of $5/2 = 2.5$ m.
	∴ U.D.L. $= 2.9 \times 2.5 = 7.25$ kN/m.
	$M_{max} = \frac{7.25 \times 5^2}{8} = 22.65$ kN.m.
	$F_{vmax} = \frac{7.25 \times 5}{2} = 18.1$ kN.
	Try a 152 x 89 x 16 UB : beam 'fully restrained by detail 'c' - Fig. 16/004 Properties & classification as for 'B2'.

Ref.	
BS5950 Cl. 4.2.5	Check bending : $M_{cx} = 341.1$ kN.m. (see 'B2')
	$\therefore M_{cx} > M_{MAX}$ ✓ O.K.
	(341.1) (22.65)
	Check shear : $P_v = 116$ kN (see 'B2')
BS5950 Cl. 4.2.3.	$\therefore P_v > F_v$ ✓ O.K.
	(116) (18.1)
	Check Deflection : $q_k = 0.75 \times 2.5 = 1.875$ kN/m.
	$\Delta = \frac{5}{384} \times \frac{1.875 \times (5000)^4}{205 \times 10^3 \times 838 \times 10^4} = 8.9$ mm.
BS5950 Table 5	Allowable $= 5000/360 = 13.9$ mm ✓ O.K.
	'B3' : 152 × 89 × 16 U.B.
	D : Beam 'B4' - span 2.5 m.
	W.D.L. = 7.25 kN/m (as 'B3')
	$M_{MAX} = \frac{7.25 \times 2.5^2}{8} = 5.7$ kN.m.
	$F_{v\,MAX} = \frac{7.25 \times 2.5}{2} = 9.1$ kN.
	Try a 127 × 76 × 13 U.B. - 'fully restrained'
BS5950 Table 6	see detail 'c', Fig 16/004.
Ref. 16/1	D = 127 mm. ; t = 4.2 mm. ; T = 7.6 mm ;
	$p_y = 275$ N/mm² ; b/T = 5.01 ; d/t = 23
	$S_{xx} = 85$ cm³ ; $I_{xx} = 477$ cm⁴.
	Classification : ($\varepsilon = 1.0$)
BS5950 Table 7	Flange : b/T = 5.01 < 8.5ε : 'plastic'
	Web : d/t = 23 < 79ε : 'plastic'
	∴ Section 'plastic'
	Check bending :
BS5950 Cl. 4.2.5	$M_{cx} = \frac{275 \times 85}{10^3} = 23.4$ kN.m.
	$\therefore M_{cx} > M_{MAX}$ ✓ O.K.
	(23.4) (5.7)
BS5950 Cl. 4.2.3	Check shear :
	$P_v = 0.6 \times 275 \times 4.2 \times 127/10^3$
	$P_v = 88$ kN.

Ref.	
	∴ $P_v > F_v$ (88) (9.1) ✓ o.k.
	Check Deflection: $q_k = 1.875$ kN/m (as 'B3')
	∴ $\Delta = \frac{5}{384} \times \frac{1.875 \times (2500)^4}{205\times10^3 \times 477\times10^4} = 1.0$ mm.
BS5950 Table 5	Allowable = 2500/360 = 6.94 mm ✓ o.k.
	'B4' : 127 × 76 × 13 U.B.
	Conservatively: make roof beams 'B5' as for 'B3', and 'B6' as for 'B4'
	Design of padstones: as pressure on the existing brickwork is to be restricted, then point loads to be borne via padstones to block inner skin (f_k = 7 N/mm²)
BS 5628 Table 4	Assume γ_m = 2.5 : block 125 mm thick.
	Try concrete padstones 300 long × 125 wide × 150 high: max. factored reaction = 73.5 kN (see beams 'B1')
	∴ Stress = $\frac{73.5\times10^3}{300 \times 125}$ = 1.96 N/mm²
	Max. local design stress = $1.25 f_k/\gamma_m$ = 1.25 × 7.0 / 2.5 = 3.5 N/mm² ✓ o.k.
	∴ Use padstones 300 × 125 × 150
	Gable wall check:–
	1) Check on wall for load bearing at Section 1–1 (Fig. 16/005) (just above ground floor)

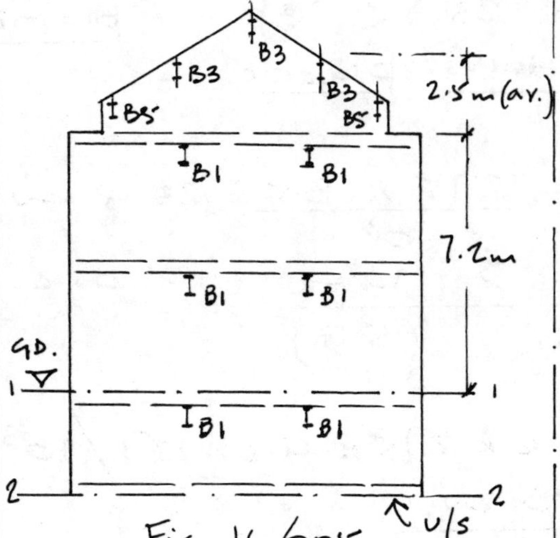

Fig. 16/005

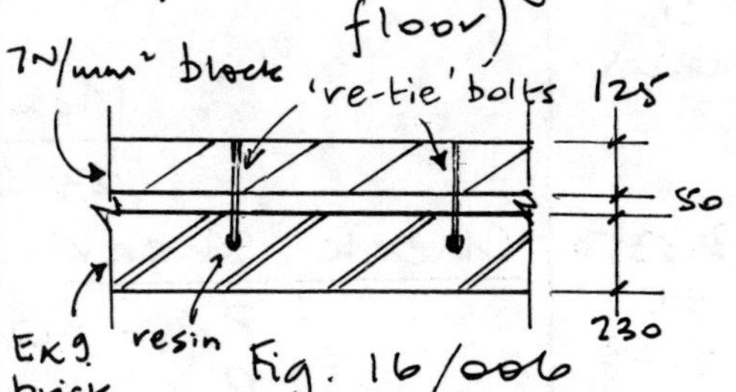

Fig. 16/006 (Wall section)

Ref.	
	Load/m. above Ground Floor Level (Level 1.1)
	Reaction from beams 'B1' = 4 × 73.5 = 294 kN
	∴ U.D.L./m. = $\frac{294}{10}$ = 29.4 kN/m.
	Reaction from roof beams 'B3' & 'B5'
	= 5 × 18.1 = 90.5 kN.
	∴ U.D.L./m. = $\frac{90.5}{10}$ = 9.0 kN/m.
	Self-weight of block inner skin (γ = 14 kN/m³):
	Factored load/sq. m. = 1.4 × 0.125 × 14
	= 2.45 kN/m²
	Average height of wall above 1-1 (Fig. 16/005)
	= 7.2 + 2.5 = 9.7 m.
	∴ U.D.L./m. = 2.45 × 9.7 = 23.8 kN/m.
	∴ Factored load/m. on inner skin
	= 29.4 + 9.0 + 23.8 = 62.2 kN/m.
BS5628 Cl. 32.2.1	Vertical Load Capacity of wall
	= $\beta . t . f_k / \gamma_m$.
	β: take simple support (lateral straps) at each floor level: Le = 3000 mm.
	tef = $\frac{2}{3}$(230 + 125) = 237 mm. (see 16/006)
	S.R = 3000/237 = 12.6
BS5628 Table 7	Padstones central to skin, e = 0
	∴ β = 0.92
	∴ Design capacity = $\frac{0.92 \times 125 \times 7}{2.5}$
	= 322 kN/m.
	Capacity > U.L.S. load ✓ o.k.
	(322) (62.2)
	2) Check on wall for underpinning safe bearing – level 2-2 (Fig 16/005)
	Self-weight of outer skin above 1-1 (Fig 16/005) (γ = 18 kN/m³)
	Factored load/sq. m. = 1.4 × 0.23 × 18 = 5.8 kN/m²

Ref.	
	Factored load/m = 5.8 × 9.7 = 56.2 kN/m.
	∴ total factored load above 1–1 = 62.2 + 56.2 = 118.4 kN/m.
	Factored load/m from gd. floor beams $= \frac{2 \times 73.5}{10} = 14.7$ kN/m.
	Self-weight of block skin (factored) = 2.45 × 2 = 4.9 kN/m.
	Self-weight of brick skin (factored) = 1.4 × 0.327 × 18 × 2 = 16.4 kN/m.
	Total factored load/m. at level 2–2 = 118.4 + 14.7 + 4.9 + 16.4 = 155 kN/m.
	'Service' load/m. $\simeq \frac{155}{1.5}$ = 103 kN/m.

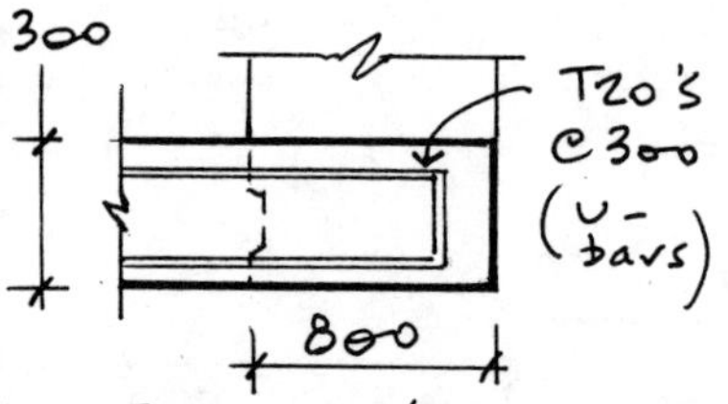

Fig. 16/007
(Detail from Fig. 16/002)

Pressure directly under wall $= \frac{103}{0.8} = 129$ kN/m²

Safe bearing = 100 kN/m² (brief)

∴ Excess pressure to be shared with raft = 129 − 100 = 29 kN/m²

or a shear of 29 × 0.8 = 23 kN/m.

This shear is transmitted to slab via U-bars (see Figs. 16/002 & 16/007)

U.L.S. shear force ≃ 1.5 × 23 = 35 kN/m.

Ref.	
BS 8110 Cl. 3.4.5.2	∴ $v = \frac{35 \times 10^3}{300 \times 10^3} = 0.117$ N/mm²
	A_s of U-bars $= 2 \times \frac{\pi \times 20^2}{4} \times \frac{1000}{300} = 2093$ mm²/m
BS 8110 Table 3.9	∴ $\frac{100 A_s}{bd} = \frac{100 \times 2093}{10^3 \times 250} = 0.84$
	∴ $v_c = 0.76$ N/mm²
	∴ $v_c > v$ ✓ o.k. (0.76) (0.117)
	∴ Underpinning & reinforcement o.k.

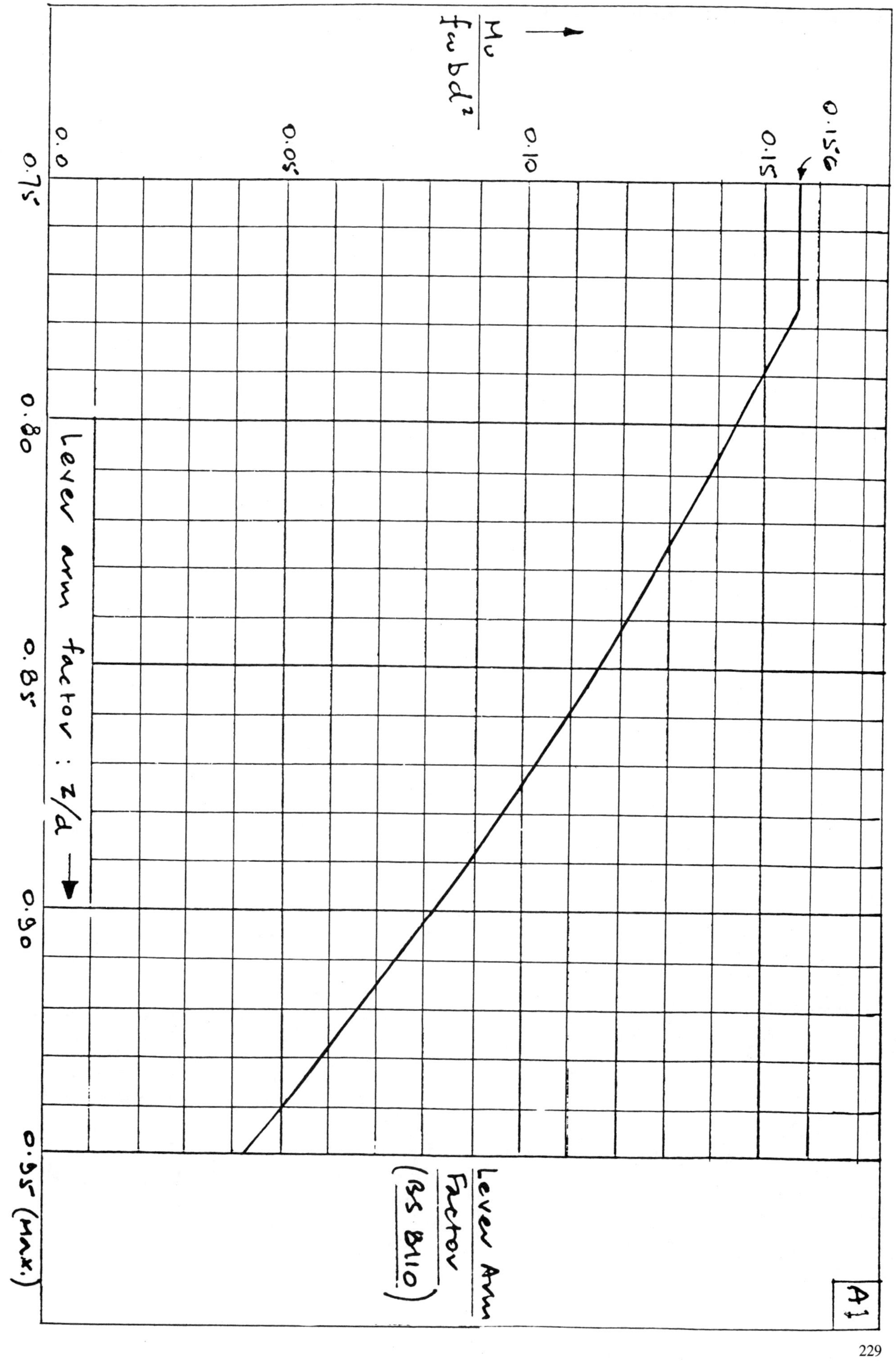
Mu / fcu bd²
0.0
0.05
0.10
0.15
0.156
0.75
0.80
0.85
0.90
0.95 (max.)
Lever arm factor : z/d
Lever Arm Factor (BS 8110)
A1

Appendix A2 Chapter references

Note! The main source of Code reference in each Chapter is stipulated at the top left of each page. Any other Code is separately stipulated. Other references are stipulated as below. The lever arm chart in Appendix A1 is referred to, throughout the reinforced concrete designs as 'App. A1'.

Chapter One

Ref 1/1 – 'Steel Designers Manual – 5th Edition' – Blackwell
Ref 1/2 – 'Design Guide for Continuous Composite Bridges: 2 – Non-Compact Sections' – Steel Construction Institute (Abbreviated below to 'S.C.I.')
Ref 1/3 – 'Simplified Version of the Steel Bridge Code, BS 5400 Pt. 3' – S.C.I.
Ref 1/4 – 'Simplified Version of the Composite Bridge Code, BS 5400 Pt. 5' – S.C.I.
Ref 1/5 – 'Reinforced Concrete Designer's Handbook – 10th Edition' – Reynolds & Steedman – E. & F. Spon

Chapter Three

Ref. 3/1 – 'Steelwork Design Guide to BS 5950: Pt. 1 – 2nd Edition' – S.C.I.
Ref. 3/2 – 'Structural Steelwork Limit State Design' – Clarke & Coverman

Chapter Four

Ref. 4/1 – 'Steel Designers Manual – 5th Edition' – Blackwell
Ref. 4/2 – 'Steelwork design Guide to BS 5950: Pt. 1 – 2nd Edition' – S.C.I.
Ref. 4/3 – 'Steel Structures – Practical Design Studies' – T.J. MacGinley – Spon
Ref. 4/4 – 'Structural Steelwork Design to BS 5950' – Morris & Plum – Longmans Scientific & Technical
Ref. 4/5 – 'Steelwork Detailers Manual' – Hayward & Weare – BSP

Chapter Five

Ref. 5/1 – 'Steelwork design Guide to BS 5950: Pt. 1 – 2nd Edition' – S.C.I.
Ref. 5/2 – 'Pile Design and Construction Practice' – M.J. Tomlinson – Viewpoint

Chapter Six

Ref. 6/1 – 'Steelwork design Guide to BS 5950: Pt. 1 – 2nd Edition' – S.C.I.

Chapter Seven

Ref. 7/1 – 'Steelwork design Guide to BS 5950: Pt. 1 – 2nd Edition' – S.C.I.

Chapter Eight

Ref. 8/1 – 'Reinforced Concrete Designer's Handbook – 10th Edition' – Reynolds & Steedman – E. & F. Spon
Ref. 8/2 – 'The Construction of new buildings behind historic facades' – David Highfield – E. & F. Spon.
Ref. 8/3 – 'Structural Foundations Manual' – Reg Atkinson – Chapman & Hall.

Chapter Nine

Ref. 9/1 – 'Structural design Guide to BS 5950: Pt. 1 – 2nd Edition' – S.C.I.

Chapter Ten

Ref. 10/1 – Design Guide on Precast Concrete Frame Buildings – Elliott & Tovey – BCA

Chapter Eleven

Ref. 11/1 – 'Reinforced Concrete Designer's Handbook – 10th Edition' – Reynolds & Steedman – E. & F. Spon

Chapter Twelve

Ref 12/1 – 'Reinforced Concrete Designer's Handbook – 10th Edition' – Reynolds & Steedman – E & F Spon
Ref 12/2 – 'Steelwork design Guide to BS 5950 Pt. 1 – 2nd Edition' – S.C.I.
Ref 12/3 – 'Structural Design in Practice – 2nd Edition' – Westbrook – Longmans Scientific & Technical Press
Ref 12/4 – 'Steel Designers Manual – 5th Edition' – Blackwell

Chapter Thirteen

Ref. 13/1 – 'Reinforced Concrete Designer's Handbook – 10th Edition' – Reynolds & Steedman – E. & F. Spon

Chapter Fourteen

Ref. 14/1 – 'Loadbearing brickwork in crosswall construction' – Curtin, Shaw, Beck & Bray – Brickwork Development Association
Ref. 14/2 – 'Designing in Reinforced Brickwork' – Curtin, Shaw, Beck & Bray – Brickwork Development Association

Chapter Fifteen

Ref. 15/1 – 'Structural Steelwork Design to BS 5950' – Morris & Plum – Longmans Scientific & Technical
Ref. 15/2 – 'Steelwork design Guide to BS 5950: Pt. 1 – 2nd Edition' – S.C.I.

Ref. 15/3 – 'The Trussed Rafter Manual' – A Gang-Nail Systems Publication
Ref. 15/4 – 'Timber Designers Manual' – Ozleton & Baird – Crosby Lockwood Staples

Chapter Sixteen

Ref. 16/1 – 'Reinforced Concrete Designer's Handbook – 10th Edition' – Reynolds & Steedman – E. & F. Spon